T0200532

Spatial Predictive Modeling
with R

Spatial Predictive Modeling with R

Jin Li

CRC Press
Taylor & Francis Group
Boca Raton London New York

CRC Press is an imprint of the
Taylor & Francis Group, an **informa** business

A CHAPMAN & HALL BOOK

First edition published 2022
by CRC Press
6000 Broken Sound Parkway NW, Suite 300, Boca Raton, FL 33487-2742

and by CRC Press
4 Park Square, Milton Park, Abingdon, Oxon, OX14 4RN

© 2022 Jin Li

CRC Press is an imprint of Taylor & Francis Group, an Informa business

ISBN: 978-0-367-55054-7 (hbk)
ISBN: 978-0-367-55056-1 (pbk)
ISBN: 978-1-003-09177-6 (ebk)

DOI: 10.1201/9781003091776

Publisher's note: This book has been prepared from camera-ready copy provided by the authors

Contents

Preface

Spatial predictive modeling

Spatially continuous data are increasingly required by various disciplines, such as ecology, agronomy, meteorology, geosciences, environmental sciences, and environmental management and conservation. Such data are, however, usually not readily available, and they are difficult and expensive to acquire, especially in areas that are difficult to access (e.g., mountainous or marine regions). Spatial data are often available as point data collected from point locations. *Spatial predictive methods* are essential tools for generating *spatial predictions* (i.e., spatially continuous data in this book) from the point data and have been widely used in various disciplines (Sanabria et al. 2013; Maier et al. 2014; Stephens and Diesing 2015; Li 2019a). *Spatial predictive modeling* (*SPM*) refers to developing optimal predictive models by applying relevant spatial predictive methods to information collected at point locations, and then generating spatial predictions with the models for the areas interested. *SPM* is an emerging discipline in applied sciences, playing a key role in the generation of spatial predictions.

Traditionally, *spatial predictive methods* may refer to a vast set of tools: (1) *non-geostatistical methods* including mathematical methods, (2) *geostatistical methods*, and (3) *combined methods* (Li and Heap 2008, 2014). These methods are usually called *spatial interpolation methods*. *Spatial predictions* generated by these methods are usually termed as *spatial interpolations*, *extrapolations* or both.

Recent advances in (1) computing technology and modeling techniques (Crawley 2007; Hastie, Tibshirani, and Friedman 2009; Kuhn and Johnson 2013), and (2) data acquisition and data processing capacity, such as remote-sensing techniques and geographic information systems, result in increasingly more information available for *SPM*. Consequently, how to select and develop optimal predictive models from a large number of predictive variables is a challenging task, particularly for geostatistical methods that are unable to handle many predictive variables, and more sophisticated *SPM* approaches are needed and have been introduced or developed to deal with a large number of predictive variables (e.g., Li, Potter, Huang, Daniell, et al. 2010; Li, Heap, Potter, and Daniell 2011b; Li, Potter, Huang, and Heap 2012b). The introduction of certain *modern statistical methods* and *machine learning methods* into *SPM* field, especially by combining modern statistics and machine learning methods with existing *spatial interpolation methods*, has resulted in a number of novel *hybrid methods* (Table 1). This development has opened an alternative source of methods and resulted in more tools for *SPM* (Li 2019a). These tools have been gradually adopted in various disciplines (Sanabria et al. 2013; Appelhans et al. 2015; Hengl et al. 2015; Ruiz-Álvarez, Alonso-Sarria, and Gomariz-Castillo 2019).

Predictive accuracy is of critical importance for spatial predictions. Improving the accuracy by choosing an appropriate method and then identifying and developing the most accurate predictive model(s) is an essential and challenging task for *SPM*. It is often difficult to select an appropriate method and develop an optimized predictive model because the *SPM* process involves many factors or components, and all relevant components need to be considered and optimized (Li and Heap 2011, 2014). Spatial predictive models are often developed in terms of predictive accuracy based on model validation methods and are fundamentally different from other modeling types (Leek and Peng 2015).

TABLE 1: Development of the *hybrid methods* for *SPM* (Li, Heap, Potter, and Daniell 2010, 2011b, 2011a; Li, Potter, Huang, Daniell, et al. 2010; Li 2011; Li, Potter, Huang, and Heap 2012a, 2012b) (modified from Li (2018a)).

No.	Hybrid method	Time
1	*Generalized linear models (GLM)* and *ordinary kriging (OK) (RKglm)*	2008 - 2009
2	*Generalized least squares (GLS)* and *OK (RKgls)*	2008 - 2009
3	*Classification and regression trees (RPART)* and *OK (RPARTOK)*	2009 - 2010
4	*RPART* and *inverse distance squared (IDS) (RPARTIDS)*	2009 - 2010
5	*Random forest (RF)* and *OK (RFOK, RKRF)*	2008 - 2009
6	*RF* and *inverse distance weighted (IDW)* or *IDS (RFIDW, RKIDS)*	2008 - 2009
7	*RFIDW* and *RFOK (RFOKRFIDW)*	2008 - 2011
8	*RF, RFIDW* and *RFOK (RFRFOKRFIDW)*	2010 - 2012
9	*Support vector machine (SVM)* and *OK (SVMOK)*	2009 - 2010
10	*SVM* and *IDW* or *IDS (SVMIDW, SVMIDS)*	2009 - 2010
11	*SVMIDW* and *SVMOK (SVMOKSVMIDW)*	2010 - 2012
12	*SVM, SVMIDW* and *SVMOK (SVMSVMOKSVMIDW)*	2010 - 2012
13	*Generalized boosted regression modeling (GBM)* and *OK (GBMOK)*	2010 - 2012
14	*GBM* and *IDW (GBMIDW)*	2010 - 2012

Predictive accuracy is the key criteria for predictive modeling, and it is used for parameter, variable and model/method selection in this book. The property of *spatial predictions* is a further criteria for predictive modeling, where professional knowledge plays its role in examining whether the predictions are scientifically sound, reasonable, and interpretable.

This book aims to introduce *SPM* as a discipline to modelers and researchers. It systematically introduces the entire process of *SPM*. The process contains the following components: data acquisition, method and variable selection, model or parameter optimization, accuracy assessment, and the generation and visualization of spatial predictions. Each of these modeling components plays an important role in model development. Incorrect or inappropriate implementation of any components may lead to less accurate or even misleading predictive model(s). This book provides tools for relevant components to improve the quality of spatial predictions in various disciplines. It also provides guidelines, suggestions, recommendations, and reproducible examples in *R* for developing the most accurate predictive model by considering these components, relevant requirements, and factors associated with each component.

This book concentrates more on the applications of predictive methods and less on the mathematical and statistical details that can be found in previous studies (e.g., Goovaerts 1997; Webster and Oliver 2001; Venables and Ripley 2002; Wackernagel 2003; Bivand, Pebesma, and Gomez-Rubio 2013; van Lieshout 2019). Since this book is specifically focusing on *SPM* with R, for one interested in machine learning and spatio-temporal statistics with *R* other relevant publications are available (e.g., Kuhn and Johnson 2013; James et al. 2017; Wikle, Zammit-Mangion, and Cressie 2019).

This book covers the whole modeling process that is not only important for *SPM*, but also provides valuable tools to other predictive modeling fields. It is expected to boost the applications of appropriate *SPM* processes, and improve the quality of spatial predictions for various disciplines. It is also expected to enhance further research in this field, and anticipated further novel and performance-improved spatial predictive methods to be developed and applied.

How this book is organized

This book introduces a number of important advances and new ideas in *SPM*, including novel spatial predictive methods, and novel accuracy-based parameter estimation and variable selection methods. Reproducible examples on applications and comparisons of various spatial predictive methods are provided for different data types including categorical data, binary data, count data, continuous data and percentage data. All examples and relevant computations are implemented in *R*. This book covers the following components for *SPM*:

1. Data acquisition (i.e., sampling design) as well as data preparation in Chapter 1;

2. Pre-selection of predictive variables, using variables in environmental sciences as examples; and exploratory analysis as an assistant tool for the pre-selection in Chapter 2;

3. Predictive accuracy assessment, model evaluation and validation in Chapter 3;

4. Spatial predictive methods:

(i) Spatial-information based interpolation methods that include mathematical methods in Chapter 4 and univariate geostatistical methods in Chapter 5;

(ii) Gradient based/detrended predictive methods that include multivariate geostatistical methods in Chapter 6, modern statistical methods in Chapter 7, and machine learning methods in Chapters 8 and 9;

(iii) Hybrids of modern statistical methods with geostatistical methods in Chapter 10; and

(iv) Hybrids of machine learning methods with geostatistical methods in Chapter 11; and

5. Further examples on the applications and comparison of spatial predictive methods in Chapter 12.

The geostatistical methods for the hybrid methods refer to the mathematical methods, univariate geostatistical methods or both. The mathematical methods are grouped in geostatistical methods because they are implemented in the `gstat` package (Gräler, Pebesma, and Heuvelink 2016) as geostatistical methods.

For each method, a brief description of how the method works and its conceptual underpinnings is provided. Then reproducible examples for ways to apply each method in *R* are provided. Parameter estimation and/or variable selection based on predictive accuracy are demonstrated for developing optimal predictive models, including novel variable selection methods, such as `steprf` and `stepgbm` in Chapter 8. And then the predictive accuracy of the optimal models are assessed based on cross-validation functions for each method. Finally, spatial predictions and prediction uncertainty are generated and visualized, using a number of novel functions that have considerably simplified the generation of the predictions.

R versions

Two versions of R were used to run R code in this book.

R version 3.6.3 (2020-02-29) Platform: x86_64-w64-mingw32/x64 (64-bit) Running under: Windows 10 x64 (build 19041)

R version 4.0.3 (2020-10-10) Platform: x86_64-w64-mingw32/x64 (64-bit) Running under: Windows 10 x64 (build 19041)

R packages

An R package, `spm` (Li 2019b), initially released on CRAN in 2017, is used as one of the core packages in this book. Three other core R packages, `spm2` (Li 2021a), `steprf` (Li 2021c), and `stepgbm` (Li 2021b), are developed for and released with this book, to cover the methods and functions that are not currently included in the `spm` package.

1. The `spm` package

The `spm` package introduces some novel accurate hybrid methods of geostatistical and machine learning methods for *SPM*. It contains two commonly used geostatistical methods (i.e., *OK* and *IDW*), two machine learning methods (*RF* and *GBM*), four hybrid methods (i.e., *RFIDW*, *RFOK*, *GBMIDW*, and *GBMOK*) and two averaging methods (*RFOKRFIDW*, and the average of *GBMOK* and *GBMIDW* (*GBMOKGBMIDW*)). For each method, two functions are provided, with one function for assessing the predictive errors and accuracy of the method based on cross-validation and the other for generating spatial predictions.

2. The `spm2` package

The `spm2` package is an extended version of `spm`, by further introducing some novel functions for statistical methods (i.e., *GLM*, *glmnet*, *GLS*), *thin plate splines*, *SVM*, kriging methods (i.e., *simple kriging, universal kriging, block kriging, kriging with an external drift*), and 228 hybrid methods plus numerous variants for *SPM*. For each method, two functions are provided, with one function for assessing the predictive errors and accuracy of the method based on cross-validation and the other for generating spatial predictions if needed. It also contains a couple of functions for data preparation and predictive accuracy assessment.

3. The `steprf` package

The `steprf` package introduces several novel variable selection methods for *RF*. They are based on averaged variable importance (*AVI*), and knowledge informed *AVI* (*KIAVI* and *KIAVI2*) methods.

4. The `stepgbm` package

The `stepgbm` package introduces a couple of novel variable selection methods for *GBM*. They are based on relative variable influence (*RVI*) and knowledge informed *RVI* (*KIRVI* and *KIRVI2*) methods.

Data sets

All data sets used in this book are available either in the spm, spm2, and sp (Pebesma and Bivand 2020) packages or online as detailed in Appendix A.

Who should read this book

This book systematically introduces whole *SPM* process to both researchers and practitioners. It can be used as the primary textbook for geostatistics and *SPM*. It can also be used as supplementary reading for modern statistics, statistical learning, machining learning, and data sciences because it provides novel functions for, and good examples of applications of, relevant methods for these courses.

It can also be used as self-learning material for researchers, modelers, and university students from the upper years. It provides reproducible examples for readers at all levels of R proficiency. The R code and R packages provide sufficient materials for non-R users as they can be easily reprogrammed into other computing languages, such as Python. It can be used as a reference for both teaching and research.

References for R

Learning materials are available online for beginners, such as
(1) "An Introduction to R";
(2) "R Language Definition";
(3) "R Installation and Administration";
(4) "R Data Import/Export"; and
(5) "Writing R Extensions".
They can also be accessed from *Manuals (in PDF)* tab under *Help* tab in *RGui*.

Caveats

Although in this book I attempt to cover relevant components, which contribute to the improvement of predictive accuracy, as comprehensively as possible, the *SPM* field is too broad to allow that to be done completely in this book. This is because different disciplines have their own specific features and requirements. Therefore, further work is needed to identify factors in relevant components or additional components that can further improve predictive accuracy in each discipline.

Furthermore, the data sets and software packages are provided in good faith, but none of the author, publishers, and distributors warrant their accuracy and are responsible for the consequences of their use.

Contact details

The author may be contacted by email at

jldata2action@gmail.com

and would be grateful for being informed of errors and improvements to the contents of this book. Errata and updates are available at `https://github.com/jinli22`.

Acknowledgment

This book would not be possible without:
(1) R, a free software environment for statistical computing and graphics (R Core Team 2020);
(2) RStudio, an integrated development environment (IDE) for R (RStudio Team 2020); and
(3) R bookdown, a powerful tool for combining analysis and reporting into the same document (Xie 2016).

I would like to acknowledge the contribution of many people to the conception and completion of this book. Over the years, many people have greatly influenced my career, but a special recognition must be given to Bob Murison, University of New England, who, with enthusiasm and patience, helped me in modern statistics and statistical computing in 1990s. Tony Arthur and Steve Henry kindly helped to get the bee data set released from CSIRO for this book. I am also grateful to Yanchang Zhao for helpful discussion and suggestion in the early stage of the preparation of the book. I am indebted to Xiufu Zhang for proofreading and critical comments on the preliminary draft. I would like to thank three reviewers for valuable comments on the proposal of this book. I am greatly appreciative to Rob Calver at Chapman & Hall/CRC for seeking a book idea back in 2015 and his team for continuous support, patience, and help at each phase of the preparation of this book. Finally, I would like to thank my family and my parents for their love and support.

Jin Li
July, 2021

Author Bio

Dr Jin Li works at Data2action, Australia as a Founder. He has research experience in spatial predictive modeling, statistical computing, ecological and environmental modeling, and ecology. As a scientist, he worked in the Chinese Academy of Sciences, University of New England, CSIRO, and Geoscience Australia. He was an Associate Editor (Jul 2008-Dec 2015) and an editorial board member (Jan 2016-April 2020) of Acta Oecologica, and a Guest Academic Editor (Mar 2018) and an Academic Editor (May 2018-Apr 2020) of PLOS ONE. He has produced over 100 various publications, developed a number of hybrid methods for spatial predictive modeling, and published four R packages for variable selections and spatial predictive modeling.

For further information see https://www.researchgate.net/profile/Jin-Li-74, https://scholar.google.com/citations?user=Jeot53EAAAAJ&hl=en and https://www.linkedin.com/in/jin-li-01421a68/.

1

Data acquisition, data quality control, and spatial reference systems

This chapter introduces relevant *sampling designs* for *spatial data*, factors to be considered for *data quality control* (*QC*), and *spatial data types* and *spatial reference systems* to be used for spatial predictive modeling.

1.1 Acquiring data for spatial predictive modeling

For spatial predictive modeling, samples are sometimes collected, stored, and ready to use. If this is the case, please go to Section 1.2 for data quality control. However, if samples need to be collected, a *sampling design* needs to be produced. For *spatial predictive modeling*, sampling designs need to focus on a survey area over space. A good sampling design ensures that data collected from a survey are capable of answering relevant research questions and fulfill survey purpose, and also is as precise and efficient as possible (Benedetti, Piersimoni, and Postigione 2017; Foster et al. 2017). Many methods have been developed to generate a *sampling design* (Stevens and Olsen 2004; Diggle and Ribeiro Jr. 2010; Wang et al. 2012; Benedetti and Piersimoni 2017). They largely fall into four categories (Li 2019a): (1) *non-random sampling*, (2) *unstratified random sampling*, (3) *stratified random sampling*, and (4) *stratified random sampling with prior information*.

1.1.1 Non-random sampling

Non-random sampling is often *ad-hoc sampling* based on expert knowledge or purely opportunistic when a certain type of environmental condition becomes available. This type of sampling design can be seen in many previous surveys (e.g., Przeslawski et al. 2011; Li, Alvarez, et al. 2017; Radke et al. 2015, 2017).

The sediment samples in the Petrel area in Appendix A were collected from several surveys (see Section 1.2 for further information) using non-random sampling designs, and will be used as an example of ad-hoc sampling below.

```
pps.df <- read.csv("./data/Sample for Petrel.csv") # petrel point sample
```

```
library(raster)
pb.df <- as.data.frame(raster("./data/bathy_1km_Petrel.tif"), xy = TRUE, na.rm =
    TRUE) # petrel bathymetry raster data imported as data-frame
names(pb.df) <- c("long", "lat", "bathy")
```

The data sets, `pps.df` and `pb.df`, are in dataframe format. They need to be converted to spatial objects as below according to Bivand, Pebesma, and Gomez-Rubio (2013) so that they can be visualized.

DOI: 10.1201/9781003091776-1

```
library(sp)

pb <- pb.df
gridded(pb) = ~long+lat # grid spatial data into SpatialPixelsDataFrame

pps <- pps.df
coordinates(pps) = ~long+lat # set spatial coordinates to create SpatialPoints
    data
proj4string(pps) <- CRS("+proj=longlat +datum=WGS84 +no_defs +ellps=WGS84 +towgs84
    =0,0,0") # set WGS84 on the SpatialPoints
```

Spatial distribution of sediment samples in the Petrel area is illustrated in Figure 1.1 by

```
par(font.axis = 2, font.lab = 2)
lab.palette <- colorRampPalette(c("dark blue", "blue", "light blue"), space = "Lab
    ")
image(pb, axes = T, xlab = "Longitude", ylab = "Latitude", col = lab.palette(255))
plot(pps, add = TRUE, pch = 1, col = "red", cex = 0.5)
```

It is apparent that the samples are non-randomly distributed and even clustered.

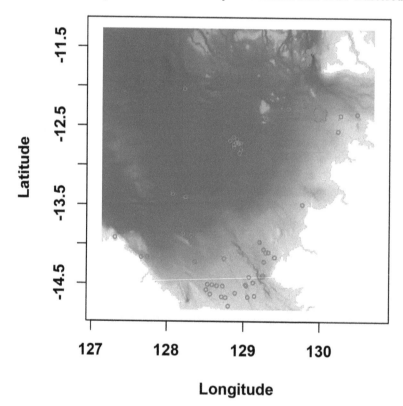

FIGURE 1.1: Samples selected (red circle) by ad-hoc sampling designs; and the background shows the spatial patterns of bathymetry in the Petrel area.

Non-random sampling can also be *systematic sampling*. This is demonstrated by applying function spsample in the sp package (Bivand, Pebesma, and Gomez-Rubio 2013; Pebesma and Bivand 2005, 2020). It samples point locations in planar coordinates using regular or random sampling methods (Pebesma and Bivand 2020). The descriptions of relevant arguments of the function are detailed in its help file, which can be accessed by ?spsample.

For sampling point locations in an area using `spsample`, the following arguments need to be specified:

(1) `x`, a spatial object;

(2) `n`, sample size; and

(3) `type`, a character to specify a sample method (e.g., "random", "regular").

We apply `spsample` to the Petrel area and produce 100 samples as below.

```
systsps <- spsample(pb, n = 100, "regular")
```

The samples selected, `systsps`, are illustrated in Figure 1.2. As expected for systematic sampling design, the samples are evenly distributed over space.

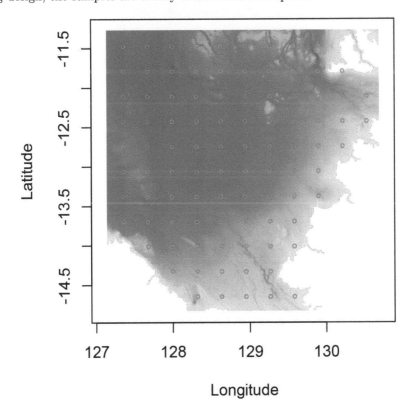

FIGURE 1.2: Samples selected (red circle) by systematic sampling design; and the background shows the spatial patterns of bathymetry in the Petrel area.

For spatial predictive modeling, non-random sampling methods are not recommended. However, non-random sampling designs were compared previously (Diggle and Ribeiro Jr. 2010), which provides some useful clues for lattice plus close pairs sampling for spatial predictive modeling.

1.1.2 Unstratified random sampling

Unstratified random sampling (or *random sampling*) is that sampling locations are randomly selected from the entire survey area, with no further information being used for stratification. This can be: (1) an *unstratified equal probability design*, or (2) an *unstratified unequal probability design* (Kincaid, Olsen, and Weber 2020).

We apply `spsample` to `pb` and produce 100 samples using unstratified equal probability design as an example below. To make the results reproducible, we need to use function `set.seed` that takes an (arbitrary) integer argument.

```
set.seed(1234)
unstran <- spsample(pb, n = 100, "random")
```

The samples resulted, `unstran`, are illustrated in Figure 1.3.

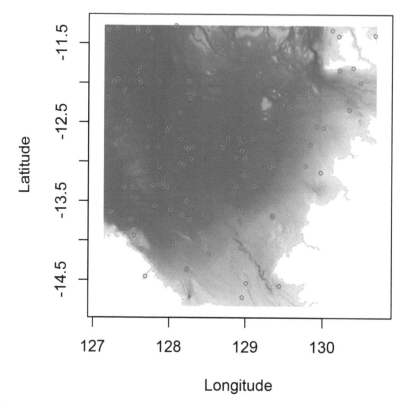

FIGURE 1.3: Samples selected (red circle) by unstratified random sampling design, and the background shows the spatial patterns of bathymetry in the Petrel area.

The unstratified random sampling is not recommended to collect data for spatial predictive modeling, so for the unstratified unequal probability design, no example will be provided. This is because (1) spatial information is available for sampling design, which would lead to spatially stratified sampling as discussed below in Section 1.1.3, and (2) unstratified random sampling design may even be overperformed by lattice sampling, a kind of non-random design (Diggle and Ribeiro Jr. 2010).

1.1.3 Stratified random sampling design

Stratified random sampling means that sampling locations are randomly selected from sub-areas known as strata. The sub-areas are formed from partitioning the entire survey area based on further information. Stratified random sampling is often used when additional information, such as spatial location information, elevation, bathymetry, soil type or geomorphological data, is available. For spatial predictive modeling, such information is important, often available (particularly the spatial information), and should be considered when designing a survey for a region. Some recently developed randomized spatial sampling procedures were reviewed and compared using *simple random sampling without replacement* as a benchmark for comparison, and the guidance has been also provided for choosing an appropriate spatial sampling method (Benedetti, Piersimoni, and Postigione 2017). Furthermore, some applications of stratified random sampling with spatial information (i.e., spatially stratified random sampling) and some *R* packages for stratified random sampling have been reviewed for spatial predictive modeling (Li 2019a).

Spatially stratified random sampling design can be generated in several ways in R (Li 2019a) using additional information. Three examples using different functions are provided below.

1. Function `spsample`

The first is to apply `spsample` to `pb` with `stratified` sampling method.

```
set.seed(1234)
stran1 <- spsample(pb, n = 100, "stratified")
```

The samples resulted, `stran1`, are shown in Figure 1.4.

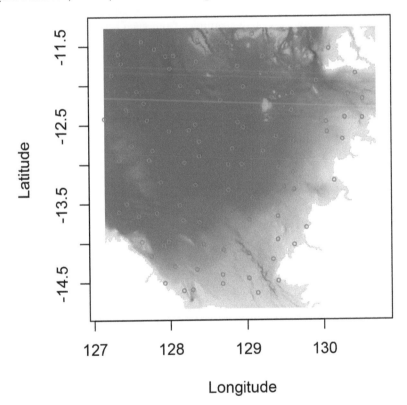

FIGURE 1.4: Samples selected (red circle) by spatially stratified random sampling design using the function `spsample` in the sp package, and the background shows the spatial patterns of bathymetry in the Petrel area.

2. Functions `stratify` and `spsample`

The second example applies the functions `stratify` and `spsample` in the `spcosa` package (Walvoort, Brus, and de Gruijter 2020) to `pb`.

The function `stratify` is to partition a spatial object into compact strata by means of k-means (Walvoort, Brus, and de Gruijter 2020). The descriptions of relevant arguments of `stratify` are detailed in its help file, which can be accessed by `?stratify`.

For sampling point locations in an area using `stratify`, the following arguments need to be specified:
(1) `x`, a spatial object;
(2) `nStrata`, number of strata; and
(3) `equalArea`, if `FALSE` the algorithm results in compact strata, and if `TRUE` the algorithm results in compact strata of equal size.

We apply `stratify` to `pb` and produce 20 strata as below.

```
library(spcosa)
```

```
# Spatially stratify the sampling area using stratify function in spcosa.
set.seed(1234)
strata20 <- stratify(pb, nStrata = 20, equalArea = TRUE) # this is time consuming!
```

The arguments of `spsample` in spcosa are the same as the arguments of `spsample` in sp. We apply `spsample` to `strata20` and select five samples from each stratum as below.

```
# select 100 samples (i.e., 5 samples from each stratum)
set.seed(1234)
sp20 <- spsample(strata20, n = 5)
sp20.100 <- as(sp20, 'data.frame')
```

Locations of the samples selected, sp20, are shown in Figure 1.5 by

```
plot(strata20, sp20)
```

3. Function `clhs`

The last example uses the function `clhs` in the `clhs` package (Roudier 2011, 2020). The function `clhs` is to implement the conditioned Latin hypercube sampling (Roudier 2011). The descriptions of relevant arguments of `clhs` are detailed in its help file, which can be accessed by `?clhs`.

For sampling point locations in an area using `clhs`, the following arguments need to be specified:
(1) `x`, a data.frame or a spatial object; and
(2) `szie`, sample size.

We apply `clhs` to `pb.df`, a `data.frame`. This sampling design uses both location information and bathymetry data to spatially stratify the sampling area and randomly select 100 samples as below.

```
library(clhs)
set.seed(1234)
sample100 <- clhs(pb.df, size = 100)
sample100.sp <- pb.df[sample100,]
sp100 <- sample100.sp[,-3]
```

Locations of the samples selected, sp100, are shown in Figure 1.6.

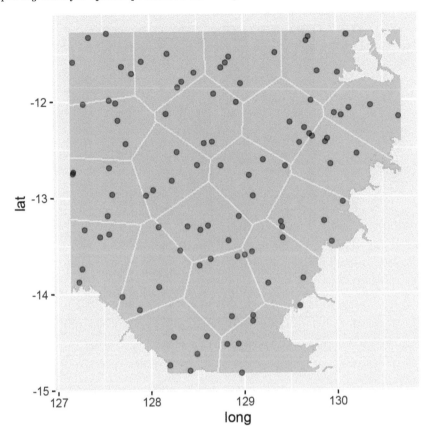

FIGURE 1.5: Samples selected (gray dot) by spatially stratified random sampling using the functions `stratify` and `spsample` in the `spcosa` package, where the region to be sampled is divided into 20 equal areas (green).

1.1.4 Stratified random sampling design with prior information

Stratified random sampling with prior information incorporates the locations of *legacy sites* into new spatially balanced sampling designs to ensure spatial balance among all sample locations (Foster et al. 2017). It is a stratified unequal probability design and can be produced using the functions `alterInclProbs` and `quasiSamp` in the `MBHdesign` package (Foster 2019). The descriptions of relevant arguments of these functions are detailed in their help files, which can be accessed by `?alterInclProbs` and `?quasiSamp`.

Stratified random sampling design with legacy sites is to be demonstrated by applying the functions `alterInclProbs` and `quasiSamp` to `pb.df`. Relevant arguments that need to be specified are also prepared and noted in the *R* code below.

```
library(MBHdesign)
set.seed( 1234)
n <- 20 # number of samples to be selected
X <- pb.df[,-3] # survey area
N <- dim(X)[1] # number of points to sample from

legacySites <- as.matrix(read.csv("./data/Sample for Petrel.csv")) # the legacy
    sites, that is, Petrel point sample
colnames(legacySites) <- c("long", "lat")
```

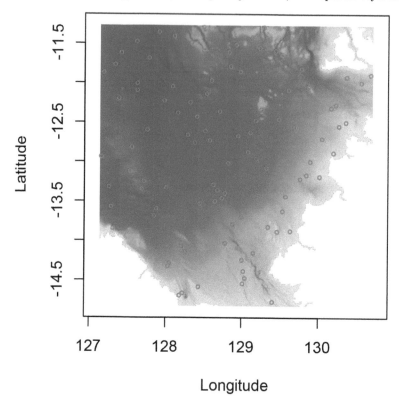

FIGURE 1.6: Samples selected (red circle) by spatially stratified random sampling design using the function clhs; and the background shows the spatial patterns of bathymetry in the Petrel area.

```
inclProbs <- rep( n/N, times = N) # uniform inclusion probabilities
altInclProbs <- alterInclProbs(legacy.sites = legacySites,
        potential.sites = X, inclusion.probs = inclProbs, mc.cores = 6) # alter
            inclusion probabilities so that new samples should be well-spaced from
            legacy sites

set.seed( 1234)
samp <- quasiSamp(n = n, dimension = 2,  study.area = NULL, potential.sites = X,
    inclusion.probs = altInclProbs) # generate the design according to the altered
        inclusion probabilities.
```

Since this sampling method uses the spatial extent of survey area (i.e., potential sites) instead of its actual spatial domain, some samples may locate outside the domain. If this occurs, sample size (i.e., the number of samples to be selected) may need to be increased to ensure sufficient samples fall within the domain.

We can visualize the adjusted inclusion probabilities (white to blue areas to show the inclusion probability changing from low to high), legacy sites (red circle) and sample locations as in Figure 1.7 by

```
X1 <- X
X1$prob <- altInclProbs
gridded(X1) = ~long+lat

image(X1,  axes = T, xlab = "Longitude", ylab = "Latitude", col = hcl.colors(10, "
    blues", rev = TRUE)) # Adjusted Inclusion Probabilities
```

```
points(legacySites, pch = 1, col = "red", cex = 0.5) # the legacy locations
points(samp[, 1:2], pch = 5, col = "red",  cex = 1) # sample locations.
```

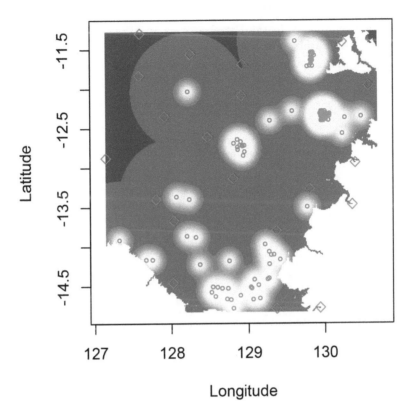

FIGURE 1.7: Samples selected (red diamond) by spatially stratified random sampling with prior information (i.e., legacy sites, red circle), with the adjusted inclusion probabilities increasing from the legacy sites (white to blue areas show the inclusion probability changing from low to high).

1.2   Data quality control

Samples may be collected, stored, and ready to use for spatial predictive modeling. However, samples may have not been *quality controlled* (*QC*), and the quality of samples can be affected by many factors in various aspects, which may also be data-specific (Pipino, Lee, and Wang 2002; Cai and Zhu 2015). Sample quality is important as samples provide the fundamental information for spatial predictive modeling. Prior to undertaking the predictive modeling, samples need to be quality controlled. That is, relevant factors need to be identified, relevant *data QC criteria* need to be developed for data *QC*, and then samples need to be cleaned accordingly. We will take Geoscience Australia's Marine Samples Database (*MARS*; at [http://dbforms.ga.gov.au/pls/www/npm.mars.search]) as an example to demonstrate the whole data *QC* process.

Sediment samples in *MARS* database were initially quality controlled prior to and after entering into the database according to various criteria, but the quality of the samples was still affected by many factors, such as data credibility, data accuracy, and completeness (Li, Potter, Huang, Daniell, et al. 2010; Li, Potter, Huang, and Heap 2012b). Data *QC* process is still required for the sediment samples.

We will use sediment samples in the Petrel area as an example, and the samples will be extracted from MARS database as below.

```
sed1 <- read.csv("./data/MARS_Grain_size_as_mud_sand_gravel_mean_texture_20200524_
    092819.csv")
# names(sed1)
dim(sed1)
```

```
[1] 385   20
```

```
# extract sediment related information
sed1 <- sed1[, c(7:10, 12:14, 16:18)]
names(sed1) <- c("sample.type", "date", "top.depth", "base.depth", "lat", "long",
    "bathy", "mud", "sand", "gravel")
```

The data structure and first three rows of sediment samples are

```
str(sed1, digits.d = 6, width = 65, strict.width = "cut")
```

```
'data.frame':    385 obs. of  10 variables:
 $ sample.type: chr  "DREDGE UNSPECIFIED" "DREDGE UNSPECIFIED""..
 $ date       : chr  "1993-10-01" "1993-10-01" "1993-10-01" "1"..
 $ top.depth  : num  NA NA NA NA NA NA NA NA NA NA ...
 $ base.depth : num  NA NA NA NA NA NA NA NA NA NA ...
 $ lat        : num  -13.494 -13.466 -13.4667 -13.468 -13.465 -..
 $ long       : num  129.985 129.935 129.983 130.034 130.082 12..
 $ bathy      : num  -4 -7 -5 -7 -5 -15 -13 -16 -11 -12 ...
 $ mud        : num  3.6 6.5 4.2 14.8 25.7 35.9 6.3 14.5 7.3 21..
 $ sand       : num  96.4 88.7 95.9 66.2 45 64.1 62.6 83 90.7 7..
 $ gravel     : num  NA 4.6 NA 18.9 29.2 NA 30.9 2.5 1.9 4.1 ...
```

```
head(sed1, 3)
```

```
          sample.type       date top.depth base.depth    lat   long bathy mud
1 DREDGE UNSPECIFIED 1993-10-01        NA         NA -13.49 130.0    -4 3.6
2 DREDGE UNSPECIFIED 1993-10-01        NA         NA -13.47 129.9    -7 6.5
3 DREDGE UNSPECIFIED 1993-10-01        NA         NA -13.47 130.0    -5 4.2
   sand gravel
1 96.4     NA
2 88.7    4.6
3 95.9     NA
```

The samples in sed1 can be quality controlled based on the *data QC criteria* developed previously (Li, Potter, Huang, Daniell, et al. 2010; Li, Potter, Huang, and Heap 2012b). In total, six data QC criteria are considered below to demonstrate how to conduct data QC.

1.2.1   Accuracy of location information

The *accuracy of location information* is important as certain spatial predictive methods are based on distance information, and location information (i.e., longitude and latitude) can also be used as predictive variables. Identification and removal of samples with less accurate location information ensure the *accuracy of spatial information*. Longitude and latitude with three digits after the decimal point are assumed to be accurate for the resolution used (i.e., 0.0025 degrees for *MARS* sediment data).

We will derive the digit number after the decimal point using the function decimaldigit in the spm2 package (Li 2021a) as below.

```
library(spm2)
sed1$lat.digit <- decimaldigit(sed1$lat)
sed1$long.digit <- decimaldigit(sed1$long)
```

Then we remove samples with two or less digits after the decimal point in either longitude or latitude.

```
sed1 <- subset(sed1, sed1$lat.digit >= 3 | sed1$long.digit >= 3)
sed1 <- sed1[, -c(11:12)]
dim(sed1)
```

```
[1] 374  10
```

This shows that sample size has been reduced from 385 to 374, that is, 11 samples have been removed.

1.2.2 Sampling methods

Data quality is often affected by the methods used to collect samples. Information of *sampling methods* for this example is stored in the column of `sample.type` as shown below.

```
sample.type <- as.data.frame(table(sed1$sample.type))
names(sample.type) <- c("Sample.type", "Frequency")
sample.type
```

```
            Sample.type Frequency
1                CORE BOX         5
2            CORE GRAVITY       130
3             DREDGE PIPE        45
4      DREDGE UNSPECIFIED        92
5             GRAB SHIPEK        31
6    GRAB SMITH MCINTYRE        68
7        GRAB UNSPECIFIED         3
```

Among these seven sampling methods, dredge methods (i.e., dredge pipe and dredge unspecified) are assumed to be unreliable, and samples collected using the methods are less accurate. So, we need to remove these dredged samples.

```
sed1 <- subset(sed1, sed1$sample.type != "DREDGE PIPE" & sed1$sample.type != "
    DREDGE UNSPECIFIED")
dim(sed1)
```

```
[1] 237  10
```

It is clear that the sample size has been reduced from 374 to 237, that is, in total, 137 samples were collected using dredge methods and have been removed.

1.2.3 Sample duplications at the same location

For distance-based prediction methods, samples are assumed to be collected from different locations (although the function `autoKrige` in the `automap` package can cope with duplicate observations (Hiemstra 2013)). We need to check if there are any *duplicates* at the same location as below.

```
sed1 <- sed1[order(sed1$mud, decreasing = TRUE),] # order the data set using mud
sed1$location.id <- paste(sed1$long, "&", sed1$lat) # create a location id
table(duplicated(sed1$location.id))
```

```
FALSE   TRUE
  230      7
```

It shows that there are seven locations with *duplicated samples*. Since the data set is sorted using mud variable in descending order, it ensures the samples with less mud content be labeled as duplicated. These duplicated samples need to be removed, and the duplicated samples with less mud content are removed below.

```
sed1$best_loc <- duplicated(sed1$location.id)
sed1 <- subset(sed1, best_loc == FALSE)
sed1 <- sed1[, -c(11:12)]
dim(sed1)
```

```
[1] 230  10
```

This shows that sample size has been reduced from 237 to 230, that is, seven duplicated samples have been removed.

1.2.4 Sample quality

Samples with a base depth more than 5 cm are assumed to be inaccurate and need to be excluded.

```
sed1 <- subset(sed1, base.depth <= 5)
dim(sed1)
```

```
[1] 125  10
```

This shows that sample size has been reduced from 230 to 125, that is, 105 samples with a base depth more than 5 cm have been removed.

1.2.5 Samples with missing values

Samples with *missing values* need to be identified and imputed. Here we identify and remove samples with missing values for sediment variables (i.e., mud, sand, and gravel) as below.

```
summary(sed1[, 8:10])
```

```
      mud             sand           gravel
Min.   : 3.0    Min.   : 1.0    Min.   :  1.0
1st Qu.:18.5    1st Qu.:37.8    1st Qu.:  6.0
Median :33.0    Median :56.0    Median : 10.0
Mean   :37.5    Mean   :52.0    Mean   : 12.7
3rd Qu.:49.0    3rd Qu.:67.0    3rd Qu.: 15.0
Max.   :99.0    Max.   :89.0    Max.   :100.0
NA's   :2       NA's   :1       NA's   :12
```

```
sed1 <- subset(sed1, mud >= 0 & sand >= 0 & gravel >= 0)
dim(sed1)
```

```
[1] 111  10
```

It shows that sample size has been reduced from 125 to 111, that is, 14 samples are with NA and have been removed.

In fact, with a further inspection, the sums of sediment for these 14 samples are all 100% if the NAs were replaced with 0s. This suggests that 0s were ignored for these samples in the MARS database, resulting in these missing values. If one wishes, it is pretty reasonable to include these 14 samples by simply replacing the NAs with 0s.

1.2.6 Data accuracy

1. Data range

For a certain data type, its values need to be within a certain *range*. For sediment data, values of all three sediment types (i.e., mud, sand, and gravel) need to be within 0 to 100% range as expected for percentage data. This has been confirmed from `summary(sed1[, 8:10])` above. The sums of these three sediment types should also be 100% exactly for each sample.

```
range(sed1$mud + sed1$sand + sed1$gravel)
```

```
[1]   99 102
```

```
table(sed1$mud + sed1$sand + sed1$gravel)
```

```
 99 100 102
  1 109   1
```

This shows that: (1) the sum of three sediment types are 100% for 109 samples, and (2) for the remaining two samples their sum is not 100%, but close to 100%. The differences could result from rounding errors in their recordings, so no further samples are removed.

2. Data limits

For marine samples, bathymetry data are stored as negative, while for terrestrial samples elevation data usually are positive. Samples with positive bathymetry are assumed to be incorrect and need to be removed.

```
sed1b <- subset(sed1, bathy <= 0)
dim(sed1b)
```

```
[1]   0 10
```

This suggests that no sample is selected. This is because bathymetry data are actually missing for all these remaining samples, which can be shown from `summary(sed1)` or simply from `sed1$bathy`. Since we can derive `bathy` data for these samples from `bathy` data for Australian EEZ as shown in Appendix A, it would not prevent the selection of these samples.

```
sed1.1 <- sed1[, -c(1:4)]
names(sed1.1); dim(sed1.1)
```

```
[1] "lat"    "long"    "bathy"   "mud"     "sand"    "gravel"
```

```
[1] 111   6
```

In total, there are 111 samples remaining after the data QC process. In comparison with the `pps.df` data set that was generated back in 2012 (Li, Potter, Huang, Daniell, et al. 2010; Li, Potter, Huang, and Heap 2012b), there are three fewer samples in this data set. This could be due to sample removal resulted from data QC process for $MARS$ database since 2012.

This example intends to provide some clues to how to QC a data set at hand. Sometimes, data noises may be resulted from repeated measures, and certain rules may need to be developed to clean such samples based on professional knowledge (e.g., Li et al. 2009).

Furthermore, exploratory analysis can be used to further detect abnormal samples as detailed in the next chapter.

1.3 Spatial data types and spatial reference systems

1.3.1 Spatial data types

To generate spatial predictions using spatial predictive models for an area, two *types of data* are required: (1) the *point data* (or *point samples*) of response and predictive variables, and (2) the *grid data* of predictive variables with location information for the area where the predictions need to be produced. Both data sets should have location information for spatial predictive modeling. Such location information needs to be geo-referenced and is often projected using various *spatial reference systems* for relevant purposes (Jiang and Li 2014).

1. Point data of response and predictive variables

We use two point data sets `petrel` and `sponge` in the `spm` package as examples below.

```
library(spm)

data(petrel)
class(petrel)
```

```
[1] "data.frame"
```

```
head(petrel, 3)
```

```
    long    lat   mud sand gravel bathy    dist relief   slope
1 130.5 -11.18 97.45 2.22   0.33   -23 0.07778      1 0.03667
2 130.3 -10.98 97.34 2.62   0.04   -44 0.19962      1 0.00000
3 132.0 -10.90 96.86 3.14   0.00   -41 0.25106      1 0.03667
```

```
petrel[1:2, ]
```

```
    long    lat   mud sand gravel bathy    dist relief   slope
1 130.5 -11.18 97.45 2.22   0.33   -23 0.07778      1 0.03667
2 130.3 -10.98 97.34 2.62   0.04   -44 0.19962      1 0.00000
```

```
data(sponge)
class(sponge)
```

```
[1] "data.frame"
```

```
head(sponge, 3)
```

```
  easting northing sponge     tpi3    var7 entro7   bs34   bs11
1  591964  8713576      8 -0.08533 0.09163 0.6542 -10.46  -9.69
2  589988  8718310      9  0.05222 0.11578 0.8445 -18.22 -14.51
3  592676  8724080      1 -0.24756 0.27239 1.3153 -22.00 -18.09
```

The samples are point data and stored together with their associated location information. The location information in the `petrel` data set is `longitude` and `latitude`, while the location information in the `sponge` data set is `easting` and `northing`.

2. Grid data of predictive variables with location information

We use two grid data sets, `petrel.grid` and `sponge.grid`, in the `spm` package as examples below.

```
data(petrel.grid)
class(petrel.grid)
```

```
[1] "data.frame"

petrel.grid[1:2, ]

         long    lat bathy   dist relief  slope
470277  128.8  -10.6   -67  1.612      5 0.2319
470278  128.8  -10.6   -67  1.610      3 0.1512

data(sponge.grid)
class(sponge.grid)

[1] "data.frame"

sponge.grid[1:2, ]

       easting northing     tpi3  var7 entro7    bs34    bs11
56205   276028  8745720  0.04556 4.651  3.395  -35.80  -30.62
56206   276028  8745730  0.13556 4.162  3.374  -35.22  -30.03
```

The examples show that location information in the `petrel.grid` data set is `longitude` and `latitude`, while the location information in the `sponge.grid` data set is `easting` and `northing`.

The spatial reference system for the location information can be changed when required as demonstrated below.

1.3.2   Spatial reference systems

There are many *spatial reference systems* (Jiang and Li 2014) or *coordinate reference systems* (Bivand, Keitt, and Rowlingson 2019). The spatial references are in a number of formats that can be found at *Spatial Reference* [https://spatialreference.org/]. For spatial predictive modeling, the spatial information or location information is often stored using *WGS84* or *utm zones*. For example, the spatial information for data sets `petrel` and `petrel.grid` in the `spm` package are stored in *WGS84*, while the spatial information in the `sponge` and `sponge.grid` data sets in the `spm` package is projected in *utm* zone 52 south. Sometimes, we may need to reproject the location information from one spatial reference system to another one.

Here we use the `sponge` and `petrel.grid` data sets to demonstrate how to reproject the location information using the function `spTransform` in the `rgdal` package (Bivand, Keitt, and Rowlingson 2019; Pebesma and Bivand 2020) and the function `st_transform` in the `sf` package (Pebesma 2018).

1. Reprojection of point data from *utm* zones to *WGS84*

Function `spTransform`

The spatial information, `easting` and `northing`, in the `sponge` data set is stored in *utm* zone 52 south. Since the `sponge` data set is in dataframe format, it needs to be converted to `SpatialPoints` format with its associated coordinate reference system prior to reprojecting as below (Bivand, Keitt, and Rowlingson 2019; Pebesma and Bivand 2020; Hijmans 2020).

```
library(sp)
library(raster)
library(rgdal)
```

Given that the data format needs to be changed from dataframe to SpatialPoints format as below, it is a good idea to reassign the dataframe `sponge` to a different object `spng` that will be reformatted. Thus the format of `sponge` will remain unchanged for future use.

```
spng <- sponge

spng[1:2, ]
```

```
    easting northing sponge     tpi3    var7 entro7   bs34   bs11
1   591964 8713576       8 -0.08533 0.09163 0.6542 -10.46  -9.69
2   589988 8718310       9  0.05222 0.11578 0.8445 -18.22 -14.51
```

```
coordinates(spng) = ~ easting + northing # set spatial coordinates to create
    SpatialPoints data
proj4string(spng) <- CRS("+proj=utm +zone=52 +south +units=m +no_defs +ellps=WGS84
    +towgs84=0,0,0")
class(spng)
```

```
[1] "SpatialPointsDataFrame"
attr(,"package")
[1] "sp"
```

```
crs(spng)
```

```
CRS arguments:
 +proj=utm +zone=52 +south +ellps=WGS84 +units=m +no_defs
```

Now the `spng` data set can be reprojected from easting and northing in *utm* zone 52 south to longitude and latitude in *WGS84* using `spTransform` as demonstrated below.

```
spng.wgs84 <- spTransform(spng, CRS("+proj=longlat +datum=WGS84 +no_defs +ellps=
    WGS84 +towgs84=0,0,0"))
class(spng.wgs84)
```

```
[1] "SpatialPointsDataFrame"
attr(,"package")
[1] "sp"
```

```
crs(spng.wgs84)
```

```
CRS arguments: +proj=longlat +datum=WGS84 +no_defs
```

```
spng.wgs84 <- as.data.frame(spng.wgs84)
names(spng.wgs84) <- c("sponge", "tpi3", "var7", "entro7", "bs34", "bs11", "long",
    "lat" )
spng.wgs84[1:2, ]
```

```
  sponge     tpi3    var7 entro7   bs34   bs11  long    lat
1      8 -0.08533 0.09163 0.6542 -10.46  -9.69 129.8 -11.64
2      9  0.05222 0.11578 0.8445 -18.22 -14.51 129.8 -11.59
```

```
# windows()
# spplot(spng.wgs84, "sponge")
write.csv(spng.wgs84, "./data/spongelonglat.csv", row.names = FALSE)
```

Function `st_transform`

Alternatively, the `spng` data set can be reprojected from `easting` and `northing` in *utm* zone 52 south to `longitude` and `latitude` in WGS84 using `st_transform` as shown below.

```
library(sf)
spng2 <- st_as_sf(sponge, coords=c("easting", "northing"))
st_crs(spng2) <- 32752
# crs(spng2)
spng.wgs84.2 <- st_transform(spng2, st_crs(4326))
class(spng.wgs84.2)
```

```
[1] "sf"          "data.frame"

crs(spng.wgs84.2)

CRS arguments: +proj=longlat +datum=WGS84 +no_defs
```

For spatial predictive modeling, coordinate information may be used as predictive variable(s). Thus `spng.wgs84.2` needs to be converted into a dataframe format as follows.

```
spng.wgs84.3 <- cbind((st_coordinates(spng.wgs84.2)), st_set_geometry(spng.wgs84
    .2, NULL))
class(spng.wgs84.3)

[1] "data.frame"

spng.wgs84.3[1:2, ]

        X      Y sponge    tpi3    var7 entro7    bs34    bs11
1 129.8 -11.64      8 -0.08533 0.09163 0.6542  -10.46   -9.69
2 129.8 -11.59      9  0.05222 0.11578 0.8445  -18.22  -14.51
```

2. Reprojection of grid data from WGS84 to utm

The spatial information, `longitude` and `latitude`, in `petrel.grid` is stored in *WGS84*. Prior to reprojecting, the `petrel.grid` data set that is in dataframe format needs to be converted to SpatialPoints format with its associated coordinate reference system.

We reassign the dataframe `petrel.grid` to a different object `pg` that will be reformatted to SpatialPoints data.

```
pg <- petrel.grid

pg[1:2, ]

          long   lat bathy  dist relief  slope
470277   128.8 -10.6   -67 1.612      5 0.2319
470278   128.8 -10.6   -67 1.610      3 0.1512

coordinates(pg) = ~ long + lat # set to create SpatialPoints data
proj4string(pg) <- CRS("+proj=longlat +datum=WGS84 +no_defs +ellps=WGS84 +towgs84
    =0,0,0") # assign coordinate reference system
class(pg); crs(pg)

[1] "SpatialPointsDataFrame"
attr(,"package")
[1] "sp"

CRS arguments: +proj=longlat +datum=WGS84 +no_defs
```

Now the `pg` data set can be reprojected from `longitude` and `latitude` in *WGS84* to `easting` and `northing` in *utm* zone 52 south as below.

```
pg.utm52s <- spTransform(pg, CRS("+proj=utm +zone=52 +south +units=m +no_defs +
    ellps=WGS84 +towgs84=0,0,0"))
class(pg.utm52s); crs(pg.utm52s)

[1] "SpatialPointsDataFrame"
attr(,"package")
[1] "sp"
```

```
CRS arguments:
 +proj=utm +zone=52 +south +ellps=WGS84 +units=m +no_defs

pg.utm52s.2 <- as.data.frame(pg.utm52s)
names(pg.utm52s.2)<- c("bathy", "dist", "relief", "slope", "easting", "northing")
pg.utm52s.2[1:2, ]


        bathy  dist relief   slope easting northing
470277    -67 1.612      5  0.2319  478366  8827977
470278    -67 1.610      3  0.1512  478640  8827977
```

Similar to the reprojection of point data spng above, st_transform can be used to reproject the grid data set from *WGS84* to *utm* zone 52 south.

3. Selection of spatial reference systems

Spatial reference system used to project spatial information is often assumed to have certain effects on the performance of predictive models, thus in practice various spatial reference systems have been developed to minimize such effects (Jiang and Li 2014). For spatial predictive modeling, a spatial reference system that can minimize distortion in distance over space is ideal. When a study area is relatively small and located within one *utm* zone, spatial data are often projected using the *utm* zone or an appropriate projection system. When the study area is spanning over two or more *utm* zones, the existing geographic coordinate system (i.e., *WGS84*) could be used. This is because the effects of spatial reference systems on the predictive accuracy of spatial predictive methods (i.e., *IDW* and *OK*) could be negligible for areas at various latitudinal locations (up to 70 decimal degrees) and spatial scales (Jiang and Li 2013, 2014; Turner, Li, and Jiang 2017). Hence, without reprojecting the spatial data (most likely it is in *WGS84*), the spatial data can be used for spatial predictive modeling for areas with latitude less than 70 decimal degrees.

Furthermore, since the predictive accuracy is the key for predictive modeling, an optimal spatial reference system should be selected to maximize predictive accuracy. The *spatial reference system* that can minimize the distortion in distance should be identified and used. The *selection* of an optimal spatial reference system can also be determined based on its effect on predictive accuracy for relevant predictive method . This can be achieved with cross-validation function that is to be introduced later for relevant predictive methods in this book.

2

Predictive variables and exploratory analysis

This chapter introduces *data preparation* of *predictive variables* and *exploratory analysis* for predictive relevant methods, including (1) principles for pre-selection of predictive variables and limitations, (2) predictive variables, and (3) role and limitations of exploratory analysis in variable pre-selection.

2.1 Principles for pre-selection of predictive variables and limitations

2.1.1 Principles

To *pre-select predictive variables*, certain principles need to be followed. The *principles* may change with relevant disciplines. For spatial predictive modeling, predictive variables should be pre-selected based on professional knowledge and the following principles (Austin 2007; Elith and Leathwick 2009; Li 2019a).

1. Causal variables

Causal variables are variables that are directly causing changes in response variable. For instance, in terrestrial environmental sciences, causal variables may include temperature, precipitation, soil nutrients, and so on for plant species and vegetation types. Predictive variables should be causal variables if they can be identified and available.

2. Variables directly caused by response variable

Variables directly caused by response variable are useful predictive variables. Such predictive variables are informative because any changes in these variables are directly resulted from the changes in the response variable. For instance, in environmental sciences, variables directly caused by response variable may include: (1) optical reflectance related variables for vegetation types, and (2) acoustic backscatters and their derived variables for seabed substrates.

3. Correlated variables

Correlated variables are variables that should be closely related to the response variable, although they may not be the causal variables nor the variables directly caused by response variable. Correlated variables may also be useful predictive variables.

2.1.2 Availability of causal variables

Causal variables are sometimes hard to identify or are even unknown; and the information of causal variables may not be available even if they are known. In such cases, proxy variables

DOI: 10.1201/9781003091776-2

(i.e., *surrogate variables*) are used instead for spatial predictive modeling. *Proxy variables* are usually variables directly caused by response variable and/or correlated variables. They can be identified based on expert or professional knowledge (e.g., McArthur et al. 2010). Certainly, predictive models can use causal variables, proxy variables, or both if causal variables are not all available.

2.1.3 Hidden predictive variables

For a predictive model, its accuracy is a key criterion for variable selection and parameter optimization and is critical for subsequent spatial predictions. When the accuracy of a predictive model resulted is unexpectedly low, and if a right predictive method has been used, it may indicate that certain important predictive variables may have been missed (i.e., *hidden variables*), and for which we may have no knowledge or even awareness (Huston 1997). Further actions are then required to identify such hidden predictive variables and relevant professional knowledge pool needs to be expanded for such actions.

2.1.4 Limitations

For spatial predictive modeling, how to select potential predictive variables can be *limited or constrained by certain factors*. For example, (1) they need to be continuously available for the entire area to be predicted; (2) spatial resolution of various predictive variables needs to meet desired resolution for the final predictions, although they can be re-scaled or aggregated. Sometimes, even though we know certain possible predictive variables, they may not meet these requirements and cannot be used for spatial predictive modeling. This is particularly true in mountainous and deep sea areas for spatial predictive modeling in the environmental sciences.

2.2 Predictive variables

Spatial predictive modeling is used in various disciplines, and *predictive variables* used for modeling may change with relevant disciplines. We take the predictive variables used for spatial predictive modeling in terrestrial and marine environmental sciences as examples below.

2.2.1 Predictive variables in terrestrial environmental sciences

For *terrestrial environmental modeling*, a number of predictive variables are available. Many previous applications provide examples of variables being used for spatial predictive modeling in terrestrial environmental sciences (Elith et al. 2006; Austin 2007; Li et al. 2009; Arthur et al. 2010; Sanabria et al. 2013; Appelhans et al. 2015; Hengl et al. 2015; Seo, Kim, and Singh 2015; Zhang et al. 2017). These variables largely fall into the following seven groups:

1. Location data

Location data refer to the coordinates of point samples and grid data. It is usually presented in `longitude` and `latitude` in *WGS84* or `easting` and `northing` in *utm*.

2. Climatic variables

Climatic variables may include temperature, precipitation, wind speed, and various derived variables (e.g., humidity, seasonality).

3. Topographical variables

Topographical variables may include elevation, slope, aspect, distance-to-sea, and relevant derived variables such as topographic position index.

4. Optical remote sensing data

Optical remote sensing data may include various reflectance bands and relevant derived variables (e.g., NDVI, EVI).

5. Vegetation information

Vegetation information may include variables like vegetation types, abundance, and coverage.

6. Substrate data

Substrate data may include soil type, soil nutrients, organic matter, and soil moisture.

7. Disturbance information

Disturbance information may include grazing, fertilization, burning, and so on.

2.2.2 Predictive variables in marine environmental sciences

For *marine environmental modeling*, the information of predictive variables is often scarce, especially for large areas (Li 2019a). In many cases, proxy variables are used for predictive modeling (McArthur et al. 2010; Miller et al. 2016). For small areas, quite often more predictive variables become available at a desired resolution such as those summarized in previous studies (Li 2019a; Diesing, Thorsnes, and Bjarnadóttir 2021). These variables largely fall into the following five groups:

1. Location data

The *location data* are the same as those for terrestrial environmental modeling.

2. Oceanographic variables

Oceanographic variables may include sea surface temperature, light availability, current velocity, suspended particulate matter, sea surface primary production, salinity, oxygen concentration, and bottom sheer stress.

3. Topographical variables

The *topographical variables* are similar to those for terrestrial environmental modeling but under water. They may include geomorphic features, bathymetry, relief, curvatures, distance-to-coast, and bathymetric position index.

4. Acoustic remote sensing variables

Acoustic remote sensing variables may include *backscatter data* and various derived variables (e.g., roughness, p-rock).

5. Seabed substrate variables

Seabed substrate variables may include seabed hardness, mud content, gravel content, carbonate, and so forth.

2.3 Exploratory analysis

Exploratory analysis for spatial predictive modeling may change with the predictive methods used. According to Li (2019a), the methods can be classified into three groups: (1) non-machine learning methods, (2) machine learning methods, and (3) hybrid methods. We are going to show relevant exploratory analysis for each group below.

2.3.1 Exploratory analysis for non-machine learning methods

For *non-machine learning methods*, such as *GLM*, exploratory analysis is often used to detect the relationships between response variable and predictive variables. By applying such analysis, one intends to find data nature and structure (Zuur, Leno, and Elphick 2010), and then certain actions can be taken to deal with relevant samples or variables. Such exploratory analysis is also used to identify various issues relating to data nature and structure for spatial predictive modeling (Li 2019a). The following examples are for how to identify and deal with some of these issues.

1. Outliers

An *outlier* is an observed value that lies at an abnormal distance from other sample values in a random sample. Outliers are usually determined by plotting relevant variables (e.g., scatter plots, box plots) (Zuur, Leno, and Elphick 2010) as shown in Figure 2.1 by

```
par(font.axis = 2, font.lab = 2)
hbee1<-read.csv("./data/bee1sub.csv", sep = ",", header = T)
hbee <- hbee1$hbee
boxplot(hbee, xlab = "Number of honey bees", horizontal = T)
```

This is based on `hbee1` data set from a previous study (Arthur et al. 2010, 2020).

Then some outliers identified could be excluded from further modeling and analysis.

Although identification of outliers is important for predictive modeling, the outliers identified may be caused by certain conditions (e.g., an optimal environmental condition) (Arthur et al. 2010, 2020). Thus they could be false outliers as shown in Figure 2.2 (Li 2008; Arthur et al. 2010) for model `glmmpql1` after fitting a `glmmPQL` model below.

```
library(MASS)
glmmpql1 <- glmmPQL(hbee ~ inf + c500 + I(inf^2) + I(links300^2) + inf:I(links300
    ^2), random = ~1|paddock/plot, data = hbee1, family = quasi(var = "mu^2", link
    = "log"), maxit = 1000)

par(font.axis = 2, font.lab = 2, las = 1)
plot(hbee, predict(glmmpql1, type = "response"), xlab = "Observed values", ylab =
    "Fitted values")
lines(hbee, hbee)
```

It is apparent that one outlier identified in Figure 2.1 (i.e., the observation with the maximal bee count number) could be no longer classified as an outlier in Figure 2.2.

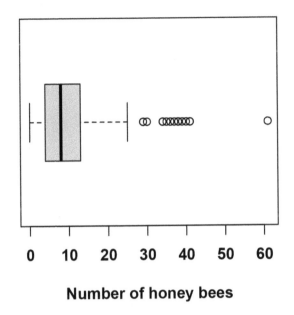

Number of honey bees

FIGURE 2.1: Boxplot of the outliers of honey bee count data.

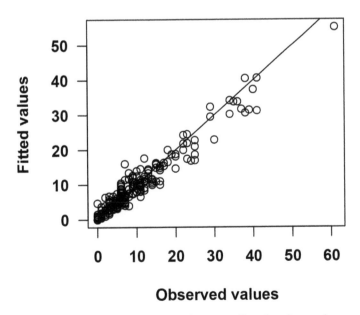

Observed values

FIGURE 2.2: Observed honey bee count data vs. fitted values of an optimal model: showing a false outlier (i.e., the sample with a count value > 60).

Outliers may also change with predictive models developed, that is, a false outlier could also be produced by a sub-optimal model, e.g, model `glmmpql2` below.

```
glmmpql2<- glmmPQL(hbee ~ inf + c500 + w2000 + w300 + I(inf^2) + I(c500^2) + I(
    w2000^2) + I(w300^2) + inf:w2000 + inf:w300 + c500:w2000 + c500:w300 + I(inf
    ^2):w2000 + I(inf^2):w300, data = hbee1, random = ~ 1|paddock/plot, family =
    quasi(link = log, var = mu^2), maxit = 1000)
```

The results are shown in Figure 2.3 (Li 2008; Arthur et al. 2010) by

```
par(font.axis = 2, font.lab = 2, las = 1)
plot(hbee, predict(glmmpql2, type = "response"), xlab = "Observed values", ylab =
    "Fitted values")
lines(hbee, hbee)
```

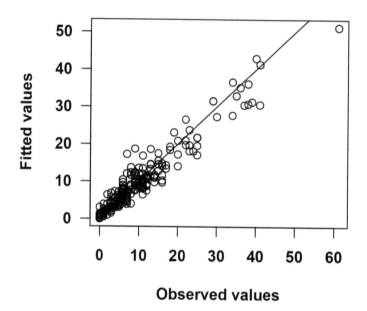

FIGURE 2.3: Observed honey bee count data vs. fitted values of a sub-optimal model: showing an outlier (i.e., the observation with a count value > 60).

It is obvious that the observation with the maximal bee count number becomes an outlier. Therefore, caution should be taken in dealing with outliers.

2. Homogeneity of variance

Variance of response variable (or depend variable) can be either *homogeneity* (*homoscedasticity*) or *heterogeneity* (*heteroscedasticity*). For spatial predictive modeling using regression methods (e.g., GLM), we need to consider the variance of response variable in relation with its mean as shown in Figure 2.4 that is based on `hbee1` data set by

```
mu <- with(hbee1, tapply(hbee, list(paddock, obs), mean))
vars <- with(hbee1, tapply(hbee, list(paddock, obs), var))

par(mfrow = c(1,2), font.axis = 2, font.lab = 2)
plot(mu, vars)
plot(mu, sqrt(vars))
```

Apparently, the variance changes with sample mean and is heterogeneous. This can be further examined based on the residuals, for details see (McCullagh and Nelder 1999; Crawley 2007).

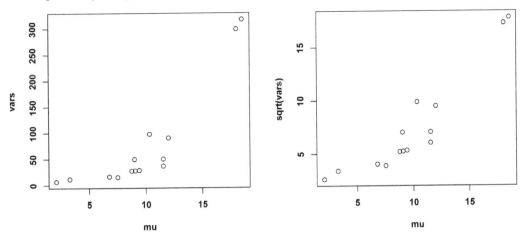

FIGURE 2.4: Relationship between variance (i.e., `vars`) and mean (i.e., `mu`) of honey bee count data: (left) variance, and (right) square-root transformed variance.

When variance is heterogeneous, relevant data transformation should be taken, or methods that can cope with variance heterogeneity should be used as demonstrated in Chapter 7.

3. Data distribution

For spatial predictive modeling, response variables can be in any *data types*, including continuous numerical data, count data, percentage data, binary data, or even categorical data. *Data distribution* of response variable changes with data types. Even for the same data type the distribution may also change. For geostatistical methods (e.g., *OK*, *KED*) and linear models (*LM*), data are assumed to be *normally distributed*.

Data normality of response variables needs to be examined prior to modeling. If data are non-normal, then data transformation is required or predictive methods that can deal with non-normality should be used. Data normality can be examined using such as QQ plots (Figure 2.5) or `histgram` of the response variable and its various transformed data (Figure 2.6), which are based on `gravel` in the `petrel` data set in the `spm` package (Li 2019b) and produced by

```
library(spm)
data(petrel)
par(font.axis = 2, font.lab = 2)
gravel <- petrel$gravel
qqnorm(gravel, main = NULL)
qqline(gravel)

par(mfrow=c(2,2), font.axis = 2, font.lab = 2)
hist(gravel, breaks=c(seq(0, 100, 5)), main = NULL)
hist(sqrt(gravel), main = NULL)
hist(log(gravel+1), main = NULL)
hist(asin(sqrt(gravel/100)), main = NULL)
```

It is obvious that the data are non-normal, and the `log` transformation is better than other transformations considered.

Data transformation can also be chosen based on its effect on predictive accuracy using cross-validation functions, such as `krigecv` in the `spm2` package (Li 2021a), as addressed in the later chapters.

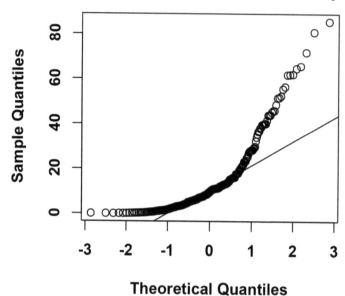

FIGURE 2.5: Normal Q-Q plot of gravel percentage data for the Petrel area.

4. Collinearity

Collinearity is about the correlations among predictive variables. It is usually determined based on variance inflation factor (*VIF*) or correlation coefficient (r) as shown in Table 2.1 that is based on the spm data set in the spm package.

```
library(spm)
data(sponge)
cor1 <- round(cor(sponge[, -3], method = "spearman"), 2)
```

In this example, no collinearity is identified. When a strong collinearity is identified between two predictive variables, one of the predictive variables is often eliminated to reduce the collinearity and also reduce computational time (Dormann et al. 2013), although caution should be taken for taking this exercise (Harrell Jr 2001; O'Brien 2007; Kuhn and Johnson 2013; Li, Alvarez, et al. 2017). If a predictive method used can deal with collinearity (e.g., *RF*), it would be better to let the variable selection process determine which variables should be removed.

5. Response curve of response variable to predictive variables

Relationships of response variable and predictive variables need to be considered to determine the *response curves* of response variable to predictive variables. The relationships can be visualized by plotting response variable against predictive variables using the lowess function in the stats package (R Core Team 2020). The sponge data set in the spm package is used as an example below.

```
library(spm)
data(sponge)
s1 <- sponge[, c(1, 2, 4:8, 3)]
par(mfrow=c(3,3), font.axis = 2, font.lab = 2)

for (i in 1:7) {
  plot(s1[, i], s1[, 8], xlab = names(s1)[i], ylab = "Species number")
  lines(lowess(s1[, 8] ~ s1[, i]), col = "blue")
}
```

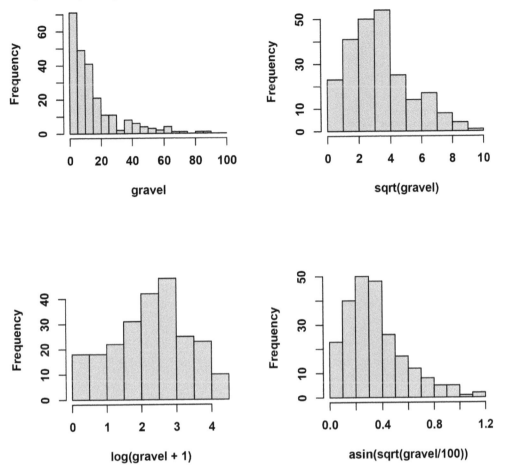

FIGURE 2.6: Distribution of: gravel, square-root transformed gravel, log transformed gravel, and arcsine transformed gravel.

TABLE 2.1: Relationships between predictive variables based on spearman correlation.

	easting	northing	tpi3	var7	entro7	bs34
easting	1.00	-0.49	-0.14	-0.30	-0.31	0.50
northing	-0.49	1.00	0.04	-0.08	-0.10	-0.25
tpi3	-0.14	0.04	1.00	0.19	0.16	-0.10
var7	-0.30	-0.08	0.19	1.00	0.98	-0.62
entro7	-0.31	-0.10	0.16	0.98	1.00	-0.60
bs34	0.50	-0.25	-0.10	-0.62	-0.60	1.00
bs11	0.54	-0.23	-0.06	-0.67	-0.65	0.95

It is apparent that for some predictors (e.g., northing and tpi3), the response curves are not linear (Figure 2.7). If a non-linear relationship is detected, relevant predictive variable may need to be specified to its second or third orders in a predictive model to capture the non-

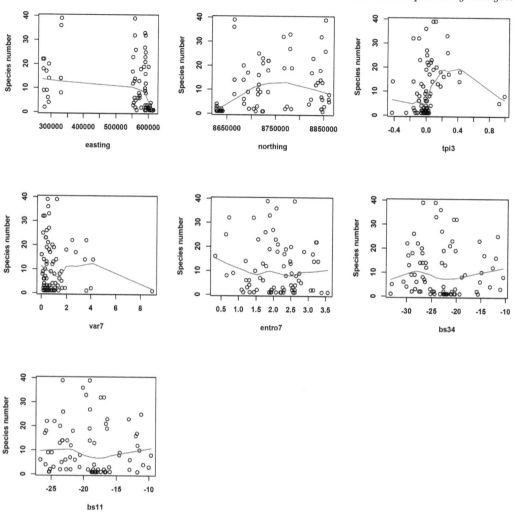

FIGURE 2.7: Relationships of species number and seven predictive variables based on the `lowess` function (blue line).

linear relationship, or a method that can automatically handle the non-linear relationship should be used (e.g., *RF*).

6. Correlations between response variable and predictive variables

The relationships between response variable and predictive variables are usually determined based on *correlation* coefficient (r). When the correlation of a predictive variable with the response variable is low, the common practice is to remove such predictive variable prior to model development. However, it would be wise to let the variable selection process determine which variables should be removed because some variables may be important predictive variables even with low correlation coefficients. That is, such correlation analysis is only indicative because it is based on the raw data instead of residuals; and the correlations based on the raw data are not necessarily the same as those based on the residuals.

7. Interactions

Interactions among predictive variables may exist and need to be identified and considered in predictive modeling. The interactions can be identified based on professional knowledge and also on visualizations with functions such as xyplot in the lattice package (Sarkar 2008), interaction.plot in the stats package, and coplot in the graphics package (R Core Team 2020). We will use coplot to show interactions of predictive variables in the hbee1 and sponge data sets.

Interactive effects of links300 and inf on honey bee number based on the hbee1 data set are shown in Figure 2.8 by

```
library(graphics)
par(font.axis = 2, font.lab = 2)
coplot(hbee ~ links300 | inf, number = 6, data = hbee1)
```

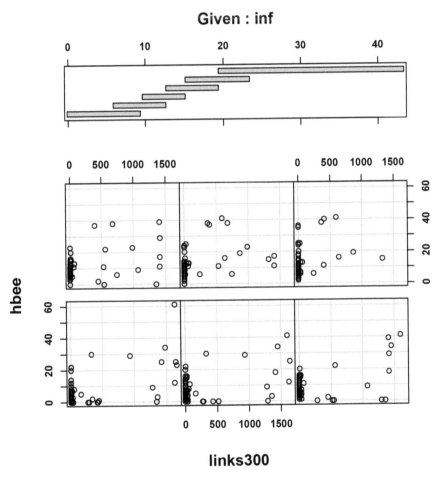

FIGURE 2.8: Interactive effects of links300 and inf on honey bee number.

In Figure 2.8, inf is divided into six intervals and the relationship between hbee and links300 changes with each of the intervals. It suggests that interactive effects of links300 and inf on honey bee number do exist.

Interactive effects of northing and easting on the sponge species number based on the sponge data set are shown in Figure 2.9 by

```
par(font.axis = 2, font.lab = 2)
coplot(sponge ~ northing | easting, data = sponge, number = 6)
```

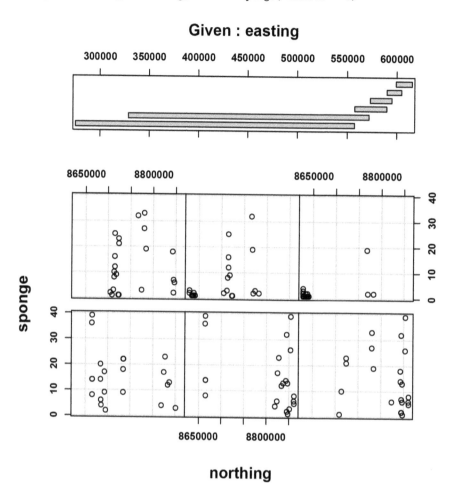

FIGURE 2.9: Interactive effects of northing and easting on sponge species number.

In Figure 2.9, easting is also divided into six intervals and the relationship between sponge and northing changes with each of the intervals. It is apparent that interactive effects of northing and easting on sponge species number also exist.

The interactions detected should be considered in predictive modeling. The interactions between latitude and longitude are used in *universal kriging (UK)* (see Section 5.4 for details) to smooth the predictions, without testing. For predictive modeling, the inclusion of interactions in a model can be tested according to their contributions to the predictive accuracy using variable selection methods that are to be introduced in the later chapters on predictive methods.

8. Other issues

There are still some other issues that are usually considered in modeling work. They may include: (1) *random errors*, (2) *spatial independence* of response variable, and (3) *temporal independence* of response variable. For spatial predictive modeling, issues like sources of random errors can be directly considered as predictive variables in the predictive modeling work. If spatial independence exists, geostatistical methods should be used, or relevant spatial information can be considered directly as predictive variables. If temporal changes exist, relevant temporal variable(s) should be considered or methods that can deal with spatio-temporal data (e.g., Hengl et al. 2008; Bakar and Sahu 2015; Gräler, Pebesma, and Heuvelink 2016) should be used.

2.3.2 Exploratory analysis for machine learning methods

For *machine learning methods*, exploratory analysis is useful for modelers to understand the data, thus providing some useful information to interpret the modeling results (Li, Siwabessy, Tran, et al. 2014). However, some of above functions of exploratory analysis for modern statistical methods are no longer required for certain machine learning methods. This is because some machine learning methods, such as *RF*, are free of assumptions on data distribution and can handle non-linear relationship and interactive effects (Cutler et al. 2007; Li, Heap, Potter, and Daniell 2011a). They can also handle highly correlated predictive variables predictors (Li 2013c; Li, Alvarez, et al. 2017). Furthermore, highly correlated predictors are encouraged to be used for *RF* because they might make meaningful contributions to improving the accuracy of predictive models (Li, Alvarez, et al. 2017).

2.3.3 Exploratory analysis for hybrid methods

For *hybrid methods*, some roles of exploratory analysis can be as useful as for the aforementioned methods. This is because if kriging methods are applied to the residuals of a detrending method (e.g., *GLM*, *RF*), the residuals are assumed to be normal. Thus the residuals need to be analyzed to check if relevant assumptions are satisfied. If not satisfied, the residuals need to be transformed for the hybrid methods. Again, the transformation of residuals can be determined according to its effect on predictive accuracy using relevant cross-validations that will be introduced in the later chapters.

3

Model evaluation and validation

For spatial predictive modeling, the ultimate purpose is to generate spatial predictions for a target area by developing and applying an accurate predictive model. The predictive accuracy of the model is critical, as it determines the quality of the resulted predictions that form the base for further scientific activities and/or decision-making. Therefore, how to correctly evaluate the accuracy of predictive models is vitally important. This chapter introduces various accuracy and error measures that are used to conduct the *evaluation of predictive models* for numerical and categorical data. It will also introduce various *model validation* methods, and techniques to handle randomness associated with cross-validation methods.

3.1 Predictive errors, observational errors, and true predictive errors

Predictive accuracy is a measure to assess how accurate are the predictions generated by a predictive model. For a dependent variable, the predictions should be as close as possible to the true values of the variable. Prior to assessing the accuracy, we need to introduce the concepts of observed and predicted values.

3.1.1 Observed values and predicted values

Observed values are the values of validation samples that are to be used to assess the performance of a predictive model. Since there are errors inevitably associated with the process acquiring observed values, observed values are often not true values. Thus for a given observed value, it may be different from its corresponding true value. The difference between true value (τ_i) and observed value (y_i), $\tau_i - y_i$, is the error associated with the observed value, which is referred to as *observational error* (ϵ_{oi}) (Li 2019a).

The *observational error* ϵ_{oi} is the sum of random error associated with observed dependent variable, and sampling and measuring errors (Li 2019a). The sampling and measuring errors result from various factors, including sampling design, the position accuracy of survey vessel, equipment used for sample collection, field operation, sample storage, sample processing procedure and analysis in laboratory, and data entry, that may affect the accuracy of observation and change with the variable observed as shown using seabed sediment data as an example in Chapter 1. However, how much error can each of these factors contribute to the sampling and measuring errors is unknown in most cases. Since the true values are often unknown, we have to use observed values to assess the predictive accuracy in practice, but these errors should be taken into account when using the predictive accuracy to assess the quality of spatial predictions resulted, where professional knowledge plays its roles.

DOI: 10.1201/9781003091776-3

Predicted values can refer to different information, and consequently, the predictive accuracy resulted can refer to different concepts (Li 2016, 2017). In this book, the *predicted values* are referring to the values predicted via predictive models for non-training samples, that is, new samples or validation samples.

3.1.2   Relationships of predictive error with observational error and true predictive error

Theoretically, the *predictive error* should be the difference between the predicted value ($\widehat{y_i}$) for and the true value of a non-training sample, which we refer to as *true predictive error* (ϵ_{tpi}). It can be expressed as: $\epsilon_{tpi} = \tau_i - \widehat{y_i}$. Because the true values are mostly unknown, the observed values are used to assess the performance of predictive models. The *predictive error* resulted is: $\epsilon_{pi} = y_i - \widehat{y_i}$, used in various error and accuracy measures as detailed in the next section. It is necessary to emphasize that here the observed values are referring to the values of validation samples or new samples. We also need to clarify the difference between residuals and predictive errors because they are often confused in previous studies as reviewed in Li (2016). The residuals are the differences between the fitted values for and the observed values of training samples, while the predictive errors are the differences between the predicted values for and the observed values of validation samples or new samples.

The *predictive error* (ϵ_{pi}) can be represented in terms of the *observational error* (ϵ_{oi}) and the *true predictive error* (ϵ_{tpi}) as in Equation (3.1):

$$
\begin{aligned}
\epsilon_{pi} &= y_i - \widehat{y_i} \\
&= (\tau_i - \widehat{y_i}) - (\tau_i - y_i) \\
&= \epsilon_{tpi} - \epsilon_{oi}
\end{aligned}
\tag{3.1}
$$

To assess the predictive accuracy, we need to further explore the relationships of predictive error with observational error and true predictive error in relation to observed values, predicted values, and true values, so that we can have a better understanding of the predictive accuracy derived.

The relationships of predictive error with observational error and true predictive error are detailed in Table 3.1 for 13 possible scenarios that are derived from relative differences among observed value, predicted value, and true value (Li 2018b, 2019a).

For some model performance measures such as *mean absolute error (MAE)* and *mean square error (MSE)*, the *absolute predictive error*, the absolute value of the difference between predicted values and observed values, is used either explicitly or implicitly, that is, $\epsilon_{api} = |y_i - \widehat{y_i}|$. So the relationships of ϵ_{api} with observational error and true predictive error are also provided in Table 3.1 which shows that:

(1)   when observed value is equal to true value (i.e., scenarios 1, 2, and 3), *predictive error* is *true predictive error*, and *absolute predictive error* is absolute *true predictive error*;

(2)   when predicted value is equal to true value (i.e., scenarios 7 and 10), *true predictive error* should be zero, and *predictive error* either underestimates or overestimates *true predictive error* by one *observational error* for scenarios 7 and 10, respectively, and *absolute predictive error* overestimates *true predictive error* by one absolute *observational error*;

TABLE 3.1: Relationships of predictive error (ϵ_{pi}) and absolute predictive error (ϵ_{api}) with observational error (ϵ_{oi}) and true predictive error (ϵ_{tpi}) for 13 scenarios derived from relative differences among observed value (y_i), predicted value ($\widehat{y_i}$) and true value (τ_i).

Scenario	y_i vs. τ_i	$\widehat{y_i}$ vs. y_i and τ_i	ϵ_{pi}	ϵ_{api}	$\epsilon_{tpi} - \epsilon_{pi}$	$\epsilon_{tpi} - \epsilon_{api}$
1	$y_i = \tau_i$	$\widehat{y_i} = y_i$	0	0	0	0
2		$\widehat{y_i} < y_i$	$\epsilon_{tpi} > 0$	$|\epsilon_{tpi}|$	0	0
3		$\widehat{y_i} > y_i$	$\epsilon_{tpi} < 0$	$|\epsilon_{tpi}|$	0	0
4	$y_i < \tau_i$	$\widehat{y_i} < y_i$	$\epsilon_{tpi} - \epsilon_{oi} > 0$	$|\epsilon_{tpi}| - |\epsilon_{oi}|$	$-\epsilon_{oi}$	$-\epsilon_{oi}$
5		$\widehat{y_i} = y_i$	0	0	$-\epsilon_{oi}$	$-\epsilon_{oi}$
6		$\tau_i > \widehat{y_i} > y_i$	$\epsilon_{tpi} - \epsilon_{oi} < 0$	$||\epsilon_{tpi}| - |\epsilon_{oi}||$	$-\epsilon_{oi}$	$- < \epsilon_{oi}$ to $+ < \epsilon_{oi}$
7		$\widehat{y_i} = \tau_i$	$-\epsilon_{oi} < 0$	$|\epsilon_{oi}|$	$-\epsilon_{oi}$	$+\epsilon_{oi}$
8		$\widehat{y_i} > \tau_i$	$\epsilon_{tpi} - \epsilon_{oi} < 0$	$|\epsilon_{tpi}| + |\epsilon_{oi}|$	$+\epsilon_{oi}$	$+\epsilon_{oi}$
9	$y_i > \tau_i$	$\widehat{y_i} < \tau_i$	$\epsilon_{tpi} - \epsilon_{oi} > 0$	$|\epsilon_{tpi}| + |\epsilon_{oi}|$	$+|\epsilon_{oi}|$	$+|\epsilon_{oi}|$
10		$\widehat{y_i} = \tau_i$	$-\epsilon_{oi} > 0$	$|\epsilon_{oi}|$	$+|\epsilon_{oi}|$	$+|\epsilon_{oi}|$
11		$y_i > \widehat{y_i} > \tau_i$	$\epsilon_{tpi} - \epsilon_{oi} > 0$	$||\epsilon_{tpi}| - |\epsilon_{oi}||$	$+|\epsilon_{oi}|$	$+ < |\epsilon_{oi}|$ to $- < |\epsilon_{oi}|$
12		$\widehat{y_i} = y_i$	0	0	$+|\epsilon_{oi}|$	$-|\epsilon_{oi}|$
13		$\widehat{y_i} > y_i$	$\epsilon_{tpi} - \epsilon_{oi} < 0$	$|\epsilon_{tpi}| - |\epsilon_{oi}|$	$-|\epsilon_{oi}|$	$-|\epsilon_{oi}|$

(3) when predicted value is equal to observed value (i.e., scenarios 5 and 12), *predictive error* underestimates or overestimates *true predictive error* by one absolute *observational error* for scenarios 5 and 12, respectively, and *absolute predictive error* underestimates *true predictive error* by one absolute *observational error*;

(4) under scenarios 4, 6, and 13, *predictive error* underestimates *true predictive error* by one absolute *observational error*, while under remaining scenarios 8, 9, and 11, *predictive error* overestimates *true predictive error* by one absolute *observational error*; and

(5) *absolute predictive error* overestimates *true predictive error* by one absolute *observational error* under scenarios 8 and 9, underestimates by one absolute *observational error* under scenarios 4 and 13, and may be underestimated or overestimated by less than one *absolute observational error* under scenarios 6 and 11 depending on relative difference of predictive values with observed and true values.

In summary, of the 13 scenarios in Table 3.1, the *predictive error* (e.g., mean error (*ME*) in Table 3.2) is exactly the same as *true predictive error* under three scenarios, underestimates and overestimates *true predictive error* by one absolute *observational error* under five scenarios, respectively. The *absolute predictive error* is exactly the same as *true predictive error* under three scenarios, underestimates and overestimates *true predictive error* by one *absolute observational error* under four scenarios, respectively, and either underestimates or overestimates *true predictive error* by less than one *observational error* under two scenarios. The number of underestimation scenarios and the number of overestimation scenarios are the same, and the quantity of the underestimation, and the quantity of the overestimation are also the same. Hence, if *observational error* is randomly distributed with a mean of zero, then *predictive error* and *absolute predictive error* largely represent *true predictive error*.

According to Equation (3.1), *true predictive error* (ϵ_{tpi}) is the sum of *predictive error* (ϵ_{pi}) and *observational error* (ϵ_{oi}), that is,

$$\epsilon_{tpi} = \epsilon_{oi} + \epsilon_{pi}$$

Therefore, to improve the quality of predictions, we need to reduce:
(1) *observational error* that can be achieved by following relevant principles and methods for data acquisition as introduced in Chapter 1; and
(2) *predictive error* that can be achieved by following relevant principles and methods for data preparation as introduced in Chapter 2 and for predictive modeling as to be introduced and discussed in the following chapters.

3.2 Accuracy and error measures for predictive models

The accuracy and error measures will be introduced for numerical data and categorical data, respectively.

3.2.1 Accuracy and error measures for numerical data

On the basis of predicted and observed values, many *accuracy and error measures* have been developed to assess the accuracy of predictive models for *numerical data* (Han and Kamber 2006; Moriasi et al. 2007; Li and Heap 2008, 2011; Bennett et al. 2013; Li 2017).

Some of these measures are provided in Table 3.2 based on observed values (y, a vector of n values) and predicted values ($\hat{y}$, a vector of n values). These measures include accuracy measures such as *variance explained* (*VEcv*) (Li 2016) and *Legates and McCabe's efficiency* (E_1) (Legates and McCabe 2013), as well as the most commonly used error measures such as *MAE* and *root mean square error* (*RMSE*) (Li and Heap 2008).

All measures in Table 3.2 are related to either ϵ_{pi} or ϵ_{api}. Of these measures, *ME* and *RME* are derived from ϵ_{pi}, while *MAE*, *RMAE*, *MSE*, *RMSE*, *RRMSE*, *SRMSE*, and *MSRE* are derived from ϵ_{api}. *MSE2* and *RMSSE* are also based on ϵ_{pi} and ϵ_{api}, respectively, where the ϵ_{pi} and ϵ_{api} are however derived from standardized predicted values and standardized observed values. *VEcv* and E_1 are slightly more complex but still related to ϵ_{api}.

The advantages and disadvantages of these measures have been discussed in previous studies (Han and Kamber 2006; Moriasi et al. 2007; Li and Heap 2008; Bennett et al. 2013; Li 2016, 2017). Several commonly used error and accuracy measures have been assessed and summarized in Table 3.3. Of these error and accuracy measures, *VEcv* and E_1 measure how accurate a predictive model is. *VEcv* has been proven to be independent of unit, scale, data mean, and variance (Li 2016, 2017) and is recommended for assessing the accuracy of numerical predictions.

When a negative *VEcv* is obtained for a predictive model, it suggests that the predictions of the model are less accurate than the grand mean of validation samples being used as predictions (Li 2016). In this case, one should consider to optimize the model or use an alternative method to develop a more accurate predictive model instead of using an error measure to hide the truth.

TABLE 3.2: Measures for assessing the performance of predictive models based on observed values (y, a vector of n values, i.e., y_1, y_2,..., y_n) and predicted values ($\hat{y}$, a vector of n values, i.e., $\widehat{y_1}$, $\widehat{y_2}$,..., $\widehat{y_n}$) for numerical data (modified from Li and Heap (2008)).

Measure	Definition
Mean error (ME) or mean bias error (MBE)	$ME = mean(y - \hat{y})$
Relative ME (RME)	$RME = mean[(y - \hat{y})/mean(y)]100$
Mean absolute error (MAE)	$MAE = mean(abs(y - \hat{y}))$
Relative MAE (RMAE)	$RMAE = [mean(abs(y - \hat{y}))/mean(y)]100$
Mean square error (MSE)	$MSE = mean((y - \hat{y})^2)$
Root MSE (RMSE)	$RMSE = [mean((y - \hat{y})^2)]^{1/2}$
Relative RMSE (RRMSE)	$RRMSE = [(mean((y - \hat{y})^2))^{1/2}/mean(y)]100$
Standardized RMSE (SRMSE)	$SRMSE = [mean((y - \hat{y})^2)]^{1/2}/sd(y)$
Mean square reduced error (MSRE)	$MSRE = [mean((y - \hat{y})^2)] \, / \, var(y)$
Mean standardized error (MSE2)	$MSE2 = mean(scale(y) - scale(\hat{y}))$
Root mean square standardized error (RMSSE)	$RMSSE = [mean((scale(y) - scale(\hat{y}))^2)]^{1/2}$
Variance explained by predictive models based on cross-validation (VEcv)	$VEcv = [1 - sum((y - \hat{y})^2)/sum((y - mean(y))^2)]100$
Legates and McCabe's efficiency (E_1)	$E_1 = [1 - sum(abs(y - \hat{y}))/sum(abs(y - mean(y)))]100$

TABLE 3.3: *Limitations* of various error measures and accuracy measures (modified from Li (2013a) and Li (2017)).

Error / accuracy measure	Unit-/scale-independent	Variance-independent	Predictive accuracy
Mean error (ME)	No	No	Unknown
Mean absolute error (MAE)	No	No	Unknown
Mean squared error (MSE)	No	No	Unknown
Root MSE (RMSE)	No	No	Unknown
Relative ME (RME)	Yes	No	Unknown
Relative MAE (RMAE)	Yes	No	Unknown
Relative RMSE (RRMSE)	Yes	No	Unknown
Standardized RMSE (SRMSE)	Yes	Yes	Unknown
Mean square reduced error (MSRE)	Yes	Yes	Unknown
Legates and McCabe's efficiency (E_1)	Yes	No	Known
Variance explained (VEcv)	Yes	Yes	Known

Although *ME*, *MAE*, and *RMSE* measure how wrong the predictive model and the predictions resulted can be and are the commonly used error measures (Li and Heap 2008; Li 2019a), they are depending on the mean of the validation samples, making it hard to know how wrong the predictions are. Therefore, *RME*, *RMAE*, and *RRMSE* that are independent on the mean are recommended for numerical predictions if error measures are required, but it should be noted that they are variance-dependent. In addition, both *RMSE* and *RRMSE* can be used as a measure of prediction uncertainties.

Note: One commonly used measure, r or r^2, should not be used for numerical data because it is an incorrect measure of predictive accuracy as being demonstrated and clarified by Li (2017).

3.2.2   Accuracy and error measures for categorical data

For *categorical data* including presence/absence data, various *accuracy measures* have been developed (Fielding and Bell 1997; Allouche, Tsoar, and Kadmon 2006). To understand these measures, we need to introduce an error matrix for two-level categorical data as in Table 3.4.

TABLE 3.4: An error matrix for observed values (y, a vector of n values, i.e., $y_1, y_2,..., y_n$) and predicted values ($\hat{y}$, a vector of n values, i.e., $\widehat{y_1}, \widehat{y_2},..., \widehat{y_n}$) for two-class categorical data (e.g., class A/class B). a, number of observations correctly predicted for class A; b, number of observations of class B that are mispredicted as class A; c number of observations of class A that are mispredicted as class B; d number of observations correctly predicted for class B.

		Observed values (y)	
		class A	class B
Predicted values ($\hat{y}$)	class A	a	b
	class B	c	d

Measures for assessing the performance of predictive models for categorical data are listed in Table 3.5.

TABLE 3.5: Measures for assessing the performance of predictive models for categorical data, with n observed values and n predicted values based on the error matrix in Table 3.4.

Measures	Definition based on the error matrix
Sensitivity (Sens)	$a/(a + c)$
Specificity (Spec)	$d/(b + d)$
True skill statistics (TSS)	$Sens + Spec - 1$
Correct classification rate (CCR)	$[(a + d)/n]100$
Classification rate by chance (CRBC)	$[(a + c)(a + b) + (b + d)(c + d)]/n^2$
Kappa	$(CCR - CRBC)/(1 - CRBC)$

Although these measures in Table 3.5 are defined based on categorical data with two classes, they can be extended for data with multiple classes.

For binary data such as presence/absence data, if they are represented in 0/1 numerical data, all error and accuracy measures for numerical data are equally applicable (e.g., see Thibaud et al. (2014)).

For categorical data, *CCR*, *kappa*, and *TSS* are often recommended (Fielding and Bell 1997; Allouche, Tsoar, and Kadmon 2006).

Note: One commonly used accuracy measure for categorical data, area under the curve (*AUC*) (or receiver operating characteristics (*ROC*)) is misleading and should not be used (Allouche, Tsoar, and Kadmon 2006; Lobo, Jiménez-Valverde, and Real 2008).

3.3 R functions for accuracy and error measures

Accuracy and error measures for numerical and categorical data can be calculated in R. The recommended measures and some other commonly used measures are implemented in the function `pred.acc` in the `spm` package. A further function, `vecv`, is also developed for the accuracy measure *VEcv* in the `spm` package. Since some error measures can be converted into *VEcv* (Li 2016), a function, `tovecv`, is also developed in the `spm` package for such conversion.

The description, usage, arguments, and returned values of `pred.acc`, `vecv`, and `tovecv` are provided below.

3.3.1 Function `pred.acc`

The function `pred.acc` calculates the following accuracy and error measures for numerical data:
(1) *ME*,
(2) *MAE*,
(3) *MSE*,
(4) *RME*,
(5) *RMAE*,
(6) *RMSE*,
(7) *RRMSE*,
(8) *VEcv*, and
(9) E_1.

And it also calculates the following accuracy and error measures for categorical data:
(1) *CCR*,
(2) *kappa*,
(3) *Sens*,
(4) *Spec*, and
(5) *TSS*.

All these measures are based on the differences between predicted values ($\hat{y}$) for and observed values (y) of new or validation samples. For 0 and 1 data, y needs to be specified as factor in order to use accuracy measures for categorical data. Moreover, *Sens*, *Spec*, and *TSS* are for categorical data with two classes (e.g., presence and absence data). For categorical data with multiple classes, only *kappa* and *CCR* are available in `pred.acc`.

To calculate accuracy and error measures using `pred.acc`, the following arguments need to be specified:
(1) `obs`, a vector of observed values of new or validation samples; and
(2) `pred`, a vector of predicted values of predictive models for new or validation samples.

1. Implementation of `pred.acc` for numerical data

The application of `pred.acc` for numerical data can be demonstrated with simulated data below.

```
library(spm)
set.seed(1234)
y <- sample(1:20, 50, replace = T)
yhat <- y + rnorm(50, 1)
pred.acc1 <- pred.acc(y, yhat)
lapply(pred.acc1, round, 2)


$me
[1] -0.91

$rme
[1] -8.86

$mae
[1] 1.01

$rmae
[1] 9.88

$mse
[1] 1.83

$rmse
[1] 1.35

$rrmse
[1] 13.25

$vecv
[1] 94.34

$e1
[1] 80.15
```

The individual measure in `pred.acc1` can be retrieved. We use *VEcv* as an example.

```
round(pred.acc1$vecv, 2)


[1] 94.34
```

2. Implementation of `pred.acc` for categorical data

The application of `pred.acc` for categorical data can also be demonstrated with simulated data.

Data with two classes

```
set.seed(1234)
y2 <- as.factor(sample(c("A", "B"), 30, replace = T))
y2hat <- y2
y2hat[1] <- c("A"); y2hat[10] <- c("A"); y2hat[25] <- c("A"); y2hat[5] <- c("B");
    y2hat[27] <- c("B")
table(y2, y2hat) # an error matrix


    y2hat
y2   A  B
  A  5  2
  B  3 20


pred.acc2 <- pred.acc(y2, y2hat)
lapply(pred.acc2, round, 2)
```

```
$kappa
[1] 0.56

$ccr
[1] 83.33

$sens
[1] 0.71

$spec
[1] 0.87

$tss
[1] 0.58

round(pred.acc2$ccr, 2)

[1] 83.33
```

Data with multiple classes

```
set.seed(1234)
y3 <- as.factor(sample(c("A", "B", "C"), 30, replace = T))
y3hat <- y3
y3hat[1] <- c("A"); y3hat[10] <- c("C"); y3hat[25] <- c("C"); y3hat[5] <- c("B");
    y3hat[30] <- c("B")
table(y3, y3hat) # an error matrix

    y3hat
y3    A  B  C
  A   5  1  0
  B   1 11  2
  C   0  1  9

pred.acc3 <- pred.acc(y3, y3hat)
lapply(pred.acc3, round, 2)

$kappa
[1] 0.74

$ccr
[1] 83.33

round(pred.acc3$kappa, 2)

[1] 0.74
```

3.3.2 Function vecv

The function vecv calculates *VEcv* for numerical data. The vecv is based on the differences between predicted values for, and observed values of, new or validation samples. It measures the proportion of variation in new or validation data explained by the predicted values obtained from predictive models (Li 2016).

To calculate accuracy measure *VEcv* using vecv, the arguments that need to be specified are the same as those for pred.acc.

1. Implementation of vecv

The application of vecv for numerical data can be demonstrated using the simulated data that were generated for pred.acc.

```
vecv1 <- vecv(y, yhat)
round(vecv1, 2)
```

```
[1] 94.34
```

3.3.3   Tovecv

The function `tovecv` converts some existing predictive error measures to *VEcv* as detailed in Li (2016) and Li (2019b). The error measures considered in `tovecv` are *MSE*, *RMSE*, *RRMSE*, *SRMSE*, and *MSRE*.

To convert error measures to *VEcv* `tovecv`, the following arguments need to be specified:
(1) `n`, sample number of new or validation samples;
(2) `mu`, mean of new or validation samples;
(3) `s`, standard deviation of new or validation samples;
(4) `m`, a value of an error measure; and
(5) `measure`, a type of error measure (i.e., "mse", "rmse", "rrmse", "srmse" or "msre").

1. Implementation of `tovecv`

The application of `tovecv` can be demonstrated using the simulated data generated for `pred` `.acc`. We will take *MSE* and *RRMSE* as examples below.

Convert MSE to VEcv

```
vecv.mse <- tovecv(length(y), mean(y), sd(y), pred.acc1$mse, "mse")
round(vecv.mse, 2)
```

```
[1] 94.34
```

Convert RRMSE to VEcv

```
vecv.rrmse <- tovecv(length(y), mean(y), sd(y), pred.acc1$rrmse, "rrmse")
round(vecv.rrmse, 2)
```

```
[1] 94.34
```

3.4   Model validation

3.4.1   Validation methods

The accuracy of predictive models is critical, as it determines the quality of predictions resulted. For a given data set, the accuracy is often assessed based on model validation methods (Kohavi 1995; Hastie, Tibshirani, and Friedman 2009) that are detailed below.

1. Hold-out validation

For *hold-out validation*, it splits up a data set into a training sub-data set (to be used to develop a predictive model) and a validation sub-data set (to be used to test how well the model performs on unseen data). The size of the sub-data sets varies depending on sample size, usually with around 50% to 90% of data for training and the remaining samples for validation.

2. K-fold cross-validation

For *k-fold cross-validation*, it randomly splits up a data set (or a stratified data set) into k groups. One of the groups is to be used as a validation sub-data set and the remaining $k-1$ groups are to be used as a training sub-data set. A predictive model is developed based on the training sub-data set, and then it is used to make predictions based on the validation sub-data set. The process is repeated until each of the k groups has been used as the validation sub-data set. Then the model performance is assessed based on the predicted values for and the observed values of the data set.

One may assess the model performance based on the predicted values for and the observed values of each validation sub-data set and then average relevant performance measure(s) over k-folds to produce the final model performance measure(s). The results would be identical if the k groups are equal in size.

3. Leave-one-out and leave-q-out cross-validation

Leave-one-out (LOO) cross-validation is a special case of k-fold cross-validation where the number of folds equals the number of the observations in the data set.

Leave-q-out cross-validation is similar to *LOO*, but it uses q samples as the validation sub-data set instead of just one sample.

4. Bootstrapping cross-validation

For *bootstrapping cross-validation*, it splits up a data set into (1) a training sub-data set by sampling the data set with replacement (i.e., with the training sub-data set resulted containing about two-thirds of samples of the data set and one-third duplicated samples), and (2) a validation sub-data set that is remaining samples left out of the bootstrapped training sub-data set. This process may repeat many times. This method can not be used for geostatistical methods such as *inverse distance weighted (IDW)* due to duplicated samples with the zero distance that cannot be used as a denominator.

5. New samples

New samples are newly acquired samples and have not been used in the data set for model training.

3.4.2 Validation functions in *R*

Many *R* functions for evaluating the predictive accuracy of relevant predictive methods based on cross-validation have been developed. Most of these functions are available in the `spm` and `spm2` packages (Li 2019b, 2021a), and some are scattered in other packages as to be detailed and implemented in Chapters 4 to 11.

In environmental sciences, the most commonly used validation methods are hold-out and leave-one-out (Li 2016, 2019a), although five- or 10-fold cross-validation is recommended (Kohavi 1995; Hastie, Tibshirani, and Friedman 2009). We will demonstrate the applications of these three validation methods.

The functions `krigepred` and `krigecv` in the `spm2` package and the `swmud` data set in the `spm` package will be used for the demonstration.

For `krigepred`, the following arguments may need to be specified:
(1) `trainx`, a dataframe contains `longitude` (`long`) and `latitude` (`lat`) and predictive variables of point samples;
(2) `trainy`, a vector of response, must have length equal to the number of rows in trainx;

(3) `trainx2`, a dataframe contains longitude (`long`), latitude (`lat`) and predictive variables of point locations (i.e., the centers of grids) to be predicted;

(4) `nmax`, the number of nearest observations that should be used for a prediction; and

(5) `vgm.args`, arguments for `vgm`, e.g., variogram model of response variable and anisotropy parameters and see `vgm` in the `gstat` package for details; and by default, "Sph" is used. For other arguments, the defaults could be used; and for details, see `?krigepred`.

1. Hold-out validation

Random sampling

The function `sample` (R Core Team 2020) can be used to randomly generate a training sub-data set (e.g., 90% observations) and a validation sub-data set (e.g., 10% observations) as shown below.

```
library(spm)
data(swmud)
set.seed(1234)
tr <- sample(1:dim(swmud)[1], floor(dim(swmud)[1]*0.9))
training1 <- swmud[tr, ]
validation1 <- swmud[-tr, ]
```

We then use `krigepred` to generate predictions using *OK* for the validation samples.

```
library(spm2)
okpred1 <- krigepred(training1[, c(1,2)], training1[, 3], validation1, nmax = 12,
    vgm.args = ("Sph"))
names(okpred1)
```

The predictive accuracy of *OK* (i.e., `okpred1`) can be evaluated by

```
okvecvho1 <- vecv(validation1$mud, okpred1$var1.pred)
round(okvecvho1, 2)
```

```
[1] 75.27
```

Stratified random sampling

The function `datasplit` in the `spm2` package, which implements a stratified random re-sampling technique, can be used to split a data set into a training sub-data set (90% observations) and a validation sub-data set (10% observations).

```
library(spm2)
set.seed(1234)
idx1 <- datasplit(swmud[, 3], k.fold = 10)
training2 <- swmud[idx1 != c(1), , drop = FALSE]
validation2 <- swmud[idx1 == c(1), , drop = FALSE]
```

Then *OK* can be employed to generate spatial predictions for the validation data set, and `vecv` can be used to assess the predictive accuracy of the predictions resulted.

```
okpred2 <- krigepred(training2[, c(1,2)], training2[, 3], validation2, nmax = 12,
    vgm.args = ("Sph"))
names(okpred2)

okvecvho2 <- vecv(validation2$mud, okpred2$var1.pred)
```

The predictive accuracy of *OK* (i.e., `okvecvho2`) is

```
round(okvecvho2, 2)
```

```
[1] 82.03
```

The predictive accuracy for random sampling is lower than that for stratified random sampling. The difference in the predictive accuracy could be due to: (1) stratified random sampling method that is expected to produce a higher accuracy than the random sampling method, and (2) randomness associated with the sampling methods that can be confirmed using the method introduced in Section 3.4.3.

2. Leave-one-out cross-validation

The function `krigecv` will be used to perform *LOO* cross-validation for *OK* based on the `swmud` data set as below.

```
set.seed(1234)
okloo1 <- krigecv(swmud[, c(1,2)], swmud[, 3], validation = "LOO", nmax = 12, vgm.
    args = ("Sph"), predacc = "VEcv")
```

The predictive accuracy of *OK* (i.e., `okloo1`) is

```
round(okloo1, 2)
```

```
[1] 83.56
```

3. 10-fold cross-validation

The function `krigecv` will also be used to perform 10-fold cross-validation for *OK* based on the `swmud` data set as follows.

```
set.seed(1234)
ok10f <- krigecv(swmud[, c(1,2)], swmud[, 3], validation = "CV", cv.fold = 10,
    nmax = 12, vgm.args = ("Sph"), predacc = "VEcv")
```

The predictive accuracy of *OK* (i.e., `ok10f`) is

```
round(ok10f, 2)
```

```
[1] 82.34
```

3.4.3 Effects of randomness associated of cross-validation methods on predictive accuracy assessments

Although five- or 10-fold cross-validation is recommended to evaluate the performance of predictive models (Kohavi 1995), the training and validation data sets that are randomly generated for each fold of the cross-validation change when the process is repeated. Consequently, the predictive accuracy or error measures resulted also change with each iteration of the cross-validation and are not stable (Li 2013c). Thus the randomness associated with the cross-validation would affect the predictive accuracy assessments.

To reduce the influence of the randomness on predictive accuracy assessments, we need to stabilize the performance measures resulted by repeating the cross-validation a certain times (e.g., 100 times) (Li 2013c, 2013b; Li, Siwabessy, Tran, et al. 2014). The stabilization of the predictive model performance measures is to be demonstrated using *OK* with 100 repetitions of 10-fold cross-validation based on the `swmud` data set.

Prior to stabilizing the predictive accuracy, we will demonstrate: (1) the dependence of predictive accuracy or error measures on random seeds; and (2) the dependence of averaged predictive model performance measures on random seeds. We will use *VEcv* as an example.

1. Dependence of predictive accuracy measures on random seeds

First, we generate 100 random seeds.

```
set.seed(1234)
random.seed <- sample(1:9999, 100)
```

Then, we use each of the random seeds for *OK* by applying `krigecv` to the `swmud` data set below.

```
random.seed.vecv <- NULL

for (i in 1:length(random.seed)) {
  set.seed(random.seed[i])
  okcv1 <- krigecv(swmud[, c(1, 2)], swmud[, 3], nmax = 12, predacc = "VEcv")
  random.seed.vecv[i] <- okcv1
}
```

The predictive accuracy resulted, *VEcv* in `okcv1`, can be visualized against `random.seed` in Figure 3.1 by

```
par(font.axis = 2, font.lab = 2)
plot(random.seed.vecv ~ random.seed, xlab = "Random seed", ylab = "VEcv (%)", col
    = "blue")
```

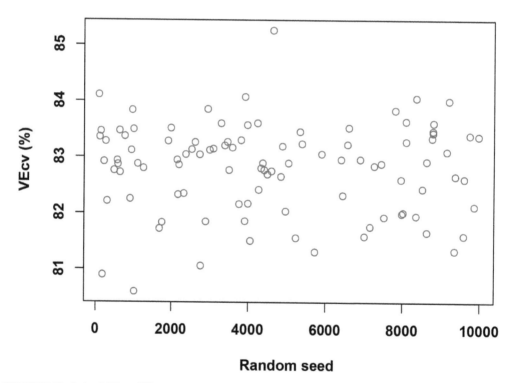

FIGURE 3.1: *VEcv* (%) of *OK* for each of 100 random seeds based on 10-fold cross-validation.

It shows that the predictive accuracy changes considerably with the random seed used, ranging from 80.59% to 85.26%.

It is clear that the predictive accuracy resulted depends on random seeds used, and we should minimize the dependence on random seed in deriving relevant predictive accuracy assessments.

2. Dependence of averaged predictive accuracy measures on random seeds

To stabilize the predictive accuracy resulted by repeating the cross-validation a certain times, we need to determine the iteration number. The choice of iteration number is data-dependent and can be determined based on the method used in the previous studies (Li 2013c, 2013b, 2019a; Li, Siwabessy, Tran, et al. 2014); and 60 to 100 times are recommended (e.g., see various cross-validation functions in spm and spm2). However, the median and average accuracy resulted may depend on the random seed used. This can be proved as follows.

We generate ten random seeds first.

```
set.seed(1234)
r.seed <- sample(1:9999, 10)
```

Then use each of the random seeds for *OK* by applying krigecv to the swmud data set 100 times below.

```
n <- 100
r.seed.vecv <- matrix(0, n, length(r.seed))
for (i in 1:length(r.seed)) {
  set.seed(r.seed[i])

  for (j in 1:n) {
  okcv1 <- krigecv(swmud[, c(1, 2)], swmud[, 3], nmax = 12, predacc = "VEcv")
  r.seed.vecv[j, i] <- okcv1
  }
}
```

The median and average predictive accuracy resulted, *VEcv* in r.seed.vecv, for the 10 random seeds are

```
library(miscTools)
colMedians(r.seed.vecv)
```

```
 [1] 82.990 82.967 82.838 82.758 82.985 82.916 83.027 83.006 82.806 82.908
```

```
colMeans(r.seed.vecv)
```

```
 [1] 82.924 82.919 82.788 82.772 82.840 82.813 82.877 82.842 82.796 82.807
```

The ranges of the median and average predictive accuracy are

```
range(colMedians(r.seed.vecv))
```

```
[1] 82.758 83.027
```

```
range(colMeans(r.seed.vecv))
```

```
[1] 82.772 82.924
```

It is apparent that the changes of median and average predictive accuracy based on 100 repetitions with random seeds are minimal, and the dependence of averaged predictive accuracy on the random seed is largely negligible. Hence, we can stabilize the predictive accuracy using any random seed for reproducible simulations.

3. Stabilization of the accuracy of predictive model

With *n* times cross-validation

We use random seed 1234 to demonstrate how to stabilize the predictive accuracy of *OK* and produce reliable averaged accuracy by repeating the cross-validation *n* (e.g., 100) times below.

```
set.seed(1234)
n <- 100
VEcv <- NULL
for (i in 1:n) {
  okcv1 <- krigecv(swmud[, c(1, 2)], swmud[, 3], nmax = 12, predacc = "VEcv")
  VEcv [i] <- okcv1
}
```

The median, mean, and range of the predictive accuracy (*VEcv* in VEcv) are

```
round(median(VEcv), 2)
```

```
[1] 82.89
```

```
round(mean(VEcv), 2)
```

```
[1] 82.85
```

```
round(range(VEcv), 2)
```

```
[1] 78.03 84.91
```

The variation of *VEcv* and their accumulative median and average with each of the iterations are calculated using the functions cummean and cummdedian in the cumstats package (Erdely and Castillo 2017) and shown in Figure 3.2 (modified from Li (2013c)) by

```
library(cumstats)
par(font.axis = 2, font.lab = 2)
plot(VEcv ~ c(1:n), xlab = "Iteration for OK", ylab = "VEcv (%)")
points(cummean(VEcv) ~ c(1:n), col = "red")
points(cummedian(VEcv) ~ c(1:n), col = "blue")
abline(h = mean(VEcv), col = "red", lwd=2)
abline(h = median(VEcv), col = "blue", lwd=2)
```

If the maximal range of the predictive accuracy (range) had been captured in VEcv, then the maximal change of accumulative average (maxdeltacummean) from n^{th} to $(n+1)^{th}$ iteration would be expected to be maxdeltacummean = range/2n. For instance, maxdeltacummean for 101th iteration, (range(VEcv)[2] - range(VEcv)[1])/(2 * 100), is 0.03% for predictive accuracy of *OK*, *VEcv* in VEcv. This maxdeltacummean is the expected maximal contribution to *VEcv* by an additional iteration.

The deviations of the accumulative median and average from the overall mean and median are illustrated in Figure 3.3 by

```
library(cumstats)
maxdeltacummean <- (range(VEcv)[2] - range(VEcv)[1]) / (2 * 100)
par(font.axis = 2, font.lab = 2)
plot((cummean(VEcv) - mean(VEcv)) ~ c(1:n), ylim = c(-0.7, 0.15), col = "red",
     xlab = "Iteration for OK", ylab = "Deviation in VEcv (%)")
points((cummedian(VEcv) - median(VEcv)) ~ c(1:n), col = "blue")
abline(h = 0, lwd = 1)
abline(h = maxdeltacummean, lty = 2)
abline(h = - maxdeltacummean, lty = 2)
```

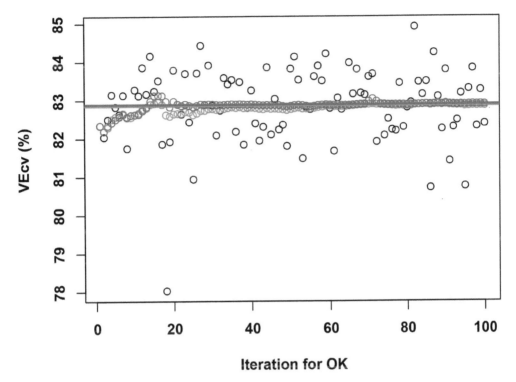

FIGURE 3.2: *VEcv* (%) of *OK* for each of the 100 times of 10-fold cross-validation (black circle), their accumulative median (blue circle), their accumulative average (red circle), overall median (blue line) and overall average (red line).

```
abline(h = 0.1, lty = 2, col = "green")
abline(h = - 0.1, lty = 2, col = "green")
```

The accumulative median is quickly stabilized as the number of iteration increases and remains largely unchanged when the iteration number is over 70 if `maxdeltacummean` is used as a threshold. Similarly, if a different threshold (e.g., 0.1%) is used, the accumulative median will be stabilized at over 70 iterations.

The accumulative average is gradually stabilized as the number of iteration increases and remains largely stable when the iteration number is over 90 if `maxdeltacummean` is used as a threshold to obtain a stabilized predictive accuracy. If a different threshold (e.g., 0.1%) is used, then the accumulative average will be stabilized at around 50 iterations.

With a pre-set threshold

If a threshold is set prior to the modeling, then it can be used to stabilize the predictive accuracy. For instance, if the threshold is set as the change in overall median no more than 0.005%. Then we can use it to obtain a stabilized predictive accuracy by

```
threshold <- 0.005
set.seed(1234)
VEcv2 <- NULL

okcv1 <- krigecv(swmud[, c(1, 2)], swmud[, 3], nmax = 12, predacc = "VEcv")
VEcv2[1] <- okcv1
```

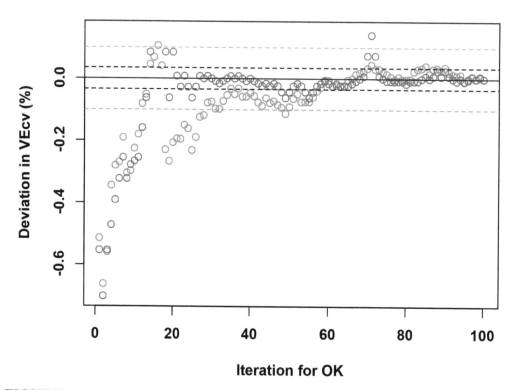

FIGURE 3.3: The deviations of the accumulative median and average from the overall mean and median, respectively, at each of the 100 iterations (black line), deviation of accumulative average (red circle), deviation of accumulative median (blue circle), the maximal change of accumulative average as threshold (black dashed lines) and 0.1% as threshold (green dashed lines).

```
vecvmedian1 <- median(VEcv2)
repeat {
  okcv1 <- krigecv(swmud[, c(1, 2)], swmud[, 3], nmax = 12, predacc = "VEcv")
  VEcv2[length(VEcv2) + 1] <- okcv1
  vecvmedian2 <- median(VEcv2)

  if (abs(vecvmedian2 - vecvmedian1) <= threshold) {
    break
  }
  vecvmedian1 <- vecvmedian2
}
```

The number of iterations and the stabilized predictive accuracy in terms of *VEcv* are

```
length(VEcv2)
```

```
[1] 64
```

```
round(vecvmedian2, 2)
```

```
[1] 82.87
```

3.4.4 Procedure for the assessment of the performance of predictive models

The performance of predictive methods can be evaluated based on the rationales described in the above sections that can be summarized for k-fold cross-validation in Table 3.6.

TABLE 3.6: Procedure for predictive accuracy assessment (modified from Li (2019b)).

Step	Description
1	Re-sample an observation data set into k sub-data sets for k-fold validation
2	Use k - 1 sub-data sets to form a training data set, the remaining one as a validation data set, which is repeated until each sub-data set has been used as the validation data set.
3	**for**
4	Each training data set and its corresponding validation data set:
5	Develop a predictive model based on the training data set
6	Validate the model using the validation data set and record the predictive values resulted
7	Repeat steps 3 to 5 for each of the k training data sets and its corresponding validation data set
8	**end for**
9	Calculate the predictive accuracy based on observed values in the observation data set and their corresponding predictive values
10	Repeat steps 1 to 8 n times to produce a stabilized predictive accuracy

4

Mathematical spatial interpolation methods

Spatial interpolation methods are an important group of *spatial predictive methods*. This chapter introduces *mathematical spatial interpolation methods*. These methods are *deterministic* and non-geostatistical spatial interpolation methods (Li and Heap 2014), which are based only on the mathematical calculations of spatial information (i.e., coordinates data) and do not involve spatial changes of data variances.

In this chapter, the mathematical methods that are available in R will be introduced, including:
(1) *inverse distance weighted (IDW)*,
(2) *nearest neighbors (NN)*, and
(3) *k nearest neighbors (KNN)*.

The `swmud` point data and `sw` grid data in the `spm` package (Li 2019b) will be used to demonstrate the applications of these methods.

Prior to introducing these methods, a general prediction formula for all spatial interpolation methods in this book will be firstly provided.

General prediction formula

The predictions of spatial interpolation methods can be represented as the weighted averages of sampled data. They all share the same *general prediction formula* (Equation (4.1)):

$$\hat{y}(x_0) = \sum_{i=1}^{n} \lambda_i \cdot y(x_i) \tag{4.1}$$

where $\hat{y}(x_0)$ is a predicted value of an attribute at the location of interest x_0, $y(x_i)$ is observed value at the sampled location x_i, λ_i is the weight assigned to the sample at x_i, and n represents the number of samples used to generate the predictions (Webster and Oliver 2001). The sample(s) are actually observation(s) and will be used interchangeably with observation(s) for relevant spatial predictive methods.

The term *attribute* is often used in geostatistics, and is also called primary variable in geostatistics. It is equivalent to *response variable* or *dependent variable* for gradient based/detrended spatial predictive methods introduced in Chapters 6 and 7.

The *spatial interpolation methods* include many methods and their features are compared in the previous reviews (Li and Heap 2008, 2014). These methods are different due to the differences in the way to determine the weights (i.e., λ_i in Equation (4.1)).

DOI: 10.1201/9781003091776-4

4.1   Inverse distance weighted

Inverse distance weighted or *inverse distance weighting* (*IDW*) predicts the value of an attribute at an unsampled location using a linear combination of values of samples weighted by an inverse function of the distance from the location of interest to the sampled locations. The *assumption* is that sampled locations closer to the unsampled location are more similar to it than those further away in their values. For *IDW*, the *weights* λ_i in Equation (4.1) can be expressed as in Equation (4.2):

$$\lambda_i = (1/d_i^p)/ \sum_{i=1}^{n} 1/d_i^p \qquad (4.2)$$

where d_i is the distance between x_0 and x_i, p is a *power parameter*, and n represents the number of samples to be used. The predictions of *IDW* are affected by two parameters (i.e., n and p). The main factor affecting predictive accuracy of *IDW* is the value of the power parameter (Isaaks and Srivastava 1989). The *number of samples used* is also an important factor for the accuracy. Weights diminish as the distance increases, especially when the value of the power parameter increases, so nearby samples have a heavier weight and thus have more influence on the predictions, and the spatial interpolation resulted is local.

4.1.1   Implementation of *IDW* in `gstat`

The predictions of *IDW* can be generated with the function `gstat` in the `gstat` package (Gräler, Pebesma, and Heuvelink 2016). The function `gstat` creates `gstat` objects that hold all the information necessary for geostatistical prediction. The descriptions of relevant arguments of `gstat` are detailed in its help file, which can be accessed by `?gstat`. For *IDW* using `gstat`, the following arguments may need to be specified:
(1) `id`, the dependent variable or the attribute to be predicted;
(2) `formula`, defines the dependent variable;
(3) `locations`, formula with only independent variables that define spatial data locations;
(4) `idp`, inverse distance power; and
(5) `nmax`, the number of nearest observations that should be used to generate predictions.

IDW is called *inverse distance squared* (*IDS*) if `idp = 2`, a commonly used option for *IDW* (Li and Heap 2008).

The application of `gstat` for *IDW* is to be demonstrated below with `idp = 2` and `nmax = 20`.

Firstly, retrieve the `swmud` data set from the `spm` package.

```
library(spm)
data(swmud)
class(swmud)
```

Then, apply `gstat` to the `swmud` data set.

```
library(gstat)
idw1 <- gstat(id = "mud", formula = mud ~ 1, locations = ~ long + lat, data =
    swmud, set = list(idp = 2), nmax = 20)
```

When `nmax` is not defined, then all samples will be used to generate predictions. The model `idw1` can be used as a predictive model to generate spatial predictions as shown in Section

4.1.4. In this example, sample data set is in data frame format. The function `gstat` can also use data set in *SpatialPointsDataFrame* to develop an *IDW* model as follows:

```
swmud2 <- swmud

library(sp)

coordinates(swmud2) = ~ long + lat
class(swmud2)

idw2 <- gstat(id = "mud", formula = mud ~ 1, data = swmud2, set = list(idp = 2),
    nmax = 20)
```

4.1.2 Parameter optimization for *IDW*

For *IDW*, only two parameters `idp` and `nmax` need to be optimized. It needs to be clarified that arguments for a function are referred to as parameters when they need to be estimated for optimization in this book. We can use the function `idwcv` in the `spm` package to estimate the optimal parameters that can maximize the predictive accuracy of *IDW*. The `idp` can also be determined using the function `estimateParameters` in the `intamap` package (Pebesma et al. 2018).

1. Implementation in `idwcv`

The descriptions of relevant arguments for `idwcv` are detailed in its help file, which can be accessed by `?idwcv`. To use `idwcv`, the following arguments may need to be specified:
(1) `longlat`, a dataframe contains longitude and latitude of point samples;
(2) `trainy`, a vector of response, must have length equal to the number of rows in `longlat`;
(3) `cv.fold`, an integer; the number of folds in cross-validation, and if > 1, then apply n-fold cross-validation; the default is 10, i.e., 10-fold cross-validation is recommended; and
(4) `predacc`, can be either "VEcv" for `vecv` or "ALL" for all measures in the function `pred.acc` as introduced in Section 3.3.1.

The remaining arguments are the same as those for `gstat` introduced above.

To estimate optimal values for `idp` and `nmax`, we need to provide relevant values that can be considered for the optimization as shown below.

```
idp <- (1:20) * 0.2
nmax <- c(5:40)
```

Then their optimal values will be estimated using `idwcv` as follows.

```
idwopt <- array(0, dim = c(length(idp), length(nmax)))

for (i in 1:length(idp)) {
  for (j in 1:length(nmax)) {
     set.seed(1234)
     idwcv1 <- idwcv(swmud[, c(1, 2)], swmud[, 3], nmax = nmax[j], idp = idp[i],
        predacc = "VEcv" )
     idwopt[i, j] <- idwcv1
  }
}
```

The predictive accuracy, *VEcv* in `idwopt`, can be visualized against with `idp` and `nmax` using the function `persp3d` in the `rgl` package (Adler and Murdoch 2020).

```
library(rgl)
```

```
persp3d(z = idwopt, x = idp, y = nmax, xlab = "idp", ylab = "nmax", zlab = " VEcv
    (%)", col = "blue")
```

It can also be illustrated in Figure 4.1 using the function `persp` in the `graphics` package (R Core Team 2020).

```
library(graphics)
```

```
persp(x = idp, y = nmax, z = idwopt, xlab = "idp", ylab = "nmax", zlab = " VEcv
    (%)", theta = 25, phi = 30, col = "green", shade = 0.3, ticktype = "detailed")
```

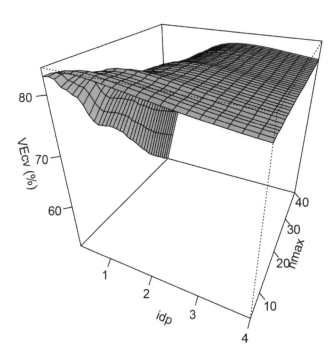

FIGURE 4.1: Predictive accuracy against `idp` and `nmax`.

The optimal values estimated for `idp` and `nmax`, which lead to the maximal predictive accuracy, are

```
para.opt <- which (idwopt == max(idwopt), arr.ind = T)
idp[para.opt[, 1]]
```

```
[1] 0.8
```

```
nmax[para.opt[, 2]]
```

```
[1] 6
```

These estimations should be used to generate spatial predictions.

2. Implementation in `estimateParameters`

The descriptions of relevant arguments for `estimateParameters` are detailed in its help file that can be accessed by `?estimateParameters`. For *IDW* using `estimateParameters`, the following arguments may need to be specified:

(1) `object`, an `intamap` object of the type;

(2) `idpRange`, range of `idp` values over which to optimize `rmse`; and

(3) `nfolds`, number of folds in n-fold cross-validation.

We need to have the data ready for `object` in `estimateParameters` first as below.

```
library(intamap)

sw2 <- sw

gridded(sw2) = ~ long + lat
proj4string(swmud2) <- CRS("+init=epsg:4326")
proj4string(sw2) <- CRS("+init=epsg:4326")

Obj = createIntamapObject(observations = swmud2,
    formulaString = as.formula(mud ~ 1),
  predictionLocations = sw2,
    class = "idw")

checkSetup(Obj)
```

Then, identify the optimal `idp` using `estimateParameters`.

```
set.seed(1234)
Obj <- estimateParameters(Obj, idpRange = seq(0.2, 4, 0.2), nfold = 10)

names(Obj)

[1] "observations"        "formulaString"       "predictionLocations"
[4] "params"              "outputWhat"          "blockWhat"
[7] "inverseDistancePower"

Obj$inverseDistancePower

[1] 3
```

It suggests that a value of 3 for `idp` should be used for *IDW*, which is different from the value estimated by `idwcv` above. This may be due to (1) the difference in the way to sample data for cross-validation, and (2) `namx` was not considered in `estimateParameters`. For `idwcv`, a stratified random sampling method is used and `nmax` was also considered together with `idp`. Hence, the results from `idwcv` will be used below.

4.1.3 Predictive accuracy of IDW with the optimal parameters

As demonstrated in Chapter 3, *predictive accuracy* changes with the `set.seed()` used. To stabilize it and get more reliable accuracy estimation, we can repeat the cross-validation n (i.e., 100 in this demonstration) times as below.

```
set.seed(1234)
n <- 100
idwvecv <- NULL

for (i in 1:n) {
  idwcv1 <- idwcv(swmud[, c(1, 2)], swmud[, 3], idp = idp[para.opt[, 1]], nmax =
      nmax[para.opt[, 2]], predacc = "VEcv")
```

```
    idwvecv[i] <- idwcv1
}
```

The median and range of predictive accuracy of *IDW* based on 100 times of 10-fold cross-validation are

```
median(idwvecv)
```

```
[1] 84.61
```

```
range(idwvecv)
```

```
[1] 81.50 87.15
```

The changes of *VEcv* with repetition times is illustrated in Figure 4.2 by

```
library(cumstats)
par(font.axis = 2, font.lab = 2)
plot(idwvecv ~ c(1:n), xlab = "Iteration for *IDW*", ylab = "VEcv (%)")
points(cummean(idwvecv) ~ c(1:n), col = "red")
points(cummedian(idwvecv) ~ c(1:n), col = "blue")
abline(h = mean(idwvecv), col = "red", lwd=2)
abline(h = median(idwvecv), col = "blue", lwd=2)
```

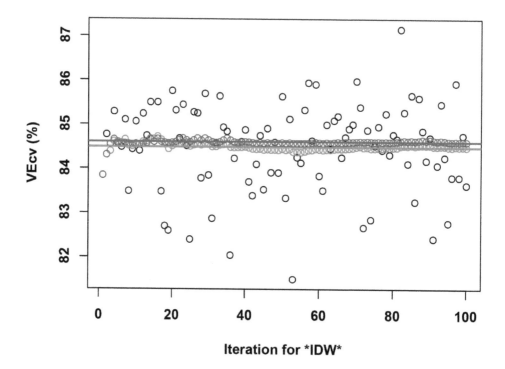

FIGURE 4.2: *VEcv* (%) of *IDW* for each of the 100 times of 10-fold cross-validation (black circle), their accumulative median (blue circle), their accumulative average (red circle), overall median (blue line) and overall average (red line).

It shows that *VEcv* changes considerably among iterations, the accumulative median and average are gradually stabilized as the number of iteration increases, and median and average remain largely unchanged when the number iteration is over 40 and 80, respectively.

4.1.4 Predictions of *IDW*

The *spatial predictions* of *IDW* can be generated using either the functions gstat and predict or the function idwpred in the spm package.

1. Functions gstat and predict

Data in dataframe format

In this example, the data sets are in dataframe format. The predictions of *IDW* can be generated using gstat and predict with the optimal parameters as shown below.

```
data(sw)
class(sw)

idw.predict <- gstat(id = "mud", formula = mud ~ 1, locations = ~ long + lat, data
    = swmud, set = list(idp = 0.8), nmax = 6)

idw.pred <- predict(idw.predict, sw)
```

Some basic properties of the predictions are

```
class(idw.pred)

[1] "data.frame"

names(idw.pred)

[1] "long"      "lat"       "mud.pred"  "mud.var"

dim(idw.pred)

[1] 500703        4

head(idw.pred, 3)

    long     lat mud.pred mud.var
1 109.4 -26.50    72.84      NA
2 109.4 -26.51    72.84      NA
3 109.4 -26.52    72.85      NA

range(idw.pred[, 3])

[1]   0.03975 97.00708

range(swmud[, 3])

[1]   0.009758 98.253112
```

Since *IDW* is a deterministic method, the variances of the predictions are not available.

Data in SpatialPointsDataFrame and SpatialPixelsDataFrame format

If the *point data* set is in *SpatialPointsDataFrame*, the *grid data* set needs to be set as *SpatialPixelsDataFrame* and then the predictions of *IDW* can be generated as below:

```
idw.predict2 <- gstat(id = "mud", formula = mud ~ 1, data = swmud2, set = list(idp
    = 0.8), nmax = 6)

idw.pred2 <- predict(idw.predict2, sw2)
```

```
class(idw.pred2)
```

```
[1] "SpatialPixelsDataFrame"
attr(,"package")
[1] "sp"
```

```
head(as.data.frame(idw.pred2), 3)
```

```
    long    lat mud.pred mud.var
1 109.4 -26.50    72.84      NA
2 109.4 -26.51    72.84      NA
3 109.4 -26.52    72.85      NA
```

```
identical(idw.pred, as.data.frame(idw.pred2))
```

```
[1] TRUE
```

2. Function `idwpred`

The descriptions of relevant arguments for `idwpred` are detailed in its help file, which can be accessed by `?idwpred`. To use `idwpred`, the following arguments may need to be specified:
(1) `longlat` and `trainy` are the same as those for `idwcv`;
(2) `longlat2`, a dataframe contains the longitude and latitude of point locations (i.e., the centers of grids) to be predicted; and
(3) `nmax` and `idp` are the same as those for `gstat`.

The predictions of *IDW* can also be generated with `idwpred` as below.

```
idwpred1 <- idwpred(swmud[, c(1,2)], swmud[, 3], sw, nmax = 6, idp = 0.8)
```

```
names(idwpred1)
```

```
[1] "LON"        "LAT"        "var1.pred" "var1.var"
```

```
identical(idw.pred$mud.pred, idwpred1$var1.pred)
```

```
[1] TRUE
```

It shows that these data sets are identical. It is clear that the function `idwpred` has simplified the application of *IDW* for generating spatial predictions and should be used.

The range of the predictions is

```
round(range(idw.pred2$mud.pred), 2)
```

```
[1]   0.04 97.01
```

The distribution of the spatial predictions of *IDW* is illustrated based on `idw.pred2` in Figure 4.3 by

```
library(raster)
```

```
names(idw.pred2) <- c('predictions', 'variances')
idw.pred2 <- brick(idw.pred2['predictions']) # create a RasterBrick for plotting.
```

```
par(font.axis = 2, font.lab = 2)
plot(idw.pred2)
```

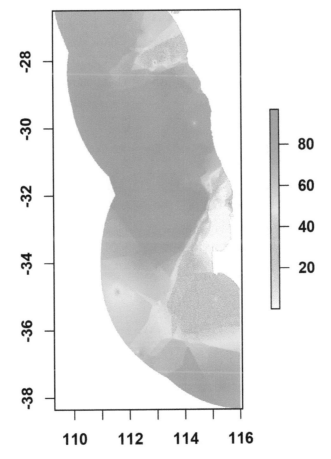

FIGURE 4.3: Spatial distribution of the *IDW* predictions.

4.2   Nearest neighbors

Nearest neighbors (*NN*) predicts the value of an attribute at an unsampled location based on the value of the nearest sample. Traditionally, for an area with samples collected from n locations, *NN* is implemented so that this area is divided into n sub-areas, known as *Thiessen* (or *Dirichlet/Voronoi*) polygons (V_i, $i = 1, 2, ... , n$), by drawing perpendicular bisectors between the sampled locations. This results in one polygon per sample and each sample is located in the center of the polygon, so that in each polygon all location points are nearer to its enclosed sample than to any other samples (Ripley 1981; Isaaks and Srivastava 1989; Webster and Oliver 2001). The predictions of the attribute at unsampled locations within polygon V_i are the measured value at the nearest single sample, x_i, that is $\hat{y}(x_0) = y(x_i)$. The *weights* in Equation (4.1) are determined according to Equation (4.3):

$$\lambda_i = \begin{cases} 1 & \text{if } x_i \text{ is the nearest sample or } \in V_i \\ 0 & \text{otherwise} \end{cases} \qquad (4.3)$$

4.2.1 Implementation of *NN* in gstat

NN can be implemented with the function `gstat` and data sets in *SpatialPointsDataFrame*, with `nmax = 1` as below.

We will apply `gstat` to `swmud2` to demonstrate the application of *NN* below.

```
nn1 <- gstat(id = "mud", formula = mud ~ 1, data = swmud2, nmax = 1)
```

For *NN*, `nmax` must be assigned 1. The value of `idp` does not affect the results according to Equation (4.2), so the default value is used. Thus, for *NN*, *parameter optimization* is not required.

4.2.2 Predictive accuracy of *NN*

The *predictive accuracy* of *NN* can be estimated by repeating the cross-validation 100 times as below.

```
set.seed(1234)
n <- 100
nnvecv <- NULL

for (i in 1:n) {
  nncv1 <- idwcv(swmud[, c(1, 2)], swmud[, 3], nmax = 1, predacc = "VEcv")
  nnvecv[i] <- nncv1
}
```

The median and range of predictive accuracy of *NN* based on 100 repetitions of 10-fold cross-validation using `idwcv` are

```
median(nnvecv)
```

```
[1] 79.28
```

```
range(nnvecv)
```

```
[1] 74.18 85.28
```

4.2.3 Predictions of *NN*

The *spatial predictions* of *NN* can be generated using `idwpred` as below.

```
nnpred1 <- idwpred(swmud[, c(1,2)], swmud[, 3], sw, nmax = 1)
```

Some basic properties of the predictions in `nnpred1` are

```
names(nnpred1)
```

```
[1] "LON"       "LAT"       "var1.pred" "var1.var"
```

```
class(nnpred1)
```

```
[1] "data.frame"
```

```
dim(nnpred1)
```

```
[1] 500703      4
```

```
head(nnpred1, 3)
```

```
     LON     LAT var1.pred var1.var
1 109.4 -26.50      69.77       NA
2 109.4 -26.51      69.77       NA
3 109.4 -26.52      69.77       NA
```

The range of the predictions of *NN* is

```
round(range(nnpred1$var1.pred), 2)
```

```
[1]   0.01 98.25
```

The distribution of the spatial predictions of *NN* is illustrated in Figure 4.4 by

```
nnpred1.1 <- nnpred1
gridded(nnpred1.1) <- ~ LON + LAT
names(nnpred1.1) <- c('predictions', 'variances')

par(font.axis = 2, font.lab = 2)
plot(brick(nnpred1.1['predictions']))
```

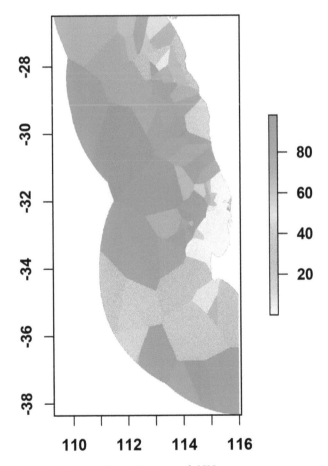

FIGURE 4.4: Spatial patterns of predictions of *NN*.

The spatial distribution of the predictions displays the patterns of *Thiessen* (or *Dirichlet/Voronoi*) polygons as expected for the predictions generated from *NN* (Li and Heap (2008)).

4.3 K nearest neighbors

In `gstat` for *NN*, if `nmax` is k and k > 1, and `idp = 0`, then it is called *k nearest neighbors* (*KNN*). For *KNN*, `idp` needs to be 0 to remove the effects of distance in Equation (4.2) and the parameter `k` needs to be optimized.

4.3.1 Parameter optimization for *KNN*

For *KNN*, only one parameter `k` needs to be estimated to maximize its predictive accuracy. We still use `idwcv` to estimate optimal value for `k` that is `nmax`.

```
knn.nmax <- c(1:40)
knnopt <- NULL

for (i in 1:length(knn.nmax)) {
  set.seed(1234)
  knncv1 <- idwcv(swmud[, c(1, 2)], swmud[, 3], idp = 0, nmax = knn.nmax[i],
      predacc = "VEcv" )
  knnopt[i] <- knncv1
}
```

The predictive accuracy, *VEcv* in `knnopt`, can be visualized against `nmax` in Figure 4.5 by

```
par(font.axis = 2, font.lab = 2)
plot(knnopt ~ knn.nmax, xlab = "nmax", ylab = "VEcv (%)", col = "blue")
```

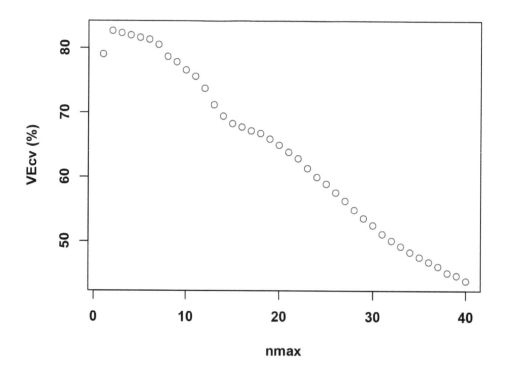

FIGURE 4.5: *VEcv* (%) of *KNN* for each `nmax` based on 10-fold cross-validation.

The optimal value estimated for `nmax` that results in the maximal predictive accuracy can be retrieved as below.

```
knn.opt <- which (knnopt == max(knnopt))
knn.nmax[knn.opt]
```

```
[1] 2
```

The estimation for `nmax` is 2, which should be used for *KNN*.

4.3.2 Predictive accuracy of `KNN` with the optimal parameter

The *predictive accuracy* of *KNN* can be stabilized by repeating the cross-validation 100 times as follows.

```
set.seed(1234)
n <- 100
knnvecv <- NULL

for (i in 1:n) {
   knncv1 <- idwcv(swmud[, c(1, 2)], swmud[, 3], idp = 0, nmax = knn.nmax[knn.opt],
       predacc = "VEcv")
   knnvecv[i] <- knncv1
}
```

The median and range of predictive accuracy of *KNN* based on 100 repetitions of 10-fold cross-validation using `idwcv` are

```
median(knnvecv)
```

```
[1] 82.93
```

```
range(knnvecv)
```

```
[1] 77.68 86.66
```

4.3.3 Predictions of *KNN*

The *spatial predictions* of *KNN* can be generated using `idwpred` as below.

```
knnpred1 <- idwpred(swmud[, c(1,2)], swmud[, 3], sw, idp = 0, nmax = knn.nmax[knn.
   opt])
```

Some basic properties of the predictions in `knnpred1` are

```
names(knnpred1)
```

```
[1] "LON"       "LAT"       "var1.pred" "var1.var"
```

```
class(knnpred1)
```

```
[1] "data.frame"
```

```
names(knnpred1)
```

```
[1] "LON"       "LAT"       "var1.pred" "var1.var"
```

```
dim(knnpred1)
```

```
[1] 500703       4
```

```
head(knnpred1, 3)
```

```
    LON    LAT var1.pred var1.var
1 109.4 -26.50    72.78       NA
2 109.4 -26.51    72.78       NA
3 109.4 -26.52    72.78       NA
```

The range of *KNN* predictions is

```
round(range(knnpred1$var1.pred), 2)
```

```
[1]   0.03 95.17
```

The spatial distribution of *KNN* predictions is illustrated in Figure 4.6 by

```
knnpred1.1 <- knnpred1
gridded(knnpred1.1) <- ~ LON + LAT
names(knnpred1.1) <- c('predictions', 'variances')

par(font.axis = 2, font.lab = 2)
plot(brick(knnpred1.1['predictions']))
```

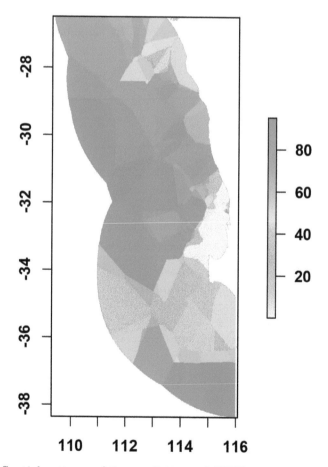

FIGURE 4.6: Spatial patterns of the predictions of *KNN*.

5

Univariate geostatistical methods

Geostatistical methods are usually grouped into *spatial interpolation methods* and are *stochastic* methods. There are many geostatistical methods (Li and Heap 2008, 2014). Geostatistics is usually believed to have originated from the work in geology and mining by Krige (1951), but it can be traced back to the early 1910s in agronomy and 1930s in meteorology (Webster and Oliver 2001). It was developed by Matheron (1963) with the theory of regionalized variables (Wackernagel 2003) and includes several methods that use *kriging* algorithms for generating spatial predictions. *Kriging* is a generic name for a family of generalized least-squares regression algorithms used in recognition of the pioneering work of Danie Krige (1951). These *kriging methods* fall into two main groups: (1) *univariate geostatistical methods*, and (2) *multivariate geostatistical methods*.

This chapter introduces *univariate geostatistical methods*. Similar to the mathematical methods in Chapter 4, univariate geostatistical methods are also based only on *spatial information* (i.e., coordinates data).

In geostatistics, methods accounting for a single variable, such as *simple kriging (SK)* and *ordinary kriging (OK)*, are univariate. Although *universal kriging (UK)* uses coordinate information, it is classified as a method accounting for a single variable in this book as also in Goovaerts (1997).

The *features* of the univariate geostatistical methods, such as *exactness*, are compared in the previous reviews (Li and Heap 2008, 2014).

All *kriging estimators* are variants of the basic Equation (5.1) which is a modification of Equation (4.1).

$$\hat{y}(x_0) - \mu = \sum_{i=1}^{n} \lambda_i \cdot [y(x_i) - \mu(x_0)] \tag{5.1}$$

where μ is a known *stationary mean*, assumed to be constant over the whole domain and calculated as the average of sample data (Wackernagel 2003); λ_i is *kriging* weight; n is the *number of samples used* to make the estimation, changing with the size of pre-defined *search window*; and $\mu(x_0)$ is the mean of samples within the search window (Li and Heap 2008).

The *kriging* weights are estimated by minimizing the variance (Equation (5.2)), as follows:

$$
\begin{aligned}
var[\widehat{Y}(x_0)] &= E[(\widehat{Y}(x_0) - Y(x_0))^2] \\
&= E[(\widehat{Y}(x_0))^2 + (Y(x_0))^2 - 2\widehat{Y}(x_0)Y(x_0))] \\
&= \sum_{i=1}^{n} \lambda_i \lambda_j C(x_i - x_j) + C(x_0 - x_0) - 2 \sum_{i=1}^{n} \lambda_i C(x_i - x_0)
\end{aligned}
\tag{5.2}
$$

where $Y(x_0)$ is the true value expected at point x_0, n represents the number of observations

DOI: 10.1201/9781003091776-5

to be included in the estimation, and $C(x_i - x_j) = Cov[Y(x_i), Y(x_j)]$ (Isaaks and Srivastava 1989). Step-by-step procedures for finding Equation (5.2) and linking it to γ can be found in Clark and Harper (2001).

In this chapter, the following methods will be introduced:
(1) *simple kriging (SK)*,
(2) *ordinary kriging (OK)*,
(3) *universal kriging (UK)*, and
(4) *block kriging (BK)*.

For these *kriging methods*, the `gstat` package will be used to demonstrate their applications and a demonstration for various kriging methods is available via `demo(examples)` (Gräler, Pebesma, and Heuvelink 2016).

The `gravel` in the `petrel` point data set in the `spm` package (Li 2019b) and the `pb.df` grid data set in Chapter 1 will be used to demonstrate the applications of the univariate geostatistical methods in *R*.

Prior to introducing the univariate geostatistical methods, variogram modeling that is an essential component of all geostatistical methods will be firstly introduced.

5.1 Variogram modeling

For variogram modeling, a few important concepts need to be introduced first. Then variogram modeling and variogram model selection will be detailed.

5.1.1 Concepts for variogram modeling

The important concepts are: (1) *semivariance* and *semivariogram* or *variogram*, (2) *variogram models (vgm)*, and (3) *anisotropy*.

1. Semivariance and semivariogram or variogram

Semivariance (γ) of Y between two data points, an important concept in geostatistics, is defined as in Equation (5.3):

$$\gamma(x_i, x_0) = \gamma(h) = var[Y(x_i) - Y(x_0)]/2 \qquad (5.3)$$

where Y is an attribute, h is the distance between data point x_i and x_0, and $\gamma(h)$ is the *semivariogram* (commonly referred to as *variogram*) (Webster and Oliver 2001).

The *semivariance* can be estimated from the data, as in Equation (5.4):

$$\hat{\gamma}(h) = \sum_{i=1}^{n} ([y(x_i) - y(x_i + h)])^2 / 2n \qquad (5.4)$$

where n is the number of sample pairs separated by distance h.

A plot of $\hat{\gamma}(h)$ against h is called *experimental variogram* (Figure 5.1). It displays the following important features (Burrough and McDonnell 1998):

(1) the *nugget*, a positive value of $\hat{\gamma}(h)$ at h close to 0, is the residual, reflecting the variance of sampling errors and the spatial variance at shorter distance than the minimum sample spacing; and

(2) the *range*, a value of distance at which the *sill* is reached, and samples separated by a distance larger than the range are assumed to be spatially independent. The range provides information about the size of a search window used in geostatistical methods.

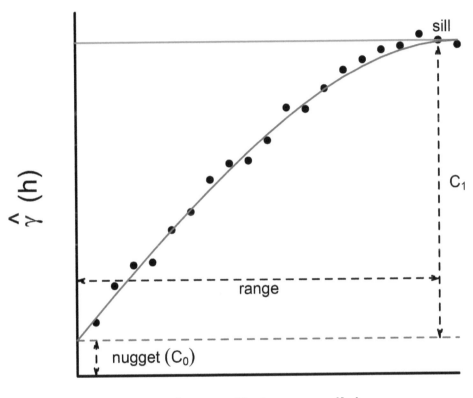

FIGURE 5.1: An example of a semivariogram as illustrated by a spherical model, with range, nugget (C_0) and sill ($C_0 + C_1$) (based on the equation in Burrough and McDonnell (1998)).

2. Variogram models

The *variogram models* may consist of simple models, including: Nugget, *Spherical (Sph)*, *Exponential (Exp)*, *Gaussian (Gau)*, and Linear or the nested sum of one or more simple models (Burrough and McDonnell 1998; Webster and Oliver 2001; Pebesma 2004). There are 20 variogram model types that can be seen via the function vgm in the gstat package below.

```
library(gstat)
vgm()
```

```
      short                                      long
1      Nug                              Nug (nugget)
2      Exp                         Exp (exponential)
3      Sph                           Sph (spherical)
4      Gau                            Gau (gaussian)
5      Exc         Exclass (Exponential class/stable)
6      Mat                              Mat (Matern)
7      Ste  Mat (Matern, M. Stein's parameterization)
8      Cir                             Cir (circular)
9      Lin                               Lin (linear)
10     Bes                              Bes (bessel)
11     Pen                      Pen (pentaspherical)
12     Per                             Per (periodic)
13     Wav                                Wav (wave)
14     Hol                                Hol (hole)
15     Log                         Log (logarithmic)
16     Pow                               Pow (power)
17     Spl                              Spl (spline)
18     Leg                           Leg (Legendre)
19     Err                   Err (Measurement error)
20     Int                           Int (Intercept)
```

A range of *variogram models* can be visualized using the function show.vgms in the gstat package. Four commonly used variogram models are illustrated according to the equations in Burrough and McDonnell (1998) in a previous study (Li and Heap 2008).

3. Anisotropy

Regarding *anisotropy* (or *geometric anisotropy*), it implies that the iso-level contours of the covariance function are elliptical (Pebesma et al. 2018). For two-dimensional spatial predictive modeling, the anisotropy ellipse is defined by two parameters:
(1) the main axis direction or the orientation angle of the ellipse, and
(2) the anisotropic ratio, the ratio of the minor range to the major range (a value between 0 and 1) (Gräler, Pebesma, and Heuvelink 2016).

5.1.2 Variogram modeling and variogram model selection

For *variogram modeling*, two steps are required. The first is to calculate the variogram from sample data, or residuals if a trend is considered, and the second is to fit nuggets, ranges and/or sills from a simple or nested variogram model to the variogram. *Variogram diagnostics* and *variogram model selection* can be carried out with the gstat, geoR (Ribeiro Jr et al. 2020) and vardiag (Glatzer 2015) packages, which can lead to the right choice of a variogram model for the data.

In this section, we will introduce:
(1) variogram modeling without anisotropy;
(2) auto-selection of variogram model;
(3) anisotropy and estimation of anisotropy parameters; and
(4) variogram modeling with anisotropy.

1. Variogram modeling without anisotropy

The *variogram modeling* can be demonstrated using the functions variogram and fit.variogram in the gstat package. The function variogram calculates the sample variogram from data or residuals, with options for directional, robust, and pooled variogram, and for irregular distance intervals. The function `fit.variogram` fits nugget(s), range(s) and/or sill(s) from a simple or nested variogram model to a sample variogram (Gräler, Pebesma, and Heuvelink 2016). There are many arguments for these functions and the descriptions of relevant arguments in the functions are detailed in their help files. These functions will be applied to the `gravel` data in the `petrel` data set to demonstrate variogram modeling.

For `variogram`, the default values are used for all arguments, and the `gravel` data will be `log` transformed according to exploratory analysis in Chapter 2. In addition, data transformation can be selected based on its effect on predictive accuracy for various kriging methods as shown in Sections 5.2.2, 5.3.3 and 5.4.4.

For `fit.variogram`, the default values are used also for all arguments except `vgm()`. For `vgm` the following arguments are specified:
(1) `psill`, (partial) sill, specified as `(max(vgm1$gamma)+ median(vgm1$gamma))/2`;
(2) `model`, model type, with five models tested (i.e., "*Exp*", "*Sph*", "*Gau*", "*Matern (Mat)*", "*Matern, M. Stein's parameterization (Ste)*"));
(3) `range`, range parameter, specified as `0.1 * diag`; and
(4) `nugget`, nugget, specified as `min(vgm1$gamma)`.

The specifications are chosen according to Hiemstra (2013) and to ensure the results of variogram modeling comparable with that from `autofitVariogram` below. The specifications for `vgm` are quantified, and relevant variogram modeling is conducted by

```
library(gstat)
library(spm)
library(sp)

data(petrel)
gravel <- petrel[, c(1, 2, 5)]

diag <- sqrt((diff(range(petrel$long)))^2+(diff(range(petrel$lat)))^2)

coordinates(gravel) = ~long+lat

gravel$loggravel <- log(gravel$gravel + 1)

vgm1 <- variogram(loggravel ~ 1, gravel)

psill1 = (max(vgm1$gamma) + median(vgm1$gamma))/2
range1 = 0.1 * diag #max(vgm1$dist)*0.5
nugget1 = min(vgm1$gamma)

model.1 <- fit.variogram(vgm1,vgm(psill1, "Exp", range1, nugget1))
model.2 <- fit.variogram(vgm1,vgm(psill1, "Gau", range1, nugget1))
model.3 <- fit.variogram(vgm1,vgm(psill1, "Sph", range1, nugget1))
model.4 <- fit.variogram(vgm1,vgm(psill1, "Mat", range1, nugget1))
model.5 <- fit.variogram(vgm1,vgm(psill1, "Ste", range1, nugget1))
```

In `variogram()` above, the specification of `loggravel ~ 1` suggests that no trend is considered, as only spatial data are considered. In `fit.variogram`, the default values (i.e., 0 ,1) are used for anisotropy parameters in `anis()`, suggesting no directional changes being considered either.

The parameters derived from the above models are listed below.

```
model.1
```

```
  model psill  range
1   Nug  0.00 0.0000
2   Exp  1.38 0.1465
```

```
model.2
```

	model	psill	range
1	Nug	0.2504	0.00000
2	Gau	1.1933	0.07359

```
model.3
```

	model	psill	range
1	Nug	0.000	0.00000
2	Sph	1.201	0.08337

```
model.4
```

	model	psill	range	kappa
1	Nug	0.00	0.0000	0.0
2	Mat	1.38	0.1465	0.5

```
model.5
```

	model	psill	range	kappa
1	Nug	0.00	0.0000	0.0
2	Ste	1.38	0.2073	0.5

The above fitted variograms are visualized in Figure 5.2. It shows that *Exp*, *Mat*, and *Ste* fit the data similarly, with a better fitting to the data than other two models (i.e., *Gau* and *Sph*) in terms of nugget, partial sill, and range.

2. Auto-selection of variogram model

The function `autofitVariogram` in the `automap` package can be used to automatically fit a variogram through `fit.variogram` to the data and select the model that has the smallest residual sum of squares with the sample variogram (Hiemstra 2013). The function `autofitVariogram` selects a variogram model from four candidate models (i.e., "Sph", "Exp", "Gau", and "Ste") as shown for `loggravel` below.

```
library(automap)

model.6 <- autofitVariogram(loggravel ~ 1, gravel)

model.6$var_model
```

	model	psill	range	kappa
1	Nug	0.4726	0.0000	0
2	Ste	0.8879	0.2154	10

The results show that the model *Ste* is selected by `autofitVariogram`.

Variogram models `model.1` to `model.6` are illustrated in Figure 5.2 by

```
library(lattice)
mypanel = function(x,y,...) {
 vgm.panel.xyplot(x,y,...)
 panel.lines(variogramLine(model.2, max(vgm1$dist)), col = 'blue')
 panel.lines(variogramLine(model.3, max(vgm1$dist)), col = 'green')
 panel.lines(variogramLine(model.4, max(vgm1$dist)), lty = "longdash", col = '
     orange')
 panel.lines(variogramLine(model.5, max(vgm1$dist)), lty = "dotted", col = 'black'
     )
 panel.lines(variogramLine(model.6$var_model, max(vgm1$dist)), col = 'purple')
}

par(font.axis = 2, font.lab = 2)
plot(vgm1, model = model.1, col = 'red', panel = mypanel, ylim = c(0, 1.8))
```

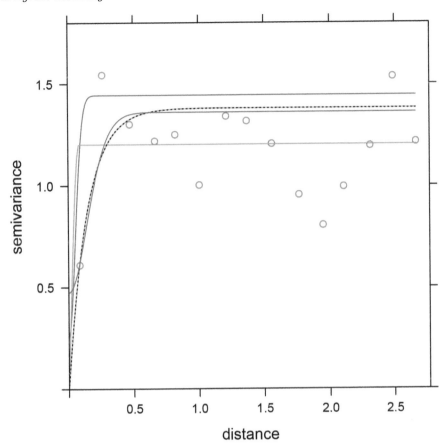

FIGURE 5.2: Variogram models (Exp (Exponential): red line, Gau (Gaussian): blue line, Sph (Spherical): green line, Mat (Matern): orange longdash, and Ste (Matern, M. Stein's parameterization): black dotted-line, Ste (selected by `autofitVariogram`): purple) for `loggravel` data.

It should be noted that `model.5` in Figure 5.2 is slightly different from `model.6`, although they all use the variogram model `Ste`; and `model.6` fits the data with a much higher nugget than `model.5`.

The models selected so far are not based on predictive accuracy. However, the choice of variogram model can be and should be determined according to predictive accuracy. How to select variogram model based on the predictive accuracy is demonstrated in Sections 5.2.2, 5.3.3 and 5.4.4.

The models selected so far are based on the assumption that the data used (i.e., `loggravel`) is isotropic. If it is suspected that such assumption may be not valid, then it should be examined. This will be demonstrated below.

3. Anisotropy and estimation of anisotropy parameters

Anisotropy can be examined visually and quantified as follows.

Visual examination of anisotropy

For *variogram modeling* with *anisotropy*, *anisotropy parameters* need to be estimated. This can be demonstrated using `loggravel` data derived from the `gravel` data set as below.

```
vgm2 <-variogram(loggravel ~ 1, gravel, alpha = c(0, 45, 90, 135))
psill2 = (max(vgm2$gamma) + median(vgm2$gamma))/2
nugget2 = min(vgm2$gamma)

model.7 <- fit.variogram(vgm2, vgm(psill2, "Ste", range1, nugget2))
```

The parameters fitted for `model.7` are

```
model.7
```

```
  model psill  range kappa
1   Nug 0.000 0.0000   0.0
2   Ste 1.382 0.2075   0.5
```

The results of `model.7` can be visually examined in Figure 5.3 by

```
par(font.axis = 2, font.lab = 2)
plot(vgm2, model = model.7)
```

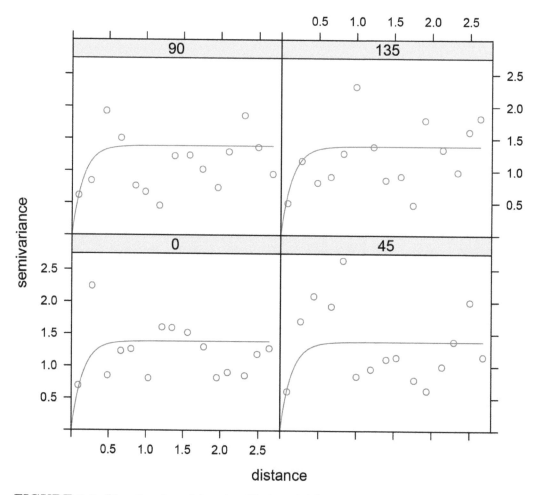

FIGURE 5.3: Directional models using 'Ste' model for `loggravel`.

The results show that the parameters are similar to those for model.5 that is without *directional changes*, i.e., suggesting that loggravel data is isotropic.

The directional changes can be further illustrated in Figure 5.4 by

```
# hist(gravel$loggravel, breaks = c(seq(0, 5, 0.5)))
g = gstat(NULL, "loggravel <= 1.5", I(loggravel <= 1.5) ~ 1, gravel)
g = gstat(g, "loggravel < 2.5", I(loggravel < 2.5) ~ 1, gravel)
g = gstat(g, "loggravel <= 3", I(loggravel <= 3) ~ 1, gravel)
    # calculate multivariable, directional variogram:
v = variogram(g, alpha=c(0,45,90,135))

par(font.axis = 2, font.lab = 2)
plot(v, group.id = TRUE, auto.key = TRUE)  # direction panels
```

The *directional changes* can also be further visually examined in *variogram-map* in Figure 5.5 by

```
# variogram maps:
par (font.axis = 2, font.lab = 2)
plot(variogram(g, cutoff=5, width=1,  map=TRUE))

# plot(variogram(g,  cutoff=2, width=0.5, map=TRUE), np=TRUE)
```

The modeling of anisotropy in Figures 5.4 and 5.5 show that an anisotropy ellipse is not apparent for loggravel, the default anis = c(0, 1) could be used for kriging methods.

Estimation of anisotropy parameters

For *variogram modeling* with *anisotropy*, the function estimateAnisotropy in the intamap package (Pebesma et al. 2018) can be used to determine anisotropy and to estimate anisotropy parameters. This function estimates geometric anisotropy parameters for two-dimensional point data using the covariance tensor identity method (Chorti and Hristopulos 2008; Pebesma et al. 2018). We will apply estimateAnisotropy to loggravel below.

```
library(intamap)

variogram1 <- estimateAnisotropy(gravel, "loggravel", loggravel ~ 1)
variogram1

$ratio
[1] 1.135

$direction
[1] -82.22

$Q
       Q11    Q22    Q12
[1,] 21.69 16.98 0.6544

$doRotation
[1] TRUE
```

It shows that anisotropy does exist for loggravel and directional changes need to be considered in variogram modeling, which is different from the results based on the visual examination of directional variogram modeling above.

To consider directional changes:
(1) alpha argument in variogram needs to be specified, and
(2) anis argument in fit.variogram should be specified. For anis argument, two parameters (i.e., the main axis direction and the anisotropy ratio) are required.

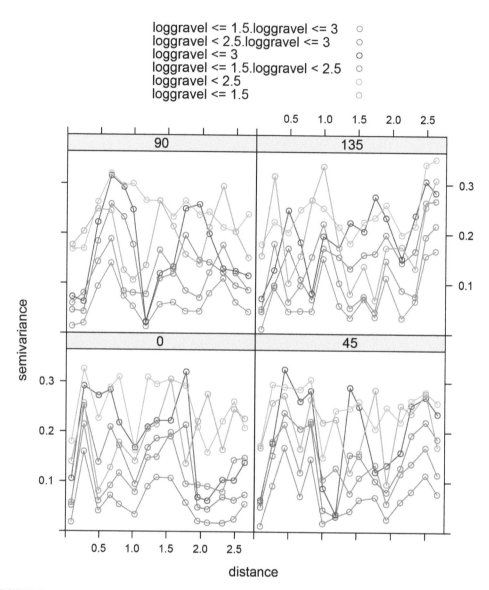

FIGURE 5.4: Directional variogram for `loggravel`.

The main *axis direction* and *anisotropy ratio* derived from `estimateAnisotropy` are different from those used in `gstat`. Since the functions `variogram` and `fit.variogram` in the `gstat` package are going to be used for kriging methods, the parameters estimated by `estimateAnisotropy` need to be converted for `variogram` and `fit.variogram` as below.

```
dir1 <- 90 - variogram1$direction
ratio1 <- 1 / variogram1$ratio
```

The parameters `dir1` and `ratio1` can now be used as the main axis direction and anisotropy ratio, respectively, for `variogram` and `fit.variogram`. Again, the choice of data transformation, alpha, variogram model and parameters for `vgm()` can be determined based on their effects

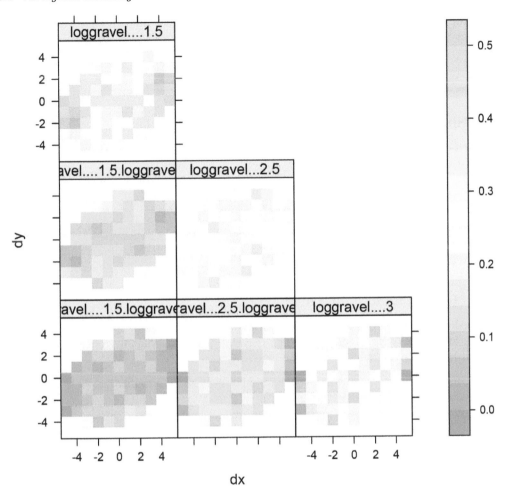

FIGURE 5.5: Variogram map for `loggravel`.

on the predictive accuracy using the functions such as `krigecv` or `okcv` that will be introduced in Sections 5.2.2, 5.3.3 and 5.4.4 for kriging methods.

4. Variogram modeling with anisotropy

Variogram modeling with anisotropy can be conducted as follows:

```
vgm3 <- variogram(log(gravel+1) ~ 1, gravel, alpha = c((0:8)*20))
psill3 = (max(vgm3$gamma) + median(vgm3$gamma))/2
nugget3 = min(vgm3$gamma)

model.8 <- fit.variogram(vgm3, vgm(psill3, "Ste", range1, nugget3,
          anis = c(dir1, ratio1)))
```

```
model.8
```

	model	psill	range	kappa	ang1	anis1
1	Nug	0.000	0.0000	0.0	0.0	1.0000
2	Ste	1.377	0.2073	0.5	172.2	0.8808

The model `model.8` should be used for kriging with anisotropy for `loggravel`.

Anisotropy and anisotropy parameters can be determined based on their effects on predictive accuracy using cross-validation functions as demonstrated in Section 5.2.2.

5.2 Simple Kriging

The predictions of *simple kriging* (*SK*) are based on Equation (5.5) that is a modified version of Equation (5.1).

$$\hat{y}(x_0) = \sum_{i=1}^{n} \lambda_i y(x_i) + [1 - \sum_{i=1}^{n} \lambda_i]\mu \qquad (5.5)$$

where μ is a known stationary mean, assumed to be constant over the whole domain and calculated as the average of sample data (Wackernagel 2003).

5.2.1 Implementation of *SK* in `krige`

The predictions of *SK* can be generated with the function `krige` in the `gstat` package. The `krige` is for various kriging methods. The descriptions of relevant arguments of `krige` are detailed in its help file, which can be accessed by `?krige`.

For *SK* using `krige`, the following arguments may need to be specified:
(1) `formula`, that defines the dependent variable;
(2) `locations`, spatial data locations (i.e., coordinates) such as ~x+y;
(3) `data`, dataframe, containing dependent variable, independent variables, and coordinates;
(4) `newdata`, prediction locations and attributes with independent variables (if present);
(5) `model`, variogram model of dependent variable;
(6) `nmax`, the number of nearest observations that should be used for generating predictions;
(7) `beta`, the mean of the dependent variable; and
(8) `block`, block size.

1. Implementation of *SK* in `krige` without anisotropy

The application of *SK* without anisotropy (i.e., *isotropy*) can be demonstrated with `nmax = 20` and the variogram model derived for `Ste` (i.e., model.5) as below.

```
library(spm)
library(gstat)

pb <- pb.df
gridded(pb) = ~long + lat
beta1 <- mean(gravel$loggravel)
# or beta1 = lm(loggravel ~ 1, gravel)$coef

sk1 = krige(loggravel ~ 1, gravel, pb, model = model.5, nmax = 20, beta = beta1)

head(sk1, 3)
```

2. Implementation of *SK* in `krige` with anisotropy

When `model.5` for *SK* above is replaced with `model.8`, the anisotropic changes are then considered in generating predictions as shown below.

```
sk2 = krige(loggravel ~ 1, gravel, pb, model = model.8, nmax = 20, beta = beta1)
head(sk2, 3)
```

It should be noted that the predictions and variances are at log scale and need to be back-transformed to original scale as below.

```
names(sk2) <- c('predictions', 'variances')
sk2.pred <- sk2
sk2.pred$predictions <- exp(sk2$predictions) - 1 # back transformation
sk2.pred$variances <- (exp(sqrt(sk2$variances)) - 1) ^ 2 # back transformation
```

The function `gstat` can also be used for *SK* in the same way as it being used for *OK* in Section 5.3.

5.2.2 Parameter optimization for *SK*

For *SK*, relevant arguments can be optimized to maximize its predictive accuracy. This can be achieved with the function `krigecv` in the `spm2` package (Li 2021a). The function `krigecv` is a cross-validation function developed for kriging methods in the `gstat` package. The descriptions of relevant arguments of `krigecv` are detailed in its help file, which can be accessed by `?krigecv`.

Using `krigecv` for *SK*, the following arguments may need to be specified:
(1) `longlat`, a dataframe contains `longitude` (`long`) and `latitude` (`lat`) of point samples;
(2) `trainy`, a vector of response, must have length equal to the number of rows in trainx;
(3) `trainpredx`, a dataframe contains predictive variables of point samples; and if `longitude` and `latitude` are going to be used as predictive variables, they should also be included but named in names other than `long` and `lat`;
(4) `validation`, validation methods, include *LOO* (i.e., leave-one-out) and *CV* (i.e., cross-validation);
(5) `cv.fold`, integer; number of folds in the cross-validation; if `cv.fold` > 1, then apply n-fold cross-validation; the default is 10, i.e., 10-fold cross-validation that is recommended;
(6) `nmaxkrige`, the number of nearest observations that should be used for a prediction;
(7) `transformation`, transform the response variable to normalize the data; can be "sqrt" for square root, "arcsine" for arcsine, "log" or "none" for non-transformation; and by default, "none" is used;
(8) `delta`, numeric; to avoid $\log(0)$ in the log transformation;
(9) `formula`, define the response vector and (possible) regressor; an object (i.e., `variogram.formula`) for `variogram` or a formula for `krige`;
(10) `vgm.args`, arguments for `vgm`, e.g., variogram model of response variable and anisotropy parameters and see `vgm` in the `gstat` package for details; and by default, "Sph" is used;
(11) `anis`, anisotropy parameters;
(12) `alpha`, direction in plane (x,y);
(13) `block`, block size and see `krige` in the `gstat` package for details;
(14) `beta`, for simple kriging; and
(15) `predacc`, can be either "VEcv" for `vecv` or "ALL" for all measures in the function `pred.acc`.

The following is to demonstrate how to find the optimal parameters for *SK* using `krigecv`. Four arguments `transformation`, `alpha`, `model` and `nmax` will be optimized in the demonstration.

1. Selection of `transformation`

The argument `transformation` can be selected by visual examination of histogram of the data as demonstrated in Chapter 2. Now we will show how to select it using `krigecv` based on the predictive accuracy of *SK* for `gravel` in the `gravel` data set.

Firstly, we need to estimate anisotropy parameters for *SK* with data under various transformation.

```
library(intamap)
```

sqrt

```
gravel$sqrtgravel <- sqrt(gravel$gravel)

variogram.sqrt <- estimateAnisotropy(gravel, "sqrtgravel", sqrtgravel ~ 1)
variogram.sqrt$doRotation
```

```
[1] TRUE
```

```
dir.sqrt <- 90 - variogram.sqrt$direction
ratio.sqrt <- 1 / variogram.sqrt$ratio
```

acrsine

```
gravel$arcsinegravel <- asin(sqrt(gravel$gravel/100))

variogram.arcsine <- estimateAnisotropy(gravel, "arcsinegravel", arcsinegravel ~
    1)
variogram.arcsine$doRotation
```

```
[1] FALSE
```

```
dir.arcsine <- 0 # as default
ratio.arcsine <- 1 # as default
```

log

For `loggravel`, the parameters estimated (i.e., `dir1` and `ratio1`) in Section 5.1.2 will be used.

none

```
variogram.none <- estimateAnisotropy(gravel, "gravel", gravel ~ 1)
variogram.none
```

```
$ratio
[1] 1.084
```

```
$direction
[1] -37.07
```

```
$Q
      Q11   Q22   Q12
[1,] 7852  8204  620.1
```

```
$doRotation
[1] FALSE
```

```
dir.none <- 0 # as default
ratio.none <- 1 # as default
```

The results show that anisotropy is weak and rotation is no longer required for `arcsinegravel` and `gravel` data. Then the parameters estimated for each transformation type will be used to determine the optimal transformation type. We will assume `vgm.args = "Ste"` and `nmax = 20`. We also need to provide a list of `alpha` for corresponding anisotropic parameters. The estimation will be conducted with `krigecv` as below.

```
library(spm2)

data(petrel)

transf <- c("sqrt", "arcsine", "log", "none")
beta.tr <- c(mean(sqrt(gravel$gravel)), mean(asin(sqrt(gravel$gravel/100))), beta1
    , mean(gravel$gravel))
anis.dir <- c(dir.sqrt, dir.arcsine, dir1, dir.none)
anis.ratio <- c(ratio.sqrt, ratio.arcsine, ratio1, ratio.none)
alpha.tr <- list (c((0:8)*20), 0, c((0:8)*20), 0)
skopt.tr <- NULL

for (i in 1:length(transf)) {
  set.seed(1234)
  skcv1 <- krigecv(petrel[, c(1, 2)], petrel[, 5], nmax = 20, transformation =
      transf[i], vgm.args = "Ste", anis = c(anis.dir[i], anis.ratio[i]), alpha =
      alpha.tr[[i]], beta = beta.tr[i], predacc = "VEcv")
  skopt.tr[i] <- skcv1
}
```

For cross-validation, the default, 10-fold, is used. The predictive accuracy for each data transformation and the optimal `transformation` are:

```
skopt.tr # predictive accuracy
```

```
[1] 24.05 23.23 15.87 29.53
```

```
max(skopt.tr) # maximum predictive accuracy
```

```
[1] 29.53
```

The optimal `transformation` that leads to the maximal predictive accuracy of *SK* is

```
skopt.transf <- which (skopt.tr == max(skopt.tr))
transf[skopt.transf] # optimal data transformation
```

```
[1] "none"
```

The results suggest that `gravel` data is isotropic, and data transformation is not required for *SK*.

2. Estimation of `alpha`

In the above demonstration for anisotropic data, `alpha` is specified as (0:8)*20, which may be not optimal. We can test various options for `alpha`. The `sqrt` will be used for `transformation` to show the estimation of `alpha` below.

```
library(spm2)

data(petrel)
beta.sqrt <- mean(sqrt(gravel$gravel))
alpha.sqrt <- list (0, c((0:17)*10), c((0:11)*15), c((0:8)*20), c((0:5)*30), c
    ((0:3)*45))
skopt.alpha <- NULL

for (i in 1:length(alpha.sqrt)) {
  set.seed(1234)
  skcv1 <- krigecv(petrel[, c(1, 2)], petrel[, 5], nmax = 20, transformation = "
      sqrt", vgm.args = "Ste", anis = c(dir.sqrt, ratio.sqrt), alpha = alpha.sqrt
      [[i]], beta = beta.sqrt, predacc = "VEcv")
  skopt.alpha[i] <- skcv1
}
```

The predictive accuracy for each option and the optimal `alpha` are

```
skopt.alpha # predictive accuracy

[1] 23.41 23.03 24.05 24.05 24.26 24.06

max(skopt.alpha) # maximum predictive accuracy

[1] 24.26
```

The optimal `alpha` that leads to the maximal predictive accuracy of *SK* is

```
skopt.alpha1 <- which (skopt.alpha == max(skopt.alpha))
alpha.sqrt[skopt.alpha1] # optimal option for `alpha`

[[1]]
[1]   0  30  60  90 120 150
```

The accuracy, *VEcv*, for the best choice of `alpha` is 24.26% that is less than the above observed predictive accuracy, 29.53%, for *SK* with `anis = c(0, 1)` and `transformation = "none"`. It suggests that although the choice of optimal `alpha` improves the predictive accuracy of *SK* for square root transformed `gravel` data, it would not affect the selection of data transformation. That is, for `gravel` data, *SK* with `transformation = "none"` and `anis = c(0, 1)` is preferred.

3. Selection of `model` and estimation of `nmax`

For *SK*, two further arguments `model` and `nmax` need to be determined by maximizing the predictive accuracy. This can be done with `krigecv`. For `model` argument, all models in `vgm()` can be tested, and we will only use a portion of the models as below for demonstration.

```
library(spm2)

data(petrel)
beta.opt <- mean(petrel[, 5])
nmax.sk <- c(5:40)
vgm.args <- c("Exp", "Gau", "Sph", "Exc", "Mat", "Ste", "Lin")
skopt <- matrix(0, length(nmax.sk), length(vgm.args))

for (i in 1:length(nmax.sk)) {
  for (j in 1:length(vgm.args)) {
    set.seed(1234)
    skcv1 <- krigecv(petrel[, c(1, 2)], petrel[, 5], nmax = nmax.sk[i],
        transformation = "none", vgm.args = vgm.args[j], beta = beta.opt, predacc
        = "VEcv")
    skopt[i, j] <- skcv1
  }
}
```

If anisotropy is detected, then the `anis` argument with relevant `direction` and `ratio` values, such as `anis = c(dir1, ratio1)` for `loggravel` above, should be considered in the application of `krigecv`.

The optimal `nmax` and `model` (i.e., `vgm.args`) that maximize the predictive accuracy are

```
sk.opt <- which(skopt == max(skopt), arr.ind = T)
nmax.sk[sk.opt[, 1]]

[1] 31

vgm.args[sk.opt[, 2]]
```

```
[1] "Exc"
```

For a thorough check, one should repeat the steps above with newly identified parameters, until the optimal `transformation`, `nmax`, `vgm.args`, and relevant `anis` parameters remain constant.

5.2.3 Predictive accuracy of *SK* with the optimal parameters

To stabilize it and get a more reliable accuracy estimation, we can repeat the cross-validation by 100 times as below.

```
set.seed(1234)
n <- 100
skvecv <- NULL

for (i in 1:n) {
  skcv1 <- krigecv(petrel[, c(1, 2)], petrel[, 5], nmax = nmax[sk.opt[, 1]],
      transformation = transf[skopt.transf], vgm.args = vgm.args[sk.opt[, 2]],
      beta = beta.opt, predacc = "VEcv")
  skvecv[i] <- skcv1
}
```

The median and range of the predictive accuracy of *SK* based on 100 repetitions of 10-fold cross-validation using `krigecv` are

```
median(skvecv)
```

```
[1] 31.41
```

```
range(skvecv)
```

```
[1] 21.84 38.53
```

The changes of *VEcv* with repetition times are illustrated in Figure 5.6.

5.2.4 *SK* predictions and variances

We can use `krigepred` in the `spm2` package with the optimal parameters for *SK* to demonstrate the generation of *spatial predictions*. This demonstration will be based on `gravel` in the `petrel` data set and `pb.df` as below.

```
skpred1 <- krigepred(petrel[, c(1,2)], petrel[, 5], pb.df[, -3], nmax = nmax[sk.
    opt[, 1]], transformation = "none", vgm.args = "Exc", beta = beta.opt)
```

Some basic properties of the predictions in `skpred1` are

```
names(skpred1)
```

```
[1] "long"      "lat"        "var1.pred" "var1.var"
```

```
class(skpred1)
```

```
[1] "data.frame"
```

```
dim(skpred1)
```

```
[1] 100828      4
```

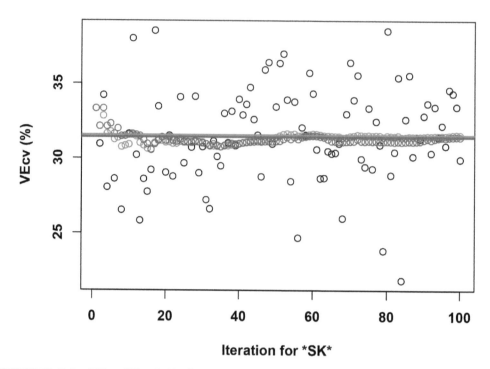

FIGURE 5.6: *VEcv* (%) of *SK* for each of 100 times of 10-fold cross-validation (black circle), their accumulative average (red circle), their accumulative median (blue circle), overall average (red line) and overall median (blue line).

```
head(skpred1, 3)
```

```
     long    lat var1.pred var1.var
1 127.1 -11.28     21.16    391.3
2 127.2 -11.28     21.24    391.3
3 127.2 -11.28     21.24    391.3
```

```
round(range(skpred1$var1.pred), 2)
```

```
[1]  0.85 68.66
```

The spatial distribution of predictions and variances of *SK* are illustrated in Figure 5.7 by

```
skpred1.1 <- skpred1
gridded(skpred1.1) <- ~ long + lat
names(skpred1.1) <- c('predictions', 'variances')

library(raster)

par(font.axis = 2, font.lab = 2)
plot(brick(skpred1.1))
```

The spatial predictions can also be visualized using the following *R* code.

```
# 1. use `spplot` on gridded data
spplot(skpred1.1, c("predictions"), xlab = expression("Longitude"^o), ylab =
    expression("Latitude"^o), key.space = list(x=0.1,y=.95, corner = c(-1.2,2.8)),
```

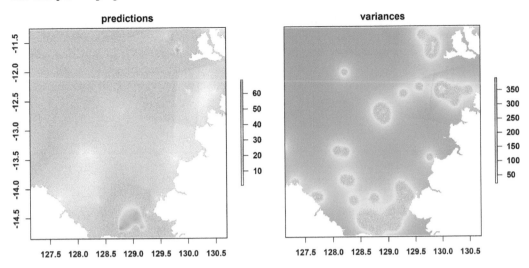

FIGURE 5.7: Spatial patterns of predictions and variances of *SK*.

```
col.regions = rev(bpy.colors(100)), scales=list(draw=T), colorkey = list(at = c(
    seq(0,70,5)), space="right", labels = c("0%","","","","20%","","","","40%","",
    "","","60%","","")), at=c(seq(0, 70, 5)))

# 2. use `spplot` on gridded data with self-define colors for `col.regions`
library(lattice)
spplot(skpred1.1, c("predictions"), xlab=expression("Longitude"^o), ylab=
    expression("Latitude"^o), key.space=list(x=0.1,y=.95, corner=c(-1.2,2.8)), col
    .regions = trellis.par.set(sp.theme(regions = list(col = colorRampPalette(c("
    lightyellow", "saddlebrown"))))), scales=list(draw=T), colorkey = list(at = c(
    seq(0,70,5)), space="right", labels = c("0%","","","","20%","","","","40%","",
    "","","60%","","")), at=c(seq(0,80, 5)))

# 3. use `spplot` on gridded data. This is only to provide an example for
    visualization of predictions and it works for variables that have about the
    same data range.
spplot(skpred1.1, col.regions = bpy.colors(100), as.table=TRUE)

# 4. use `spplot` on gridded data with point data added
petrel.points <- petrel[, 1:2]
coordinates(petrel.points) <- ~ long + lat
petrel.pts = list("sp.points", petrel.points, pch = 1, col = "black", cex = 0.1)
spplot(skpred1.1, c("predictions"), sp.layout = list(petrel.pts), col.regions =
    rev(bpy.colors()))

# 5. use `spplot` on a RasterLayer
r1 <- raster(skpred1.1)
spplot(r1, col.regions = rev(bpy.colors(256)))

# 6. use `plot` on a RasterLayer
plot(r1)
```

SK assumes second-order stationary, that is, sample mean, variance, and covariance are constant over the domain or the region of interest (Webster and Oliver 2001; Wackernagel 2003). If such assumption cannot be satisfied, then other methods, such as *OK*, need to be used (Burrough and McDonnell 1998; Li and Heap 2008). In practice, *OK* is often preferred to *SK* due to several features detailed in Li and Heap (2014).

It needs to be clarified that the *variances* associated with the predictions by the kriging

methods introduced in this book are independent of the data values and **do not** reflect
the uncertainties of the predictions (Goovaerts 1997). Therefore, the variances cannot be
used as an *uncertainty* measure of kriging predictions (Goovaerts 1997; Li and Heap 2008).
Such variances, however, reflect the variations in spatial departures among samples, so they
are good indicators where samples are sparse and, thus, may provide useful information for
selecting further sampling locations (Li 2019a).

5.3   Ordinary kriging

Ordinary kriging (OK) is similar to *SK*, but with one difference that is *OK* predicting the
value of the attribute using Equations (5.4) and (5.1) by replacing μ with a local mean
$\mu(x_0)$ (i.e., local constant mean, or local stationary mean) and forcing $[1 - \sum_{i=1}^{n} \lambda_i] = 0$
(Goovaerts 1997; Clark and Harper 2001). The *local mean* is the mean of samples within
the search window. Therefore, *OK* essentially uses equations (5.4) and (4.1) to make the
predictions. *OK* estimates the local constant mean, then performs *SK* on the corresponding
residuals, so *OK* only requires a local stationary mean within the *search window* (Goovaerts
1997).

The predictions of *OK* can be generated using the functions `gstat` and `krige`.

5.3.1   Implementation of *OK* in `gstat`

For *OK* using `gstat`, the following arguments may need to be specified:
(1) `formula`, defines dependent variable;
(2) `locations`, spatial data locations (i.e., coordinates) such as ~ x + y;
(3) `data`, dataframe, containing dependent variable, independent variables, and coordinates;
(4) `model`, variogram model of dependent variable; and
(5) `nmax`, the number of nearest observations that should be used for generating predictions.

The application of *OK* will be demonstrated for `loggravel` with `nmax = 20` and `model = model.5`
as below, where `model.5` is from Section 5.1.2.

```
library(sp)
library(gstat)

gstat1 <- gstat(formula = loggravel ~ 1, data = gravel, model = model.5, nmax =
    20)
ok1 <- predict(gstat1, pb)
```

The spatial predictions and variances in `ok1` are

```
head(ok1, 3)

  var1.pred var1.var
1     2.644    1.531
2     2.644    1.531
3     2.644    1.531


summary(ok1)


Object of class SpatialPixelsDataFrame
Coordinates:
        min     max
```

```
long 127.14 130.70
lat  -14.83 -11.27
Is projected: NA
proj4string : [NA]
Number of points: 100828
Grid attributes:
     cellcentre.offset cellsize cells.dim
long            127.15     0.01       356
lat             -14.83     0.01       356
Data attributes:
   var1.pred        var1.var
 Min.    :0.04   Min.    :0.0094
 1st Qu.:1.94    1st Qu.:1.3093
 Median :2.31    Median :1.4836
 Mean   :2.26    Mean    :1.3881
 3rd Qu.:2.62    3rd Qu.:1.5447
 Max.   :4.32    Max.    :2.2420
```

5.3.2 Implementation of *OK* in `krige`

For *OK* using `krige`, the arguments are similar to those used for *SK* above but with the default `beta`.

We also use `loggravel` with `nmax = 20` and `model = model.5` for `krige` as follows.

```
ok2 = krige(loggravel ~ 1, gravel, pb, model = model.5, nmax = 20)
head(ok2, 3)

summary(ok2)

Object of class SpatialPixelsDataFrame
Coordinates:
        min     max
long 127.14 130.70
lat   -14.83 -11.27
Is projected: NA
proj4string : [NA]
Number of points: 100828
Grid attributes:
     cellcentre.offset cellsize cells.dim
long            127.15     0.01       356
lat             -14.83     0.01       356
Data attributes:
   var1.pred        var1.var
 Min.    :0.04   Min.    :0.0094
 1st Qu.:1.94    1st Qu.:1.3093
 Median :2.31    Median :1.4836
 Mean   :2.26    Mean    :1.3881
 3rd Qu.:2.62    3rd Qu.:1.5447
 Max.   :4.32    Max.    :2.2420

identical(ok1, ok2)

[1] TRUE
```

The results show that `gstat` and `krige` produce the same results. Either of them can be used for *OK* modeling.

If `model.5` in `ok1` or `ok2` is replaced with `model.8` (see Section 5.1.2 for details), then anisotropic changes will be considered in *OK* modeling.

5.3.3   Parameter optimization for *OK*

Similar to *SK*, relevant arguments of *OK* need to be optimized to maximize its predictive accuracy. The *parameter optimization* can be conducted using:
(1) the function okcv in the spm package or krigecv,
(2) the function autoKrige in the automap package (Hiemstra 2013), and
(3) the function estimateParameters in the intamap package.

There are four arguments transformation, alpha, model, and nmax that need to be determined to maximize the predictive accuracy of *OK*.

1. Implementation of *OK* in okcv and krigecv

The functions okcv and krigecv can be used to estimate relevant parameters to maximize the predictive accuracy of *OK*. Since okcv and krigecv will produce the same results for *OK*, we can use either of them. The following is to demonstrate how to seek optimal transformation, alpha, model, and nmax using okcv or krigecv.

Selection of transformation

The same as what has been stated for *SK* in Section 5.2.2, the argument transformation was initially selected according to the visual examination of histograms of the data. Now we can select it using krigecv based on the predictive accuracy of *OK*. For cross-validation, the default, 10-fold, is used. For *OK*, anis.dir, anis.ratio, and alpha.tr should be the same as those for *SK* as detailed in Section 5.2.2.

```
library(spm2)

data(petrel)
transf <- c("sqrt", "arcsine", "log", "none")
anis.dir <- c(dir.sqrt, dir.arcsine, dir1, dir.none)
anis.ratio <- c(ratio.sqrt, ratio.arcsine, ratio1, ratio.none)
alpha.tr <- list (c((0:8)*20), 0, c((0:8)*20), 0)
okopt.tr <- NULL

for (i in 1:length(transf)) {
  set.seed(1234)
  okcv1 <- krigecv(petrel[, c(1, 2)], petrel[, 5], nmax = 20, transformation =
      transf[i], vgm.args = "Ste", anis = c(anis.dir[i], anis.ratio[i]), alpha =
      alpha.tr[[i]], predacc = "VEcv")
  okopt.tr[i] <- okcv1
}
```

The optimal transformation type that results in the maximal predictive accuracy is

```
okopt.transf <- which (okopt.tr == max(okopt.tr))
transf[okopt.transf] # optimal `transformation` type
```

```
[1] "none"
```

The results suggest that for *OK*, to maximize its predictive accuracy, the gravel data is isotropic and should not be transformed.

Estimation of alpha

The optimal alpha for *OK* can be estimated in the same way as for *SK* in Section 5.2.2, which will not be presented here.

Selection of model and estimation of nmax

The optimal model and nmax can be chosen with okcv to maximize the predictive accuracy of

OK as below. The okcv takes the same arguments as krigecv and produces the same results as krigecv.

```
data(petrel)
nmax.ok <- c(5:40)
vgm.args.ok <- c("Exp", "Gau", "Sph", "Exc", "Mat", "Ste", "Lin")
okopt <- matrix(0, length(nmax.ok), length(vgm.args.ok))

for (i in 1:length(nmax.ok)) {
  for (j in 1:length(vgm.args.ok)) {
    set.seed(1234)
    okcv1 <- okcv(petrel[, c(1, 2)], petrel[, 5], nmax = nmax.ok[i],
        transformation = "none", vgm.args = vgm.args.ok[j], anis = c(dir.none,
        ratio.none), predacc = "VEcv")
    okopt[i, j] <- okcv1
  }
}
```

In this example, anis argument could be omitted as the values used are the same as the default values. It is kept so that one can easily modify it to model anisotropic data.

The optimal nmax and vgm that maximize the predictive accuracy are

```
ok.opt <- which (okopt == max(okopt, na.rm=T), arr.ind = T)
nmax.ok[ok.opt[, 1]]
```

```
[1] 8
```

```
vgm.args.ok[ok.opt[, 2]]
```

```
[1] "Gau"
```

Since NAs are in okopt for "EXP", na.rm=T is used. These parameters should be used to produce a stabilized predictive accuracy for and to generate spatial predictions of *OK*.

2. Implementation of *OK* in autoKrige

The function autoKrige can also be used for *OK*. The arguments for autokrige are similar to those for krige but without the model argument. This is because it performs automatic kriging based on a variogram automatically generated through autofitVariogram for a given data set as shown below.

```
library(automap)
```

```
autoKrige1 <- autoKrige(loggravel ~ 1, gravel, pb, nmax = 20)
```

```
names(autoKrige1)
```

```
[1] "krige_output" "exp_var"      "var_model"     "sserr"
```

```
autoKrige1$var_model
```

```
  model  psill  range kappa
1   Nug 0.4726 0.0000     0
2   Ste 0.8879 0.2154    10
```

Although autoKrige1$var_model and model.5 are using the same variogram model (i.e., Ste), the parameters estimated are not the same. Consequently, the predictions, ok1 and autoKrige1$krige_output are different as shown in the summary(ok1) in Section 5.3.1 and summary (autoKrige1$krige_output) below.

The function autoKrige can generate predictions of *OK* directly.

```
names(autoKrige1$krige_output)
```

```
[1] "var1.pred"  "var1.var"   "var1.stdev"
```

```
summary(autoKrige1$krige_output)
```

```
Object of class SpatialPixelsDataFrame
Coordinates:
        min    max
long 127.14 130.70
lat  -14.83 -11.27
Is projected: NA
proj4string : [NA]
Number of points: 100828
Grid attributes:
     cellcentre.offset cellsize cells.dim
long            127.15     0.01       356
lat            -14.83     0.01       356
Data attributes:
   var1.pred          var1.var          var1.stdev
 Min.   :0.401    Min.   :0.498    Min.   :0.705
 1st Qu.:1.971    1st Qu.:1.272    1st Qu.:1.128
 Median :2.304    Median :1.480    Median :1.216
 Mean   :2.257    Mean   :1.384    Mean   :1.169
 3rd Qu.:2.625    3rd Qu.:1.530    3rd Qu.:1.237
 Max.   :4.120    Max.   :2.176    Max.   :1.475
```

Despite the automatic selection of variogram model using autoKrige, other arguments still need to be estimated.

3. Implementation of *OK* in estimateParameters

For *OK* using estimateParameters, the object argument that needs to be specified is detailed in Section 4.1.2.

```
library(intamap)
```

```
krigingObj = createIntamapObject(
        observations = gravel,
        formulaString = as.formula('loggravel ~ 1'),
        class = "automap")
```

```
krigingObj = estimateParameters(krigingObj)
```

The variogram model selected and parameters estimated for the model are

```
krigingObj$variogramModel
```

```
  model  psill  range   ang1   anis1
1   Nug 0.4786 0.0000    0.0  1.0000
2   Gau 0.9117 0.2346  172.2  0.8808
```

The model selected, krigingObj$variogramModel, can then used for *OK* modeling as below.

```
intamap1  = krige(loggravel ~ 1, gravel, pb, model = krigingObj$variogramModel,
    nmax = 20)
```

The predictions intamap1 and ok1 are compared as below.

```
head(intamap1 , 3)
```

```
   var1.pred var1.var
1      2.677    1.535
2      2.677    1.535
3      2.677    1.535
```

```
identical(ok1$var1.pred, intamap1$var1.pred)
```

```
[1] FALSE
```

```
range(ok1$var1.pred - intamap1$var1.pred)
```

```
[1] -2.396  1.672
```

The function `estimateParameters` calls `autofitVariogram`, but selected *Gau* model instead of *Ste* model. As a result, the predictions `intamap1` and `ok1` are different.

Given the difference in variogram models selected, methods that can maximize the predictive accuracy should be used to select and conduct a variogram model.

5.3.4 Predictive accuracy of *OK* with the optimal parameters

To stabilize *predictive accuracy* and get a more reliable accuracy estimation, we need to repeat the cross-validation by a certain number of times. Since `anis = c(dir.none, ratio.none )` is the same as the default, `anis` is not specified in the following application.

```
set.seed(1234)
n <- 100
okvecv <- NULL

for (i in 1:n) {
  okcv1 <- krigecv(petrel[, c(1, 2)], petrel[, 5], nmax = nmax.ok[ok.opt[, 1]],
      transformation = transf[okopt.transf], vgm.args = vgm.args.ok[ok.opt[, 2]],
      predacc = "VEcv")
  okvecv[i] <- okcv1
}
```

The median and range of predictive accuracy of *OK* based on 100 repetitions of 10-fold cross-validation using `krigecv` are

```
median(okvecv)
```

```
[1] 33.98
```

```
range(okvecv)
```

```
[1] -59.53  41.60
```

The changes of *VEcv* with repetition times are illustrated in Figure 5.8. It should be noted that for one repetition, the *VEcv* is negative that is not unusual for environmental data. As a result, it dragged down the subsequent accumulative averages.

The accuracy can also be assessed using the function `krige.cv` in the `gstat` package. There is also a function `autoKrige.cv` in the `automap` package, which in fact calls `krige.cv` to perform cross-validation.

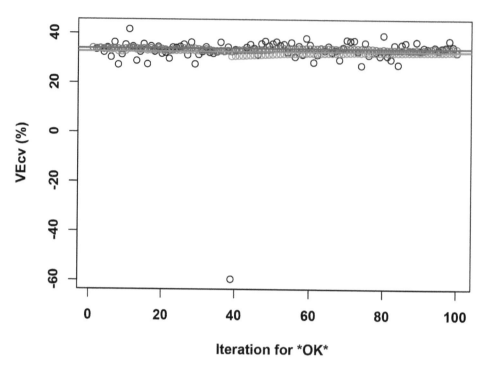

FIGURE 5.8: *VEcv* (%) of *OK* for each of 100 times of 10-fold cross-validation (black circle), their accumulative median (blue circle), their accumulative average (red circle), overall median (blue line) and overall average (red line).

5.3.5   OK predictions and variances

We will use `krigepred` with the optimal parameters for *OK* to generate *spatial predictions* for `pb.df` that is from Section 5.2.1.

```
okpred1 <- krigepred(petrel[, c(1,2)], petrel[, 5], pb.df[, -3], nmax = nmax.ok[ok
    .opt[, 1]], transformation = "none", vgm.args = "Gau")
```

Some basic properties of the predictions in `okpred1` are

```
names(okpred1)
```

```
[1] "long"      "lat"       "var1.pred" "var1.var"
```

```
class(okpred1)
```

```
[1] "data.frame"
```

```
dim(okpred1)
```

```
[1] 100828      4
```

```
head(okpred1, 3)
```

```
   long    lat var1.pred var1.var
1 127.1 -11.28    25.17    106.5
2 127.2 -11.28    25.17    106.5
3 127.2 -11.28    25.17    106.5
```

The spatial distribution of predictions and variances of *OK* are illustrated in Figure 5.9 by

```
okpred1.1 <- okpred1
gridded(okpred1.1) <- ~ long + lat
names(okpred1.1) <- c('predictions', 'variances')

library(raster)

par(font.axis = 2, font.lab = 2)
plot(brick(okpred1.1))
```

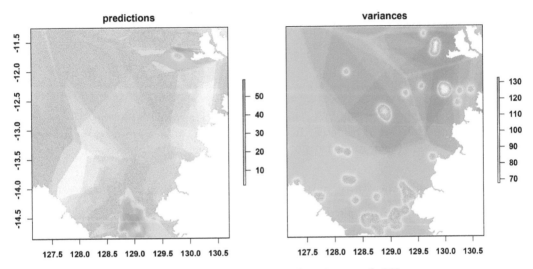

FIGURE 5.9: Spatial patterns of predictions and variances of *OK*.

The prediction range, `range(okpred1$var1.pred)`, is from 2.01% to 59%.

OK requires a stationary mean of the local search window. If the mean is not constant (or non-stationary) within the search window, detrending methods, such as *UK*, *kriging with an external drift* (*KED*) (Verfaillie, Van Lancker, and Van Meirvenne 2006), *regression kriging* (*RK*) (Knotters, Brus, and Oude Voshaar 1995), *universal cokriging* (*UCK*) (Stein, Hoogerwerf, and Bouma 1988) or other detrending methods (Li and Heap 2008, 2014), should be used.

5.4 Universal kriging

Universal kriging (*UK*) is an extension of *OK* by incorporating the local trend within the neighborhood search window as a smoothly varying function of the coordinates. It is also called *kriging with a trend* (*KT*) (Li and Heap 2008) or *external drift kriging* (Gräler, Pebesma, and Heuvelink 2016). *UK* estimates the trend components within each search window and then performs *SK* on the corresponding residuals in the same way as *OK*.

5.4.1   Variogram modeling without anisotropy for *UK*

Similar to *OK*, a *variogram model* needs to be developed for *UK*, but it is based on residuals of the detrended data. Below is an example to demonstrate how to develop a variogram model based on the residuals of `loggravel ~ long + lat` for `loggravel` data.

```
vgm.uk <- variogram(loggravel ~ long + lat, gravel)

psill.uk = (max(vgm.uk$gamma) + median(vgm.uk$gamma))/2
range.uk = 0.1 * diag
nugget.uk = min(vgm.uk$gamma)

model.uk <- fit.variogram(vgm.uk, vgm(psill.uk, "Ste", range.uk, nugget.uk))
model.uk


  model psill  range kappa
1   Nug  0.00 0.0000   0.0
2   Ste  1.35 0.2023   0.5
```

In `variogram()`, the specification of `loggravel ~ long + lat` suggests that a trend is considered but only spatial data are considered. In `vgm()`, the default values are used for anisotropy parameters in `anis()`, suggesting that no directional changes have been considered.

5.4.2   Variogram modeling with anisotropy for *UK*

Estimation of anisotropy parameters for UK

Similar to *OK*, a directional *variogram model* may also need to be developed but will be based on residuals for *UK*. *Anisotropy parameters* can be estimated by applying `estimateAnistropy` to the residuals such as the residuals of `loggravel ~ long + lat` as shown below.

```
library(intamap)

variogram.uk <- estimateAnisotropy(gravel, "loggravel", loggravel ~ long + lat)

[generalized least squares trend estimation]

variogram.uk

$ratio
[1] 1.135

$direction
[1] -80.22

$Q
        Q11   Q22    Q12
[1,] 21.49 16.93 0.8115

$doRotation
[1] TRUE
```

The results indicate that anisotropy does exist for the residuals of `loggravel ~ long + lat`, and directional changes need to be considered in variogram modeling. Since the `gstat` package is going to be used for kriging methods, the parameters estimated by `estimateAnisotropy` need to be converted as below.

```
dir.uk <- 90 - variogram.uk$direction
ratio.uk <- 1 / variogram.uk$ratio
```

Variogram modeling

Variogram modeling with anisotropy can be conducted using the functions `variogram` and `fit.variogram`.

```
vgm.uk2 <- variogram(log(gravel+1) ~ long + lat, gravel, alpha = c((0:8)*20))

model.uk2 <- fit.variogram(vgm.uk2, vgm(psill.uk, "Ste", range.uk, nugget.uk, anis
    = c(dir.uk, ratio.uk)))
model.uk2

  model psill  range kappa  ang1   anis1
1   Nug 0.000 0.0000   0.0    0.0  1.0000
2   Ste 1.347 0.2026   0.5  170.2  0.8808
```

The model `model.uk2` can be used for *UK* with anisotropy as shown below.

5.4.3 Implementation of *UK* in `krige` with anisotropy

```
uk <- krige(loggravel ~ long + lat, gravel, pb, model = model.uk2, nmax = 20)

[using universal kriging]

head(uk, 3)

  var1.pred var1.var
1     4.054    3.239
2     4.050    3.235
3     4.045    3.232
```

The predictions and variances are at log scale and need to be back-transformed to original scale in the same way as that for `sk2` in Section 5.2.1.

5.4.4 Parameter optimization for *UK*

For *UK*, four arguments `transformation`, `alpha`, `model`, and `nmax` need to be optimized to maximize its predictive accuracy.

1. Selection of `transformation`

The argument `transformation` can be chosen using `krigecv` based on the predictive accuracy of *UK* as below. For cross-validation, the default 10-fold, is also used.

Firstly, we need to estimate anisotropy parameters for *UK* with data under four types of data transformation (i.e., "sqrt", "arcsine", "log", and "none").

```
library(intamap)
```

sqrt

```
variogram.sqrt.uk <- estimateAnisotropy(gravel, "sqrtgravel", sqrtgravel ~ long +
    lat)

[generalized least squares trend estimation]

variogram.sqrt.uk
```

```
$ratio
[1] 1.107

$direction
[1] -59.88

$Q
        Q11   Q22  Q12
[1,] 92.23 83.41 7.71

$doRotation
[1] TRUE

dir.sqrt.uk <- 90 - variogram.sqrt.uk$direction
ratio.sqrt.uk <- 1 / variogram.sqrt.uk$ratio
```

arcsine

```
variogram.arcsine.uk <- estimateAnisotropy(gravel, "arcsinegravel", arcsinegravel
    ~ long + lat)

[generalized least squares trend estimation]

variogram.arcsine.uk

$ratio
[1] 1.082

$direction
[1] -57.89

$Q
      Q11   Q22     Q12
[1,] 1.23 1.148 0.08445

$doRotation
[1] FALSE

dir.arcsine.uk <- 0 # as default
ratio.arcsine.uk <- 1 # as default
```

log

For "log" transformation, the parameters estimated for `loggravel` (i.e., `dir.uk` and `ratio.uk`) in Section 5.4.2 will be used.

none

```
variogram.none.uk <- estimateAnisotropy(gravel, "gravel", gravel ~ long + lat)

[generalized least squares trend estimation]

variogram.none.uk

$ratio
[1] 1.085

$direction
[1] -37.54

$Q
      Q11   Q22    Q12
```

```
[1,] 7861 8196 629.8

$doRotation
[1] FALSE


dir.none.uk <- 0 # as default
ratio.none.uk <- 1 # as default
```

The results show that anisotropy is weak and rotation is no longer required for `arcsinegravel` and `gravel` data.

These parameters will be used to select an optimal transformation type by applying `krigecv` with `vgm.args = "Ste"` and `nmax = 20` to `petrel` data set as follows.

```
library(spm2)

data(petrel)
transf <- c("sqrt", "arcsine", "log", "none")
anis.dir.uk <- c(dir.sqrt.uk, dir.arcsine.uk, dir.uk, dir.none.uk)
anis.ratio.uk <- c(ratio.sqrt.uk, ratio.arcsine.uk, ratio.uk, ratio.none.uk)
alpha.uk <- list (c((0:8)*20), 0, c((0:8)*20), 0)
ukopt.tr <- NULL

for (i in 1:length(transf)) {
  set.seed(1234)
  ukcv1 <- krigecv(petrel[, c(1, 2)], petrel[, 5], nmax = 20, transformation =
      transf[i], formula = var1 ~ long + lat, vgm.args = "Ste", anis = c(anis.dir.
      uk[i], anis.ratio.uk[i]), alpha = alpha.uk[[i]], predacc = "VEcv")
  ukopt.tr[i] <- ukcv1
}
```

The optimal type of `transformation` that leads to the maximal predictive accuracy is

```
ukopt.transf <- which (ukopt.tr == max(ukopt.tr))
transf[ukopt.transf]

[1] "none"
```

2. Estimation of `alpha`

The optimal `alpha` can also be estimated for *UK*, which is the same as the parameter estimation of `alpha` for *SK* in Section 5.2.2 and will not be presented here.

3. Selection of `model` and estimation of `nmax`

For *UK*, we also need to seek optimal `model` and `nmax`. They can also be carried out with `krigecv`.

```
library(spm2)

data(petrel)
nmax.uk <- c(5:40)
vgm.args.uk <- c("Exp", "Gau", "Sph", "Exc", "Mat", "Ste", "Lin")
ukopt <- matrix(0, length(nmax.uk), length(vgm.args.uk))

for (i in 1:length(nmax.uk)) {
  for (j in 1:length(vgm.args.uk)) {
    set.seed(1234)
    ukcv1 <- krigecv(petrel[, c(1, 2)], petrel[, 5], nmax = nmax.uk[i],
        transformation = transf[ukopt.transf], formula = var1 ~ long + lat, vgm.
        args = vgm.args.uk[j], anis = c(dir.none.uk, ratio.none.uk), predacc = "
        VEcv")
    ukopt[i, j] <- ukcv1
```

```
    }
}
```

The optimal `nmax` and `model` (i.e., `vgm.args.uk`) that maximize the predictive accuracy are

```
uk.opt <- which (ukopt == max(ukopt, na.rm = T), arr.ind = T)
nmax.uk[uk.opt[, 1]]
```

```
[1] 31
```

```
vgm.args.uk[uk.opt[, 2]]
```

```
[1] "Ste"
```

Since `NA`s are in `ukopt` for "EXP", `na.rm=T` is used. These parameters should be used for *UK*.

5.4.5   Predictive accuracy of `UK` with the optimal parameters

Similar to *OK*, we can stabilize the *predictive accuracy* of *UK* and get a more reliable accuracy estimation by repeating the cross-validation 100 times as below.

```
set.seed(1234)
n <- 100
ukvecv <- NULL

for (i in 1:n) {
  ukcv1 <- krigecv(petrel[, c(1, 2)], petrel[, 5], nmax = nmax.uk[uk.opt[, 1]],
      transformation = transf[ukopt.transf], formula = var1 ~ long + lat, vgm.args
      = vgm.args.uk[uk.opt[, 2]], anis = c(dir.none.uk, ratio.none.uk), predacc =
      "VEcv")
  ukvecv[i] <- ukcv1
}
```

The median and range of predictive accuracy of *UK* based on 100 repetitions of 10-fold cross-validation using `krigecv` are

```
median(ukvecv)
```

```
[1] 32.72
```

```
range(ukvecv)
```

```
[1] 26.26 38.48
```

5.4.6   UK predictions and variances

We will use `krigepred` with the optimal parameters for *UK* to generate *spatial predictions* for `pb.df` that is from Section 5.2.1.

```
ukpred1 <- krigepred(petrel[, c(1,2)], petrel[, 5], pb.df[, -3], nmax = nmax.uk[uk
    .opt[, 1]], transformation = "none", formula = var1 ~ long + lat, vgm.args = "
    Ste")
```

Some basic properties of the predictions in `ukpred1` are

```
names(ukpred1)
```

```
[1] "long"      "lat"      "var1.pred" "var1.var"
```

```
class(ukpred1)
```

```
[1] "data.frame"
```

```
dim(ukpred1)
```

```
[1] 100828      4
```

```
summary(ukpred1)
```

```
      long            lat           var1.pred         var1.var
 Min.   :127    Min.   :-14.8    Min.   :-73.9    Min.   :  37
 1st Qu.:128    1st Qu.:-13.6    1st Qu.: 12.6    1st Qu.: 213
 Median :129    Median :-12.8    Median : 17.3    Median : 226
 Mean   :129    Mean   :-12.8    Mean   : 17.2    Mean   : 340
 3rd Qu.:129    3rd Qu.:-12.0    3rd Qu.: 22.3    3rd Qu.: 271
 Max.   :131    Max.   :-11.3    Max.   : 72.5    Max.   :6673
```

The results show that some predictions are negative, which are incorrect, and need to be corrected. The variances are abnormally high, and for illustration purposes we cap them to 500 as below.

```
ukpred1.1 <- ukpred1
ukpred1.1$var1.pred[ukpred1.1$var1.pred <= 0] <- 0
ukpred1.1$var1.var[ukpred1.1$var1.var >= 500] <- 500
```

The spatial distribution of predictions and variances of *UK* are illustrated in Figure 5.10 by

```
gridded(ukpred1.1) <- ~ long + lat
names(ukpred1.1) <- c('predictions', 'variances')
```

```
library(raster)
```

```
par(font.axis = 2, font.lab = 2)
plot(brick(ukpred1.1))
```

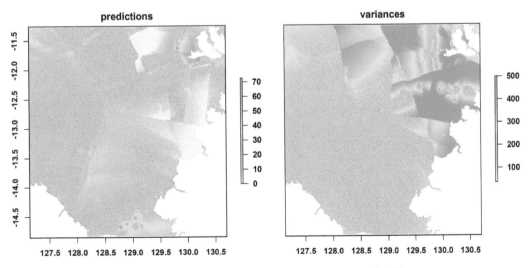

FIGURE 5.10: Spatial patterns of predictions and variances of *UK*.

The prediction range, `range(ukpred1.1$predictions)`, is from 0% to 72.55%.

5.5 Block kriging

Block kriging (*BK*) is an extension of kriging methods (e.g., *SK*, *OK*, *UK*) and predicts a block value or average value instead of a point value over a surface by replacing the point-to-point covariance with the point-to-block covariance (Goovaerts 1997; Wackernagel 2003). This can be done by specifying `block` argument in relevant function for generating predictions, which would result in *BK* methods. Taking `krigepred` as an example, if `beta` argument is also specified, a *block simple kriging* (*BSK*) will be resulted; if the defaults for `beta` and `formula` are also used, a *block ordinary kriging* (*BOK*) will be resulted; and if `formula` argument is also specified to include external trends, a *block universal kriging* (*BUK*) will be resulted.

Below is an example to demonstrate the application of *BOK* by applying `krigepred` with the optimal parameters for *OK* to `pb.df` that is from Section 5.2.1.

```
bkpred1 <- krigepred(petrel[, c(1,2)], petrel[, 5], pb.df[, -3], nmax = nmax.ok[ok
    .opt[, 1]], transformation = "none", vgm.args = "Gau", block = c(50, 50))
```

Some basic properties of the predictions in `bkpred1` are

```
names(bkpred1)
```

```
[1] "long"      "lat"      "var1.pred" "var1.var"
```

```
class(bkpred1)
```

```
[1] "data.frame"
```

```
dim(bkpred1)
```

```
[1] 100828        4
```

```
summary(bkpred1)
```

```
      long            lat           var1.pred         var1.var
 Min.    :127   Min.    :-14.8   Min.    : 2.83   Min.    :14.3
 1st Qu.:128   1st Qu.:-13.6   1st Qu.: 9.90   1st Qu.:16.1
 Median :129   Median :-12.8   Median :14.01   Median :18.6
 Mean    :129   Mean    :-12.8   Mean    :15.62   Mean    :20.6
 3rd Qu.:129   3rd Qu.:-12.0   3rd Qu.:19.94   3rd Qu.:23.5
 Max.    :131   Max.    :-11.3   Max.    :59.24   Max.    :41.9
```

The spatial distribution of predictions and variances of *BOK* are illustrated in Figure 5.11 by

```
bkpred1.1 <- bkpred1
gridded(bkpred1.1) <- ~ long + lat
names(bkpred1.1) <- c('predictions', 'variances')

library(raster)

par(font.axis = 2, font.lab = 2)
plot(brick(bkpred1.1))
```

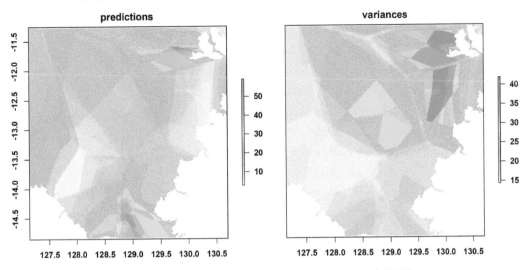

FIGURE 5.11: Spatial patterns of predictions and variances of *BOK*.

The prediction range, `range(bkpred1$var1.pred)`, is from 2.83% to 59.24%.

BOK with *anisotropy* can be implemented using `model.8` (Section 5.1.2) and specifying `block` argument (e.g., `block = c(50, 50)`) in `ok2` in Section 5.3.2.

We can seek relevant optimal *parameters* for *BOK* in the same way as for *SK*, *OK*, and *UK* and then use the parameters optimized to assess the predictive accuracy of *BOK* and generate *BOK* predictions, which will not be presented here and will be left for interested readers to explore.

Similar to *OK* in Section 5.3.3, *BOK* predictions can be produced using `autoKrige` with `block` argument specified.

6

Multivariate geostatistical methods

Multivariate geostatistical methods are *gradient based/detrended spatial predictive methods* that can use secondary information or predictive variables. In geostatistics, methods accounting for secondary information, like *kriging with an external drift (KED)* and *ordinary cokriging (OCK)*, are called *multivariate geostatistical methods*. Multivariate kriging means that there are several variables which can be either primary variables or both primary and secondary variables (Li and Heap 2014).

The predictions of multivariate geostatistical methods can be represented as the weighted averages of sampled data as in Equation (4.1).

In this chapter, the multivariate geostatistical methods that are available in R are introduced, including:
(1) *simple cokriging (SCK)*,
(2) *OCK*, and
(3) *KED*.

For multivariate geostatistical methods, the `petrel` point data in the `spm` package and the `pb.df` grid data in Chapter 1 will be used to demonstrate the applications of the multivariate geostatistical methods in R.

Prior to introducing individual multivariate geostatistical methods, a general introduction for all cokriging methods will be provided.

Cokriging

The *cokriging (CK)* uses non-exhaustive secondary information and explicitly accounts for the spatial cross-correlation between the primary and secondary variables (Goovaerts 1997; Li and Heap 2008). Equation (5.1) can be extended to incorporate the secondary information to derive the following Equation (6.1):

$$\widehat{y_1}(x_0) - \mu_1 = \sum_{i_1=1}^{n_1} \lambda_{i_1}[y_1(x_{i_1}) - \mu_1(x_{i_1})] + \sum_{j=2}^{n_v} \sum_{i_j=1}^{n_j} \lambda_{i_j}[y_j(x_{i_j}) - \mu_j(x_{i_j})] \qquad (6.1)$$

where μ_1 is the stationary mean of the primary variable, $y_1(x_{i_1})$ is the data of the primary variable at point i_1, $\mu_1(x_{i_1})$ is the mean of samples within the search window, n_1 is the *number of sampled points* within the search window used to make the prediction for point x_0, λ_{i_1} is the weight estimated to minimize the prediction variance, n_v is the number of secondary variables, n_j is the number of jth secondary variable within the *search window*, λ_{i_j} is the weight assigned to i_j^{th} point of jth secondary variable, $y_j(x_{i_j})$ is the data at i_j^{th} point of jth secondary variable, and $\mu_j(x_{i_j})$ is the mean of samples of jth secondary variable within the search window (Li and Heap 2008).

The *cross-semivariance* (or *cross-variogram*) can be estimated from data using the following Equation (6.2):

$$\hat{\gamma}_{12}(h) = \sum_{i=1}^{n} [y_1(x_i) - y_1(x_i + h)] \cdot [y_2(x_i) - y_2(x_i + h)]/2n \qquad (6.2)$$

where n is the number of pairs of sample points of variables y_1 and y_2 at point x_i, $x_1 + h$ separated by distance h (Burrough and McDonnell 1998). Cross-semivariances can increase or decrease with h depending on the correlation between the two variables and the Cauchy-Schwartz relation must be checked to ensure a positive CK prediction variance in all circumstances (Burrough and McDonnell 1998; Li and Heap 2008).

6.1 Simple cokriging

In Equation (6.1), replacing $\mu_1(x_i)$ with the stationary mean (μ_1) of the primary variable, and replacing $\mu_j(x_{i_j})$ with the stationary mean (μ_j) of the secondary variables will give a SCK estimator (Goovaerts 1997) (Equation (6.3)):

$$
\begin{aligned}
\widehat{y_1}(x_0) &= \sum_{i_1=1}^{n_1} \lambda_{i_1} [y_1(x_{i_1}) - \mu_1] + \mu_1 + \sum_{j=2}^{n_v} \sum_{i_j=1}^{n_j} \lambda_{i_j} [y_j(x_{i_j}) - \mu_j(x_j)] \\
&= \sum_{i_1=1}^{n_1} \lambda_{i_1} [y_1(x_{i_1})] + \sum_{j=2}^{n_v} \sum_{i_j=1}^{n_j} \lambda_{i_j} [y_j(x_{i_j})] + (1 - \sum_{i_1=1}^{n_1} \lambda_{i_1})\mu_1 - \sum_{j=2}^{n_v} \sum_{i_j=1}^{n_j} \lambda_{i_j}\mu_j
\end{aligned}
\qquad (6.3)
$$

The predictions of SCK are essentially the same as that of SK if all secondary variables are recorded at every sampled point or if the primary and secondary variables are not correlated (Goovaerts 1997). When the point of interest is beyond the correlation range of both the primary and secondary data, the SCK estimator then reverts to the stationary mean of the primary variable (Li and Heap 2008). The *sill* of the *cross-semivariogram* model is the correlation coefficient of the primary and secondary variables (Li and Heap 2008).

6.1.1 Data normality and correlation

For SCK, secondary variables that are correlated to the primary variable need to be selected. For example, for `gravel` variable in the `petrel` data, `sand` in the `petrel` data can be used as a secondary variable as they are correlated. The correlation between `gravel` and `sand` is

```
cor(petrel$gravel, petrel$sand)
```

```
[1] -0.5294
```

SCK assumes that both primary and secondary variables are normally distributed. We will still use log transformation of `gravel` as suggested in Chapter 2 to normalized the `gravel` data. We will then examine the *data normality* of `sand` data, and the results show that arcsine transformation of `sand` is largely normally distributed (Figure 6.1) and should be used.

The correlation between the transformed `gravel` and `sand` data is

```
cor(log(petrel$gravel + 1), asin(sqrt(petrel$sand /100)))
```

```
[1] -0.3125
```

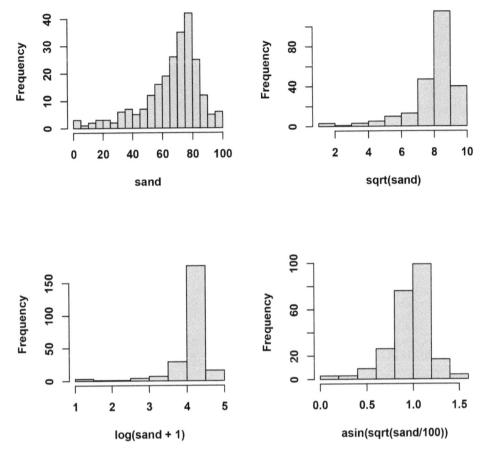

FIGURE 6.1: Distribution of (1) sand, (2) square-root transformed sand, (3) log transformed sand, and (4) arcsine transformed sand.

6.1.2 Parameter optimization for *SCK*

SCK can be performed with the gstat package. We will use the function gstat.cv in the gstat package to find the best model and nmax for *SCK*. For gstat.cv, the following arguments need to be specified:

(1) object, object of class gstat;

(2) nfold, integer; if larger than 1, then apply n-fold cross-validation; if nfold equals nrow(data) (default), apply leave-one-out cross-validation; and

(3) nmax, the number of nearest observations that should be used for a kriging prediction.

A further function fit.lmc in the gstat package is also required for fitting a linear model of coregionalization to a multivariable sample variogram. For fit.lmc, the following arguments need to be specified:

(1) v, multivariable sample variogram, output of variogram;

(2) g, gstat object, output of gstat;

(3) fit.method, fitting method, used by gstat, and see ?fit.variogram for details; and

(4) correct.diagonal, multiplicative correction factor to be applied to partial sills of direct variograms only.

The correlation of the untransformed `gravel` and `sand` is higher than that of the transformed data, suggesting that the untransformed data should be used, while the examination of data normality suggests that the transformed data should be used. Therefore, the data transformation also needs to be considered to satisfy the assumption of data normality. We will conduct parameter estimation for both transformed and untransformed data and then select the one with higher predictive accuracy.

1. For transformed data

Since outputs of `gstat.cv` are the observed values and residuals, `vecv` or `pred.acc` in the `spm` package can be used to calculate *VEcv* and other relevant accuracy/error measures (e.g., *MAE*, *RMAE*, *RMSE*, *RRMSE*). We will apply `gstat.cv` to the transformed `gravel` and `sand` to find optimal `model` and `nmax` for *SCK* as demonstrated below.

```
library(spm) # for `vecv`
library(sp)
library(gstat)

gravel2 <- petrel
coordinates(gravel2) = ~ long + lat

beta1 <- mean(log(gravel2$gravel + 1))
beta2 <- mean(asin(sqrt(gravel2$sand/100)))

nmax.sck <- c(5:40)
vgm.args.sck <- c("Exp", "Gau", "Sph", "Exc", "Mat", "Ste", "Lin")
sckopt <- matrix(0, length(nmax.sck), length(vgm.args.sck))

for (i in 1:length(nmax.sck)) {
  for (j in 1:length(vgm.args.sck)) {
    set.seed(1234)
    sck.g <- gstat(id = "gravel", formula = log(gravel + 1) ~ 1, beta = beta1,
        data = gravel2, nmax = nmax.sck[i])

    sck.g <- gstat(sck.g, id = "sand", formula = asin(sqrt(sand/100)) ~ 1, beta =
        beta2, data = gravel2, nmax = nmax.sck[i])

    sck.g.cv <- gstat(sck.g, model = vgm(psill1, vgm.args.sck[j], range1, nugget1)
        , fill.all=T)

    sck.v <- variogram(sck.g.cv)

    sck.fit <- fit.lmc(sck.v, sck.g.cv, fit.method = 6, correct.diagonal = 1.01)

    sckcv <- gstat.cv(sck.fit, nmax = nmax.sck[i], nfold = 10)

    sckvecv1 <- vecv(sckcv$observed, sckcv$observed - sckcv$residual)
    sckopt[i, j] <- sckvecv1
  }
}
```

In this demonstration, `psill1`, `range1`, and `nugget1` are from Section 5.1.2.

The optimal `nmax` and `vgm` that maximize the predictive accuracy are

```
sck.opt <- which (sckopt == max(sckopt), arr.ind = T)
nmax.sck[sck.opt[, 1]]
```

```
[1] 37
```

```
vgm.args.sck[sck.opt[, 2]]
```

```
[1] "Lin"
```

The maximal predictive accuracy is

```
max(sckopt)
```

```
[1] 33.2
```

The effects of different data transformation types on the predictive accuracy can also be tested in the same way as shown in Chapter 5.

2. For untransformed data

We will also use `gstat.cv` to select the best `model` and estimate the optimal `nmax` for *SCK* for untransformed `gravel` and `sand`. We still use the above `nmax.sck`, `vgm.args.sck` as well as `range1`.

```
beta1.2 <- mean(gravel2$gravel)
beta2.2 <- mean(gravel2$sand)

vgm.sck2 <- variogram(gravel ~ 1, gravel)
psill.sck2 = (max(vgm.sck2$gamma) + median(vgm.sck2$gamma))/2
nugget.sck2 = min(vgm.sck2$gamma)

sckopt2 <- matrix(0, length(nmax.sck), length(vgm.args.sck))

for (i in 1:length(nmax.sck))  {
for (j in 1:length(vgm.args.sck)) {
    set.seed(1234)
    sck.g <- gstat(id = "gravel", formula = gravel ~ 1, beta = beta1.2, data =
        gravel2, nmax = nmax.sck[i])
    sck.g <- gstat(sck.g, id = "sand", formula = sand ~ 1, beta = beta2.2, data =
        gravel2, nmax = nmax.sck[i])

    sck.g.cv <- gstat(sck.g, model = vgm(psill.sck2, vgm.args.sck[j], range1,
        nugget.sck2), fill.all=T)

    sck.v <- variogram(sck.g.cv)

    sck.fit <- fit.lmc(sck.v, sck.g.cv, fit.method = 6, correct.diagonal = 1.01)

    sckcv <- gstat.cv(sck.fit, nmax = nmax.sck[i], nfold = 10)

    sckvecv1 <- vecv(sckcv$observed, sckcv$observed - sckcv$residual)
    sckopt2[i, j] <- sckvecv1
  }
}
```

The optimal `nmax` and `vgm` that result in the maximal predictive accuracy are

```
sck.opt2 <- which (sckopt2 == max(sckopt2), arr.ind = T)
nmax.sck[sck.opt2[, 1]]
```

```
[1] 13
```

```
vgm.args.sck[sck.opt2[, 2]]
```

```
[1] "Lin"
```

The maximal predictive accuracy is

```
max(sckopt2)
```

```
[1] 40.19
```

The predictive accuracy of the *SCK* model resulted for the untransformed data is much higher than that for the transformed data. Hence, the parameters derived for the untransformed data should be used for *SCK*.

6.1.3   Predictive accuracy of *SCK* with the optimal parameters

We will stabilize *predictive accuracy* to get a more reliable accuracy estimation by repeating the cross-validation 100 times. Prior to deriving the reliable accuracy estimation, we need to conduct variogram modeling for *SCK* first.

1. Variogram modeling

For two variables *cokriging*, two direct variograms and a cross-variogram need to be created for the variables, such as gravel and sand. Like using fit.variogram to fit a variogram object, fit.lmc is to fit a linear model of coregionalization to a multivariable sample variogram as shown below.

```
sck.g <- gstat(id = "gravel", formula = gravel ~ 1, beta = beta1, data = gravel2,
    nmax = nmax.sck[sck.opt2[, 1]])

sck.g <- gstat(sck.g, id = "sand", formula = sand ~ 1, beta = beta2, data =
    gravel2, nmax = nmax.sck[sck.opt2[, 1]])

sck.g <- gstat(sck.g, model = vgm(psill.sck2, vgm.args.sck[sck.opt2[, 2]], range1,
    nugget.sck2), fill.all=T)

sck.v <- variogram(sck.g)

sck.fit <- fit.lmc(sck.v, sck.g, fit.method = 6, correct.diagonal = 1.01)
```

In this example, range1 is from Section 5.1.2, and nmax.sck[sck.opt2[, 1]], vgm.args.sck[sck.opt2[, 2]], psill.sck2, and nugget.sck2 are for the untransformed data.

The fitting results for two variograms and their cross-variogram for *SCK* are displayed below and shown in Figure 6.2 by

```
sck.fit
```

```
data:
gravel : formula = gravel`~`1 ; data dim = 237 x 7 nmax = 13 beta = 14.9
sand : formula = sand`~`1 ; data dim = 237 x 7 nmax = 13 beta = 66.33
variograms:
                  model    psill   range
gravel[1]           Nug  236.109  0.0000
gravel[2]           Lin   46.501  0.8309
sand[1]             Nug  155.593  0.0000
sand[2]             Lin  238.365  0.8309
gravel.sand[1]      Nug -153.225  0.0000
gravel.sand[2]      Lin   -9.401  0.8309
```

```
par(font.axis = 2, font.lab = 2)
plot(sck.v, model=sck.fit)
```

The gravel and sand are negatively and weakly correlated according to their cross-variogram for *SCK* (Figure 6.2).

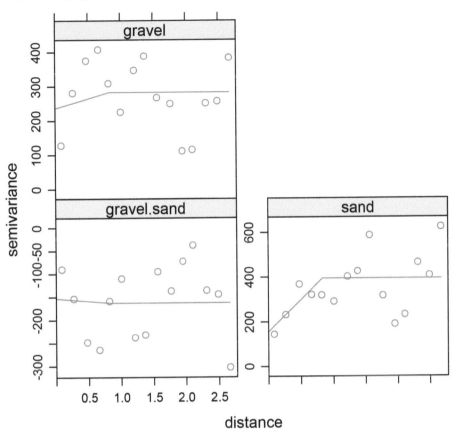

FIGURE 6.2: Two variograms and their cross-variogram of gravel and sand for *SCK*.

2. Implementation of *SCK* in `gstat.cv`

```
set.seed(1234)
n <- 100
sckvecv <- NULL
sckpredacc <- NULL

for (i in 1:n) {
  sckcv1 <- gstat.cv(sck.fit, nmax = nmax.sck[sck.opt[, 1]], nfold = 10)

  sckvecv[i] <- vecv(sckcv1$observed, sckcv1$observed - sckcv1$residual)

  sckacc <- pred.acc(sckcv1$observed, sckcv1$observed - sckcv1$residual)
  sckpredacc <- rbind(sckpredacc, sckacc)
}
```

The median and range of predictive accuracy of *SCK* based on 100 repetitions of 10-fold cross-validation using `gstat.cv` are

```
median(sckvecv)
```

```
[1] 40.84
```

```
range(sckvecv)
```

```
[1] 35.52 46.44
```

The accuracy/error measures from `pred.acc` for *SCK* are in `sckpredacc` and can be extracted by the function `colMedians` in the `miscTools` package (Henningsen and Toomet 2019) as below.

```
library(miscTools)
class(sckpredacc) # it is a matrix of lists


[1] "matrix" "array"


sckpredacc1 <- data.frame(lapply(data.frame(sckpredacc), unlist))
round(colMedians(sckpredacc1), 2)


     me    rme    mae   rmae    mse   rmse  rrmse   vecv     e1
  -0.54  -3.60   8.21  55.14 158.05  12.57  84.39  40.84  31.09
```

6.1.4 *SCK* predictions and variances

To generate *SCK* predictions and variances using the function `gstat`, first we need to conduct variogram modeling, and then we can generate *SCK* predictions and variances.

1. Variogram modeling

We will conduct variogram modeling for untransformed `gravel` data with `range1`, `beta1.2`, `beta2.2`, `psill.sck2`, and `nugget.sck2` as below.

```
sck.g <- gstat(id = "gravel", formula = gravel ~ 1, beta = beta1.2, data = gravel2
    , nmax = nmax.sck[sck.opt2[, 1]])

sck.g <- gstat(sck.g, id = "sand", formula = sand ~ 1, beta = beta2.2, data =
    gravel2, nmax = nmax.sck[sck.opt2[, 1]])

sck.g <- gstat(sck.g, model = vgm(psill.sck2, vgm.args.sck[sck.opt2[, 2]], range1,
    nugget.sck2), fill.all=T)

sck.v <- variogram(sck.g)

sck.fit <- fit.lmc(sck.v, sck.g, fit.method = 6, correct.diagonal = 1.01)
```

The results of variogram modeling are

```
sck.fit


data:
gravel : formula = gravel`~`1 ; data dim = 237 x 7 nmax = 13 beta = 14.9
sand : formula = sand`~`1 ; data dim = 237 x 7 nmax = 13 beta = 66.33
variograms:
                  model    psill   range
gravel[1]          Nug  236.109  0.0000
gravel[2]          Lin   46.501  0.8309
sand[1]            Nug  155.593  0.0000
sand[2]            Lin  238.365  0.8309
gravel.sand[1]     Nug -153.225  0.0000
gravel.sand[2]     Lin   -9.401  0.8309
```

This `sck.fit` model will be used to generate spatial predictions for *SCK*.

2. Predictions and variances

The predictions and variances of *SCK* are produced as shown below for `pb` that is from Chapter 1.

```
sck.pred<-predict(sck.fit, newdata=pb)
```

```
sck.gravel.pred<- sck.pred
sck.gravel.pred$predictions <- sck.gravel.pred$gravel.pred
sck.gravel.pred$variances <- sck.gravel.pred$gravel.var
```

The predictions and variances of *SCK* are summarized and illustrated in Figure 6.3 by

```
summary(sck.gravel.pred$predictions)
```

```
   Min. 1st Qu.  Median   Mean 3rd Qu.    Max.
   7.33   12.95   14.90  15.20   16.16   37.46
```

```
summary(sck.gravel.pred$variances)
```

```
   Min. 1st Qu.  Median   Mean 3rd Qu.    Max.
    250     269     275    274     280     283
```

```
library(raster)
```

```
par(font.axis = 2, font.lab = 2)
plot(brick(sck.gravel.pred), c("predictions", "variances"))
```

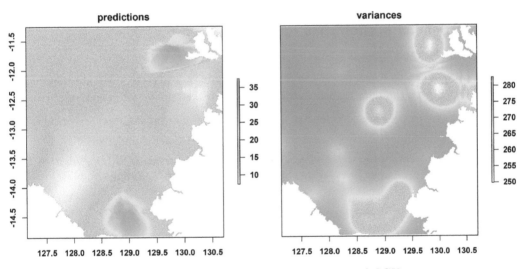

FIGURE 6.3: Spatial patterns of predictions and variances of *SCK*.

SCK and *SK* methods are expected to produce identical predictions when the primary and secondary variables are measured at the same locations and the cross-covariance is proportional to the primary auto-covariance (Goovaerts 1997). However, their predictions are not the same for `gravel` data. This could be due to the weak correlation of the secondary variable (i.e., `sand`) with the primary variable (i.e., `gravel`), as the contribution of the secondary variable to *SCK* predictions depends on a number of factors, including: (1) its correlation with the primary variables, (2) its pattern of spatial continuity, and (3) the spatial configuration of both primary and secondary sample points (Goovaerts 1997; Li and Heap 2014). The R functions for *SK* and *SCK* may also contribute to such differences, which could be tested using highly correlated and co-located samples.

Although the predictive accuracy of *SCK* is higher than other methods (e.g., *SK*, *OK*) for `gravel` data, its predictions are much lower than the observed values. In practice, it should be used with caution.

6.2 Ordinary cokriging

Ordinary cokriging (*OCK*) predicts the value of the primary variable using Equation (6.3) by replacing $\mu - 1$ and μ_j with local means $\mu_1(x_0)$ and $\mu_j(x_0)$ (i.e., the mean of samples within the search window), and forcing $\sum_{i_1=1}^{n_1} \lambda_{i_1} = 1$, and $\sum_{i_j=1}^{n_j} \lambda_{i_j} = 0$ (Goovaerts 1997; Li and Heap 2008). These two constraints may result in negative weights. To reduce the occurrence of negative weights, these two constraints are combined to form the single constraint: $\sum_{i_1=1}^{n_1} \lambda_{i_1} + \sum_{i_j=1}^{n_j} \lambda_{i_j} = 1$ (Goovaerts 1997; Li and Heap 2008).

6.2.1 Data requirements

Similar to *SCK*, secondary variable(s) that are correlated to the primary variable need to be selected for *OCK*. *OCK* can also be performed using the `gstat` package (Gräler, Pebesma, and Heuvelink 2016), and a demo for *OCK* is available via `demo(cokriging)` in the `gstat` package.

6.2.2 Parameter optimization for *OCK*

1. For transformed data

Similar to *SCK*, we can also use `gstat.cv` to select the best `model` and estimate the optimal `nmax` for *OCK*, with the data transformation types identified above for `gravel` and `sand`.

We can use `vecv` or `pred.acc` to calculate *VEcv* and other relevant accuracy/error measures based on the outputs of `gstat.cv`.

For a primary variable of `gravel` and a secondary variable of `sand`, *OCK* is similar to *SCK* but without `beta` argument in creating the `gstat` objects for *OCK* as shown below.

```
library(spm) # for `vecv`

nmax.ock <- c(5:40)
vgm.args.ock <- c("Exp", "Gau", "Sph", "Exc", "Mat", "Ste", "Lin")
ockopt <- matrix(0, length(nmax.ock), length(vgm.args.ock))

for (i in 1:length(nmax.ock)) {
  for (j in 1:length(vgm.args.ock)) {
    set.seed(1234)
    ock.g <- gstat(id = "gravel", formula = log(gravel + 1) ~ 1, data = gravel2,
        nmax = nmax.ock[i])

    ock.g <- gstat(ock.g, id = "sand", formula = asin(sqrt(sand/100)) ~ 1, data =
        gravel2, nmax = nmax.ock[i])

    ock.g.cv <- gstat(ock.g, model = vgm(psill1, vgm.args.ock[j], range1, nugget1)
        , fill.all=T)

    ock.v <- variogram(ock.g.cv)

    ock.fit <- fit.lmc(ock.v, ock.g.cv, fit.method = 6, correct.diagonal = 1.01)

    ockcv <- gstat.cv(ock.fit, nmax = nmax.ock[i], nfold = 10)

    ockvecv1 <- vecv(ockcv$observed, ockcv$observed - ockcv$residual)
    ockopt[i, j] <- ockvecv1
  }
```

```
}
```

In this example, `psill1`, `range1`, and `nugget1` are from Section 5.1.2.

The optimal `nmax` and `vgm` that maximize the predictive accuracy are

```
ock.opt <- which (ockopt == max(ockopt), arr.ind = T)
nmax.ock[ock.opt[, 1]]
```

```
[1] 5
```

```
vgm.args.ock[ock.opt[, 2]]
```

```
[1] "Gau"
```

For `namx`, 5 is the minimum value tested, the optimal number could be even smaller than 5. We will leave it as it is for this demonstration. In practice, further test of a smaller number is required to ensure an optimal value estimated for `nmax`.

2. For untransformed data

Similar to *SCK*, we can also use `gstat.cv` to select the best `model` and estimate the optimal `nmax` for *OCK* and assess the predictive accuracy of *OCK* resulted for untransformed `gravel` and `sand`. We still use the above `nmax.ock`, `vgm.args.ock` as well as `range1`.

```
vgm.ock <- variogram(gravel ~ 1, gravel)
```

```
psill.ock = (max(vgm.ock$gamma) + median(vgm.ock$gamma))/2
nugget.ock = min(vgm.ock$gamma)
```

```
ockopt2 <- matrix(0, length(nmax.ock), length(vgm.args.ock))
```

```
for (i in 1:length(nmax.ock)) {
  for (j in 1:length(vgm.args.ock)) {
    set.seed(1234)
    ock.g <- gstat(id = "gravel", formula = gravel ~ 1, data = gravel2, nmax =
        nmax.ock[i])

    ock.g <- gstat(ock.g, id = "sand", formula = sand ~ 1, data = gravel2, nmax =
        nmax.ock[i])

    ock.g.cv <- gstat(ock.g, model = vgm(psill.ock, vgm.args.ock[j], range1,
        nugget.ock), fill.all=T)

    ock.v <- variogram(ock.g.cv)

    ock.fit <- fit.lmc(ock.v, ock.g.cv, fit.method = 6, correct.diagonal = 1.01)

    ockcv <- gstat.cv(ock.fit, nmax = nmax.ock[i], nfold = 10)

    ockvecv1 <- vecv(ockcv$observed, ockcv$observed - ockcv$residual)
    ockopt2[i, j] <- ockvecv1
  }
}
```

The optimal `nmax` and `vgm` that result in the maximal predictive accuracy are

```
ock.opt2 <- which (ockopt2 == max(ockopt2), arr.ind = T)
nmax.ock[ock.opt2[, 1]]
```

```
[1] 6
```

```
vgm.args.ock[ock.opt2[, 2]]
```

```
[1] "Gau"
```

The predictive accuracy, `max(ockopt2)`, of the *OCK* model resulted for the untransformed data is 45.93% that is much higher than 37.83%, the accuracy for the transformed data. Therefore, the optimal parameters for the untransformed data should be used for *OCK*.

6.2.3   Predictive accuracy of *OCK* with the optimal parameters

We will stabilize the *predictive accuracy* to get a more reliable accuracy estimation by repeating the cross-validation 100 times. We need to conduct variogram modeling first, and then we can generate the reliable accuracy estimation.

1. Variogram modeling

To apply *OCK* to `gravel` and `sand`, we still use the above `nmax.ock` and `vgm.args.ock`.

```
ock.g <- gstat(id = "gravel", formula = gravel ~ 1, data = gravel2, nmax = nmax.
    ock[ock.opt2[, 1]])
```

```
ock.g <- gstat(ock.g, id = "sand", formula = sand ~ 1, data = gravel2, nmax = nmax
    .ock[ock.opt2[, 1]])
```

```
ock.g <- gstat(ock.g, model = vgm(psill.ock, vgm.args.ock[ock.opt2[, 2]], range1,
    nugget.ock), fill.all=T)
```

```
ock.v <- variogram(ock.g)
```

```
ock.fit <- fit.lmc(ock.v, ock.g, fit.method = 6, correct.diagonal = 1.01)
```

In this example, `range1` is from Section 5.1.2, and `psill.ock` and `nuggest.ock` are for the untransformed data.

```
ock.fit
```

```
data:
gravel : formula = gravel`~`1 ; data dim = 237 x 7 nmax = 6
sand : formula = sand`~`1 ; data dim = 237 x 7 nmax = 6
variograms:
                  model    psill   range
gravel[1]           Nug  275.721  0.0000
gravel[2]           Gau    2.374  0.8309
sand[1]             Nug  214.436  0.0000
sand[2]             Gau  196.448  0.8309
gravel.sand[1]      Nug -177.121  0.0000
gravel.sand[2]      Gau   21.381  0.8309
```

The fitting results of `fit.lmc` for two variograms and their cross-variogram for *OCK* are shown in Figure 6.4 by

```
par(font.axis = 2, font.lab = 2)
plot(ock.v, model=ock.fit)
```

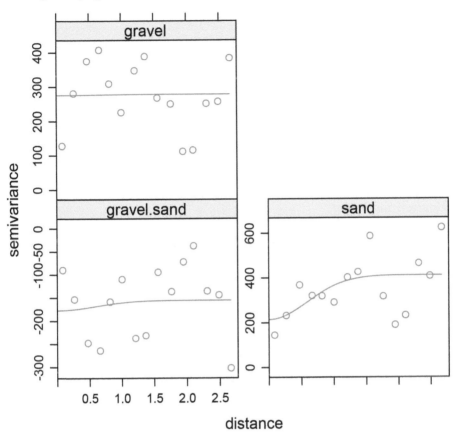

FIGURE 6.4: Two variograms and their cross-variogram of gravel and sand for *OCK*.

2. Implementation of *OCK* in `gstat.cv`

```
set.seed(1234)
n <- 100
ockvecv <- NULL
ockpredacc <- NULL

for (i in 1:n) {
  ockcv1 <- gstat.cv(ock.fit, nmax = nmax.ock[ock.opt[, 1]], nfold = 10)

  ockvecv[i] <- vecv(ockcv1$observed, ockcv1$observed - ockcv1$residual)

  ockacc <- pred.acc(ockcv1$observed, ockcv1$observed - ockcv1$residual)
  ockpredacc <- rbind(ockpredacc, ockacc)
}
```

The median and range of predictive accuracy of *OCK* based on 100 times of 10-fold cross-validation using `gstat.cv` are

```
median(ockvecv)
```

```
[1] 46.9
```

```
range(ockvecv)
```

```
[1] 41.21 52.92
```

The accuracy/error measures from pred.acc for *OCK* are

```
ockpredacc1 <- data.frame(lapply(data.frame(ockpredacc), unlist))
round(colMedians(ockpredacc1), 2)
```

```
    me     rme     mae    rmae     mse    rmse   rrmse    vecv      e1
 -0.47   -3.15    7.77   52.18  141.86   11.91   79.95   46.90   34.80
```

6.2.4 *OCK* predictions and variances

To generate *OCK* predictions using the gstat package, we need to conduct variogram modeling first, and then we can generate *OCK* predictions and variances.

1. Variogram modeling

For the untransformed gravel and sand data, we can use the above ock.fit to generate *OCK* predictions and variances.

2. Predictions and variances

The predictions and variances of *OCK* are produced as below for pb that is from Chapter 1.

```
ock.pred<-predict(ock.fit, newdata=pb)
```

```
names(ock.pred)
ock.gravel.pred<- ock.pred
ock.gravel.pred$predictions <- ock.gravel.pred$gravel.pred
ock.gravel.pred$variances <- ock.gravel.pred$gravel.var
```

The predictions and variances of *OCK* are summarized and illustrated in Figure 6.5 by

```
summary(ock.gravel.pred$predictions)
```

```
  Min. 1st Qu.  Median    Mean 3rd Qu.    Max.
  1.58   10.29   13.01   15.80   20.93   63.06
```

```
summary(ock.gravel.pred$variances)
```

```
  Min. 1st Qu.  Median    Mean 3rd Qu.    Max.
   322     322     322     322     323     325
```

```
library(raster)
```

```
par(font.axis = 2, font.lab = 2)
plot(brick(ock.gravel.pred), c("predictions", "variances"))
```

When the primary and secondary variables are all measured at the same locations, then *OCK* and *OK* will produce the same predictions (Burrough and McDonnell 1998). For the gravel, their predictions are not the same, which could be due to the same reasons mentioned for *SCK* vs. *SK* in Section 6.1.4.

6.3 Kriging with an external drift

Kriging with an external drift (*KED*) is similar to *UK* but incorporates the local trend within the neighborhood search window as a linear function of a secondary variable instead

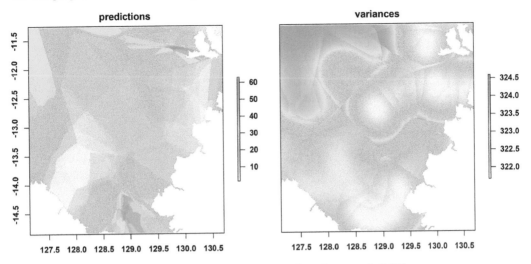

FIGURE 6.5: Spatial patterns of predictions and variances of *OCK*.

of the spatial coordinates (Goovaerts 1997). *KED* can be extended to include both secondary variables and coordinate information if the gstat package is used. Unlike *CK* that requires non-exhaustive secondary variables, *KED* requires the secondary variables that need to be available not only at all data points of primary variable but also at all points to be predicted.

6.3.1   Application of *KED*

The examples for *UK* in Chapter 5 can be used to demonstrate the application of *KED* by replacing loggravel ~ long + lat with loggravel ~ bathy in the formula if bathy is the only variable considered for local trend.

Similar to the kriging methods in Chapter 5, the function autofitVariogram in the automap package (Hiemstra 2013) can be used to select a variogram model for *KED*. The external drift is considered with such as loggravel ~ bathy in variogram().

6.3.2   Variable selection and parameter optimization for *KED*

Four arguments, transformation, alpha, model, and nmax, need to be optimized for *KED*. Similar to *UK*, they can also be determined based on their effects on the predictive accuracy using functions such as krigecv but their optimization will not be presented here. If more than one predictive variable is considered for *KED*, then variable selection is required. In addition, anisotropy parameters need to be estimated for *KED*.

1. Variable selection for *KED*

The *variable selection* can also be conducted based on the effects of relevant variables on the predictive accuracy with the function krigecv in the spm2 package. We will select an optimal set of secondary variables for *KED* by assuming that the parameters estimated for *UK* are also useful for *KED*. That is, we will use vgm.args = "Ste", nmax = 31, and anis = c(0, 1) for variable selection for *KED*.

```
library(spm)
library(spm2)

data(petrel)
```

```
formula.ked <- c(var1 ~ bathy,
                 var1 ~ bathy + dist,
                 var1 ~ bathy + relief,
                 var1 ~  bathy + slope,
                 var1 ~ bathy + dist + relief,
                 var1 ~ bathy + relief +slope,
                 var1 ~ bathy + dist + relief + slope)
kedopt <- NULL

for (i in 1:length(formula.ked)) {
  set.seed(1234)
  kedcv1 <- krigecv(petrel[, 1:2], petrel[, 5], petrel[, -c(1:5)], nmax = 31,
      transformation = "none", formula = formula.ked[[i]], vgm.args = "Ste", anis
      = c(0, 1), predacc = "VEcv")
  kedopt[i] <- kedcv1
}
```

The variables selected are

```
ked.opt <- which (kedopt == max(kedopt))
formula.ked[ked.opt]

[[1]]
var1 ~ bathy + dist + relief
```

It indicates that `var1 ~ bathy + dist + relief` should be used for *KED*.

2. Estimation of anisotropy parameters for *KED*

We use `gravel` in the `petrel` data set to demonstrate the estimation of *anisotropy parameters* for *KED*. The `gravel` data will be log-transformed according to Chapter 5.

```
gravel <- petrel[, c(1, 2, 5:8)]
coordinates(gravel) = ~long + lat
gravel$loggravel <- log(gravel$gravel + 1)
```

Similar to *UK*, a directional variogram model can be developed based on the residuals of `loggravel ~ bathy + dist + relief` for *KED*. Anisotropy parameters can be estimated with `estimateAnisotropy` for the residuals by

```
library(intamap)
anis1 <- estimateAnisotropy(gravel, "loggravel", loggravel ~ bathy + dist + relief
    )

anis1

$ratio
[1] 1.177

$direction
[1] -84.16

$Q
       Q11    Q22    Q12
[1,] 20.48 14.88 0.5782

$doRotation
[1] TRUE
```

The results show that data anisotropy needs to be considered. Since the `gstat` package is going to be used for *KED*, the parameters estimated by `estimateAnisotropy` need to be converted for the `gstat` package.

```
ked.dir1 <- 90 - anis1$direction
ked.ratio1 <- 1 / anis1$ratio
```

In practice, all relevant parameters of *KED* should be optimized as shown for *SK* in Chapter 5 and the applications of *KED* in Chapter 12 provide further examples for parameter estimation and variable selection.

6.3.3 Predictive accuracy of KED

The *predictive accuracy* of *KED* can be estimated using leave-one-out (LOO) cross-validation.

```
library(spm2)
kedcv <- krigecv(petrel[, 1:2], petrel[, 5], petrel[, -c(1:5)], nmax = 31,
    transformation = "none", formula = var1 ~ bathy + dist + relief, vgm.args = "
    Ste", anis = c(ked.dir1, ked.ratio1), validation = "LOO", predacc = "VEcv")
```

The predictive accuracy of *KED* based on *LOO* cross-validation, kedcv, is 28.73%. This example is to demonstrate that *LOO* cross-validation can also be conducted with krigecv, although 10-fold cross-validation is preferred (Kohavi 1995).

6.3.4 KED predictions and variances

We will use krigepred with the optimal parameters for *UK*, and the variables selected and anisotropy parameters estimated for *KED* to generate predictions and variances for petrel .grid data.

```
kedpred1 <- krigepred(petrel[, c(1, 2, 6:8)], petrel[, 5], petrel.grid[, -6], nmax
    = 31, transformation = "none", formula = var1 ~ bathy + dist + relief, vgm.
    args = "Ste", anis = c(ked.dir1, ked.ratio1))
```

Some basic properties of the predictions in kedpred1 are

```
names(kedpred1)
```

```
[1] "long"      "lat"      "var1.pred" "var1.var"
```

```
class(kedpred1)
```

```
[1] "data.frame"
```

```
dim(kedpred1)
```

```
[1] 248675      4
```

```
summary(kedpred1)
```

```
      long           lat           var1.pred         var1.var
 Min.   :129    Min.   :-11.8   Min.   :-17.1   Min.   :  64.6
 1st Qu.:129    1st Qu.:-11.5   1st Qu.: 12.6   1st Qu.: 219.7
 Median :129    Median :-11.2   Median : 19.7   Median : 294.1
 Mean   :129    Mean   :-11.2   Mean   : 20.8   Mean   : 347.1
 3rd Qu.:130    3rd Qu.:-10.9   3rd Qu.: 28.6   3rd Qu.: 398.7
 Max.   :130    Max.   :-10.6   Max.   :101.3   Max.   :1648.6
```

The results show that some predictions are beyond the range of percentage data (i.e., 0 to 100%) and need to be corrected. They are corrected by resetting the faulty predictions to the nearest bound of the data range. The variances are abnormally high (i.e., a small portion are higher than 300) and for illustration purposes, we cap them to 300.

```
kedpred2 <- kedpred1
kedpred2$var1.pred[kedpred2$var1.pred >= 100] <- 100
kedpred2$var1.pred[kedpred2$var1.pred <= 0] <- 0
kedpred2$var1.var[kedpred2$var1.var >= 300] <- 300
```

The spatial distribution of predictions and variances of *KED* can be illustrated in Figure 6.6 by

```
kedpred2.1 <- kedpred2
gridded(kedpred2.1) <- ~ long + lat
names(kedpred2.1) <- c('predictions', 'variances')

library(raster)

par(font.axis = 2, font.lab = 2)
plot(brick(kedpred2.1))
```

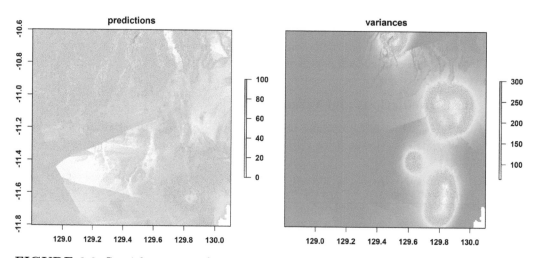

FIGURE 6.6: Spatial patterns of predictions and variances of *KED*.

7

Modern statistical methods

This chapter introduces modern statistical methods, which are also *gradient based/detrended spatial predictive methods*. *Modern statistical methods* use secondary information or predictive variables for detrended spatial predictive modeling.

In this chapter, the following *modern statistical methods* that are available in *R* are introduced:
(1) *linear models (LM)*,
(2) *trend surface analysis (TSA)*,
(3) *thin plate splines (TPS)*,
(4) *generalized linear models (GLM)*, including `glmnet`, and
(5) *generalized least squares (GLS)*.

The `petrel` point data and `petrel.grid` grid data in the `spm` package (Li 2019b) will be used to demonstrate the applications of the modern statistical methods.

7.1   Linear models

The predictions of *linear models* (or *linear regression models*) *(LM)* can be represented in Equation (7.1), in a similar way as in Equation (4.1).

$$\widehat{y_i} = \sum_{j=0}^{p} \beta_j \cdot x_{ij} \tag{7.1}$$

where $\widehat{y_i}$ is the predicted value of response variable at the location of interest (i), x_{ij} is the observed value of predictive variable (x_j) at the location i, β_j is the *regression coefficient* assigned to variable x_j, and p represents the number of predictive variables used for generating the predictions. When j is 0, β_0 is the intercept and its associated x_0 is 1. Here $\widehat{y_i}$ is equivalent to $\hat{y}(x_0)$ in Equation (4.1).

The *coefficients* (β_j) are estimated by minimizing square error loss, as follows (7.2):

$$minimize \left\{ \sum_{i=1}^{n} (y_i - \hat{y})^2 / (2n) \right\} \tag{7.2}$$

where y_i is the observed value of sample i and n represents the number of samples. For a step-by-step procedure to derive β_j, Crawley's book (2007) is recommended.

For *LM*, data are *assumed* to be independent of each other, *normally distributed*, and *homogeneous in variance*. Regression methods explore a possible functional relationship between

DOI: 10.1201/9781003091776-7   121 response variable (or primary variable) and predictive variables (or secondary variables, secondary information). A predictive model can be developed by a thorough understanding of the relationships between the response variable and predictive variables (e.g., coordinate information, elevation, bathymetry).

7.1.1 Relationships of response variable with predictive variables

For *LM*, `gravel` in `petrel` and `petrel.grid` will be used. The correlations of `gravel` with available *predictive variables* are shown below.

```
library(spm)

cor(petrel)
```

```
           long     lat     mud    sand  gravel   bathy    dist  relief   slope
long     1.0000  0.4581 -0.0657  0.1018 -0.0496  0.2590 -0.3040  0.0800  0.0740
lat      0.4581  1.0000 -0.0572  0.2169 -0.1940 -0.4913  0.4335  0.2273  0.2437
mud     -0.0657 -0.0572  1.0000 -0.6063 -0.3537 -0.0852 -0.0595 -0.0544 -0.0321
sand     0.1018  0.2169 -0.6063  1.0000 -0.5294 -0.1373  0.2433 -0.1112 -0.1272
gravel  -0.0496 -0.1940 -0.3537 -0.5294  1.0000  0.2524 -0.2227  0.1888  0.1838
bathy    0.2590 -0.4913 -0.0852 -0.1373  0.2524  1.0000 -0.6665 -0.4052 -0.3831
dist    -0.3040  0.4335 -0.0595  0.2433 -0.2227 -0.6665  1.0000 -0.0872 -0.0551
relief   0.0800  0.2273 -0.0544 -0.1112  0.1888 -0.4052 -0.0872  1.0000  0.9511
slope    0.0740  0.2437 -0.0321 -0.1272  0.1838 -0.3831 -0.0551  0.9511  1.0000
```

The relationships of `gravel` with available predictive variables are illustrated in Figure 7.1 by

```
gravel1 <- petrel[, c(1, 2, 6:9, 5)]

par(mfrow=c(3,2), font.axis = 2, font.lab = 2)
for (i in 1:6) {
  plot(gravel1[, i], gravel1[, 7], xlab = names(gravel1)[i], ylab = "Gravel (%)")
  lines(lowess(gravel1[, 7] ~ gravel1[, i]), col = "blue")
}
```

The relationships are weak non-linear, which is quite common for environmental data when causal variables are unknown and proxy variables are used as predictors instead.

7.1.2 Implementation of *LM* in `lm`

We can use the function `lm` in the `stats` package (R Core Team 2020) to demonstrate the application of *LM*. The function `lm` can be used to fit linear models. For `lm`, the following two arguments are essential:
(1) `formula`, a symbolic description of the model to be fitted; and
(2) `data`, a dataframe, containing the variables in the model.

We will use `gravel` as a response variable and all other variables in the `gravel1` data set as predictive variables. Here we consider all available predictive variables and their second- and third-order terms (i.e., quadratic, cubic polynomials) in `lm1` as below.

```
lm1 <- lm(log(gravel + 1) ~ . + I(long^2) + I(long^3) + I(lat^2) + I(lat^3) + I(
    bathy^2) + I(bathy^3) + I(dist^2)+ I(dist^3) + I(relief^2)+ I(relief^3) + I(
    slope^2) + I(slope^3), gravel1)

lm1
```

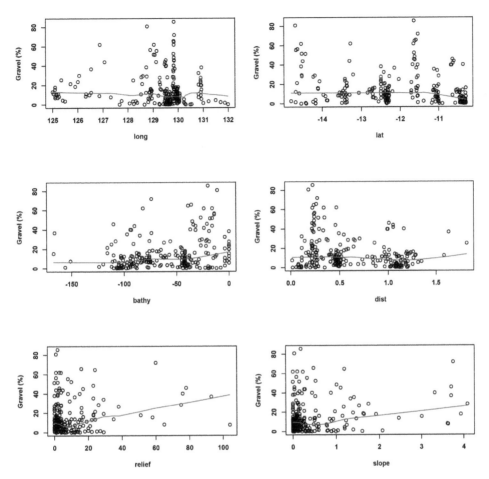

FIGURE 7.1: Relationships of gravel and predictive variables based on the lowess function (blue line).

```
lm(formula = log(gravel + 1) ~ . + I(long^2) + I(long^3) + I(lat^2) +
    I(lat^3) + I(bathy^2) + I(bathy^3) + I(dist^2) + I(dist^3) +
    I(relief^2) + I(relief^3) + I(slope^2) + I(slope^3), data = gravel3)
```

(Intercept)	long	lat	bathy	dist	relief
-17782.4208	395.9424	-116.0185	0.0685	5.2085	-0.0374
slope	I(long^2)	I(long^3)	I(lat^2)	I(lat^3)	I(bathy^2)
0.7461	-3.0142	0.0076	-9.2652	-0.2443	0.0006
I(bathy^3)	I(dist^2)	I(dist^3)	I(relief^2)	I(relief^3)	I(slope^2)
0.0000	-4.9611	1.5054	0.0022	0.0000	-0.5423
I(slope^3)					
0.0744					

The interactions among relevant predictive variables are not considered in this demonstration. In practice, they should be considered if they were believed to be important. For instance, two-way interactions could be considered by replacing log(gravel + 1)~ . with log (gravel + 1)~ . ^2 in lm1.

7.1.3   Model selection based on likelihood methods

LM are usually simplified or selected based on likelihood measures such as *Akaike information criteria* (*AIC*) or *Bayesian information criteria* (*BIC*). In fact, the term *model selection* is *variable selection* in essence. When variable selection is based on likelihood, the variable selection is usually termed as model selection.

Various functions, such as `anova`, `update`, `stepAIC`, `dropterm`, and `addterm` in the `MASS` package (Venables and Ripley 2002; Ripley 2020), as well as `regsubsets` in the `leaps` package (James et al. 2017; Lumley 2020), are available for model selection. These functions will be used for model selection in this section.

1. Functions `anova` and `update`

The function `anova` computes the analysis of variance (or deviance) tables for one or more fitted model objects. The function `update` refits a model by extracting the call stored in the object, and updating and evaluating the call. We will apply `anova` to `lm1`.

```
anova(lm1)
```

```
Analysis of Variance Table

Response: log(gravel + 1)
             Df Sum Sq Mean Sq F value        Pr(>F)
long          1      4     4.2    5.45        0.0204 *
lat           1      9     9.1   11.87        0.0007 ***
bathy         1     10    10.2   13.37        0.0003 ***
dist          1      1     1.1    1.45        0.2302
relief        1     32    32.4   42.29 0.0000000005 ***
slope         1      0     0.3    0.41        0.5219
I(long^2)     1      1     1.3    1.67        0.1978
I(long^3)     1      0     0.3    0.42        0.5153
I(lat^2)      1     11    11.2   14.61        0.0002 ***
I(lat^3)      1     14    14.2   18.51 0.0000255301 ***
I(bathy^2)    1      5     5.1    6.71        0.0102 *
I(bathy^3)    1      2     2.1    2.75        0.0987 .
I(dist^2)     1      6     6.4    8.39        0.0042 **
I(dist^3)     1      2     2.4    3.11        0.0791 .
I(relief^2)   1      0     0.1    0.09        0.7599
I(relief^3)   1      2     2.4    3.07        0.0812 .
I(slope^2)    1      0     0.4    0.52        0.4734
I(slope^3)    1      0     0.4    0.58        0.4462
Residuals   218    167     0.8
---
Signif. codes:  0 '***' 0.001 '**' 0.01 '*' 0.05 '.' 0.1 ' ' 1
```

We will use `update` to remove those non-significant (with $p < 0.05$) second- and third-order terms.

```
lm1.1 <- update(lm1, .~. - I(long^2) - I(long^3) - I(bathy^3) - I(dist^3) - I(
    relief^2) - I(relief^3) - I(slope^2) - I(slope^3))
```

```
anova(lm1.1)
```

```
Analysis of Variance Table

Response: log(gravel + 1)
           Df Sum Sq Mean Sq F value   Pr(>F)
long        1      4     4.2     5.2   0.0230 *
lat         1      9     9.1    11.4   0.0009 ***
bathy       1     10    10.2    12.8   0.0004 ***
dist        1      1     1.1     1.4   0.2395
```

```
relief       1    32    32.4    40.6 0.000000001 ***
slope        1     0     0.3     0.4        0.5302
I(lat^2)     1     8     7.8     9.7        0.0020 **
I(lat^3)     1    17    17.3    21.7 0.000005517 ***
I(bathy^2)   1     4     3.9     4.8        0.0291 *
I(dist^2)    1     4     4.1     5.2        0.0241 *
Residuals  226   180     0.8
---
Signif. codes:  0 '***' 0.001 '**' 0.01 '*' 0.05 '.' 0.1 ' ' 1
```

The results show that `slope` is not significant and should be removed. Although `dist` is also not significant, it should be kept to maintain marginality (Venables and Ripley 2002) as its second-order `I(dist^2)` is significant.

```
lm1.2 <- update(lm1.1, .~. - slope)

anova(lm1.2)

Analysis of Variance Table

Response: log(gravel + 1)
            Df Sum Sq Mean Sq F value     Pr(>F)
long         1      4     4.2     5.3     0.0228 *
lat          1      9     9.1    11.4     0.0008 ***
bathy        1     10    10.2    12.9     0.0004 ***
dist         1      1     1.1     1.4     0.2387
relief       1     32    32.4    40.8 0.000000001 ***
I(lat^2)     1      8     8.0    10.1     0.0017 **
I(lat^3)     1     17    17.3    21.7 0.000005340 ***
I(bathy^2)   1      4     3.8     4.8     0.0300 *
I(dist^2)    1      4     4.1     5.1     0.0249 *
Residuals  227    180     0.8
---
Signif. codes:  0 '***' 0.001 '**' 0.01 '*' 0.05 '.' 0.1 ' ' 1
```

Now all variables are significant except `dist`, and no further variable can be removed.

2. stepAIC

The function `stepAIC` performs stepwise model selection based on *AIC* or *BIC*. We can use `stepAIC` to select predictive variables. Relevant arguments of `stepAIC` are detailed in its help file that can be accessed via `?stepAIC`. For `stepAIC`, we may need to specify the following arguments for its application:

(1) `object`, a model of an appropriate class;
(2) `k`, the multiple of the number of degrees of freedom used for the penalty, with a default of 2; and
(3) `direction`, the mode of stepwise search, can be `both`, `forward` or `backward`, with a default of `both`.

Now we will use `stepAIC` to simplify the model, `lm1`, based on *AIC* (i.e., using the default for `k`) and the default for `direction`.

```
library(MASS)

lm2 <- stepAIC(lm1, trace = F)
```

The model selected is

```
lm2
```

```
Call:
lm(formula = log(gravel + 1) ~ long + lat + bathy + dist + I(long^2) +
    I(lat^2) + I(lat^3) + I(bathy^2) + I(bathy^3) + I(dist^2) +
    I(dist^3) + I(relief^2) + I(relief^3), data = gravel3)

Coefficients:
 (Intercept)           long            lat          bathy           dist
-1572.889785      17.715242    -112.956634       0.066670       5.208052
   I(long^2)       I(lat^2)       I(lat^3)      I(bathy^2)     I(bathy^3)
   -0.070690      -9.022039      -0.237989       0.000592       0.000002
   I(dist^2)      I(dist^3)    I(relief^2)    I(relief^3)
   -5.051983       1.576930       0.000775      -0.000006
```

We can also simplify the model based on *BIC* by specifying k as k = log(n), where n is the sample size.

```
lm3 <- stepAIC(lm1, trace = F, k = log(dim(gravel1)[1]))
```

The model resulted is

```
lm3
```

```
Call:
lm(formula = log(gravel + 1) ~ lat + bathy + dist + I(long^2) +
    I(lat^2) + I(lat^3) + I(bathy^2) + I(relief^2) + I(relief^3),
    data = gravel3)

Coefficients:
  (Intercept)            lat          bathy           dist      I(long^2)
-446.10228681  -114.26780935     0.03292989     1.12212695    -0.00137518
     I(lat^2)       I(lat^3)     I(bathy^2)    I(relief^2)    I(relief^3)
  -9.14284574    -0.24180085     0.00011227     0.00089543    -0.00000745
```

3. Functions dropterm and addterm

The function dropterm tries fitting all models that differ from the current model by dropping a single term and by maintaining marginality. The function addterm tries fitting all models that differ from the current model by adding a single term from those supplied and by maintaining marginality.

We will simplify lm3 using dropterm as below.

```
dropterm (lm3, k = log(dim(gravel1)[1]), test="F")

Single term deletions

Model:
log(gravel + 1) ~ lat + bathy + dist + I(long^2) + I(lat^2) +
    I(lat^3) + I(bathy^2) + I(relief^2) + I(relief^3)
            Df Sum of Sq  RSS    AIC  F Value        Pr(F)
<none>                    182  -8.31
lat          1     21.42  203  12.64    26.76   0.00000051 ***
bathy        1     17.13  199   7.58    21.40   0.00000626 ***
dist         1     13.78  196   3.55    17.22   0.00004707 ***
I(long^2)    1     19.27  201  10.12    24.08   0.00000176 ***
I(lat^2)     1     20.99  203  12.14    26.23   0.00000065 ***
I(lat^3)     1     20.53  202  11.59    25.65   0.00000085 ***
I(bathy^2)   1      4.56  186  -7.90     5.69   0.01785    *
I(relief^2)  1      9.57  191  -1.61    11.95   0.00065    ***
I(relief^3)  1      5.66  187  -6.50     7.07   0.00838    **
---
Signif. codes:  0 '***' 0.001 '**' 0.01 '*' 0.05 '.' 0.1 ' ' 1
```

The `dropterm` with F test is for models based on data in *Gaussian* distribution. The model resulted shows that no further variables can be removed.

Besides `dropterm`, `addterm` can also be used to update relevant models. For example,

```
lm0 <- lm(log(gravel + 1) ~ 1, gravel1)
addterm(lm0, lm1, k = log(dim(gravel1)[1]), test = "F")
```

```
Single term additions

Model:
log(gravel + 1) ~ 1
             Df Sum of Sq RSS  AIC F Value   Pr(F)
<none>                    271 36.9
long          1      4.18 266 38.7    3.69 0.05610 .
lat           1     13.08 258 30.7   11.93 0.00065 ***
bathy         1     13.80 257 30.0   12.63 0.00046 ***
dist          1      7.35 263 35.9    6.56 0.01107 *
relief        1      7.37 263 35.9    6.57 0.01097 *
slope         1      6.47 264 36.7    5.75 0.01726 *
I(long^2)     1      4.20 266 38.7    3.71 0.05537 .
I(long^3)     1      4.23 266 38.7    3.73 0.05465 .
I(lat^2)      1     11.98 259 31.7   10.89 0.00112 **
I(lat^3)      1     10.95 260 32.6    9.91 0.00186 **
I(bathy^2)    1      9.53 261 33.9    8.57 0.00375 **
I(bathy^3)    1      5.38 265 37.7    4.76 0.03009 *
I(dist^2)     1      6.80 264 36.4    6.05 0.01460 *
I(dist^3)     1      3.84 267 39.0    3.38 0.06708 .
I(relief^2)   1      7.14 264 36.1    6.37 0.01227 *
I(relief^3)   1      4.87 266 38.1    4.31 0.03902 *
I(slope^2)    1      8.90 262 34.5    7.99 0.00511 **
I(slope^3)    1      8.82 262 34.6    7.91 0.00533 **
---
Signif. codes:  0 '***' 0.001 '**' 0.01 '*' 0.05 '.' 0.1 ' ' 1
```

It appears that all `long` related terms and `dist^3` are not significant and should be removed.

4. Function `regsubsets`

The function `regsubsets` performs model selection by exhaustive search, forward or backward stepwise, or sequential replacement. We can also use `regsubsets` to select predictive variables. For `regsubsets`, we need to specify the following three arguments:
(1) `formula`, a symbolic description of the model to be fitted;
(2) `data`, a dataframe, containing the variables in the model; and
(3) `method`, use `exhaustive` search, `forward` selection, `backward` selection or `seqrep` sequential replacement to search.

We will apply `regsubsets` to `lm1` that is a model with all predictive terms.

```
library(leaps)
```

```
reg1 <- regsubsets(log(gravel + 1) ~. + I(long^2) + I(long^3) + I(lat^2) + I(lat
     ^3) + I(bathy^2) + I(bathy^3) + I(dist^2)+ I(dist^3) + I(relief^2)+ I(relief
     ^3) + I(slope^2) + I(slope^3), gravel1, method = "backward")
```

We will select a model using *BIC* below.

```
reg1.summary <- summary(reg1)
id <- which.min(reg1.summary$bic)
```

The variables selected are

```
names(which(reg1.summary$which[id, ] == TRUE))
```

```
[1] "(Intercept)" "lat"          "bathy"        "dist"         "I(long^2)"
[6] "I(lat^2)"     "I(lat^3)"     "I(relief^2)" "I(relief^3)"
```

```
coef(reg1, id)
```

```
  (Intercept)              lat           bathy            dist        I(long^2)
-404.237779569  -103.269480561     0.017634973     1.016019242     -0.001182998
     I(lat^2)         I(lat^3)     I(relief^2)     I(relief^3)
  -8.265003088     -0.218718763     0.000902304    -0.000006865
```

We can also use `reg1` and `reg1.summary` to select the model with other measures such as `rss` (i.e., residual sum of squares for each model) or `cp` (i.e., Mallows' Cp).

It should be noted that the model selection based on the likelihood methods may result in the most parsimonious models, but not necessarily the most accurate predictive models, especially when proxy variables are used as predictors instead of causal variables (Li, Alvarez, et al. 2017). Although such model selection methods are useful approaches for inferential or exploratory analyses (Leek and Peng 2015), they may not be sufficient for predictive modeling.

7.1.4   Variable selection based on predictive accuracy

The traditional model selection methods such as *AIC* and *BIC* may not necessarily result in the most accurate predictive models. For predictive modeling, predictive accuracy instead of *AIC* or *BIC* should be used for *model selection* (or *variable selection*) (Li, Alvarez, et al. 2017).

Variable selection based on predictive accuracy can be carried out with the function `bestglm` in the `bestglm` package (McLeod, Xu, and Lai 2020). The function `bestglm` performs the best subset selection using 'leaps' algorithm (Furnival and Wilson 1974) or complete enumeration that is used for the non-*Gaussian* and for the case where the input matrix contains factor variables with more than 2 levels (Morgan and Tatar 1972).

For `bestglm`, we need to provide the following arguments:
(1) `Xy`, a dataframe containing design matrix `x` and output variable `y`, with all columns named;
(2) `family`, one of the *glm* distribution functions;
(3) `IC`, information criteria to use: "AIC", "BIC", "BICg", "BICq", "LOOCV", or "CV"; and
(4) `CVArgs`, used when `IC` is set to "CV"; for k-fold cross-validation, three components need to be specified, that is, `Method` (should be "HTF"(), "K" (fold number), and "REP" (number of replications)).

We need to prepare a required data set for `bestglm` by considering all predictive terms in `lm1` below.

```
X <- data.frame(subset(gravel1, select = -gravel),
long2 = gravel1$long^2, long3 = gravel1$long^3,
lat2 = gravel1$lat^2, lat3= gravel1$lat^3,
bathy2 = gravel1$bathy^2, bathy3 = gravel1$bathy^3,
dist2 = gravel1$dist^2, dist3 = gravel1$dist^3,
relief2 = gravel1$relief^2, relief3 = gravel1$relief^3,
slope2 = gravel1$slope^2, slope3= gravel1$slope^3)
```

```
# if two-way interactions need to be considered, for example, the interaction
    between `bathy` and `dist`, then we can add it to the `X` data set like `
    bathydist = gravel1$bathy * gravel1$dist`.
```

```
y <- log(gravel1$gravel +1)
Xy <- as.data.frame(cbind(X, y))
```

Then apply `bestglm` to the data set.

```
library(bestglm)

set.seed(1234)

lm.cv <- bestglm(Xy, IC = "CV", family = gaussian)
```

The model resulted is

```
lm.cv$BestModel
```

```
Call:
lm(formula = y ~ ., data = data.frame(Xy[, c(bestset[-1], FALSE),
    drop = FALSE], y = y))

Coefficients:
  (Intercept)            lat          bathy           dist         relief
-411.69718332  -103.08785231     0.01731052     1.02637316     0.02526885
        long3           lat2           lat3
  -0.00000584    -8.23320563    -0.21746637
```

```
lm.cv$Subsets[, 20:21]
```

```
     logLikelihood        CV
0         -15.7394     1.168
1          -9.5368     1.138
2           0.7226     1.076
3           9.9143     1.021
4          12.3317     1.035
5          15.6410     1.169
6          23.2589     1.438
7*         27.3666     1.014
8          29.8458     1.044
9          32.4727     1.081
10         34.5165     1.174
11         37.1539     1.248
12         39.0122     1.827
13         40.4280    28.214
14         40.7062    50.210
15         40.9573   149.163
16         41.1102   137.387
17         41.3161   379.246
18         41.4694   446.567
```

The results show that `model 7` in `lm.cv$Subsets` is the optimal model with the lowest *MSE* (i.e., `cv`) and the optimal model is with seven predictive variables, `lat`, `bathy`, `dist`, `relief`, `long3`, `lat2`, and `lat3`. It should be noted that `bestglm` claims that it is to find the best fitted model according to log-likelihood, but if `IC = cv` it is to find the predictive model with the least predictive error.

We can also use the *LOO* method.

```
lm.loocv <- bestglm(Xy, IC = "LOOCV", family = gaussian)

lm.loocv
```

```
LOOCV
BICq equivalent for q in (0.705926243008822, 0.788893926175818)
Best Model:
                  Estimate     Std. Error t value    Pr(>|t|)
(Intercept) -1534.60062726 470.1614224176  -3.264 0.0012704987
long            17.00459935   7.2417158735   2.348 0.0197382529
lat           -114.27567249  26.5997621815  -4.296 0.0000259129
bathy            0.06722556   0.0131879731   5.097 0.0000007315
dist             5.38718398   1.5568152622   3.460 0.0006457290
relief           0.01985760   0.0055314464   3.590 0.0004063416
long2           -0.06784941   0.0283677477  -2.392 0.0175931350
lat2            -9.11876420   2.1417084839  -4.258 0.0000303908
lat3            -0.24036627   0.0570884833  -4.210 0.0000369247
bathy2           0.00061818   0.0001975219   3.130 0.0019826446
bathy3           0.00000202   0.0000008483   2.382 0.0180696759
dist2           -5.26871557   2.1485932260  -2.452 0.0149645010
dist3            1.64947137   0.8766198398   1.882 0.0611835462
```

```
round(lm.loocv$Subsets[, c("logLikelihood", "LOOCV")], 2)
```

```
     logLikelihood LOOCV
0           -15.74  1.15
1            -9.54  1.10
2             0.72  1.02
3             9.91  0.95
4            12.33  0.94
5            15.64  0.93
6            23.26  0.88
7            27.37  0.86
8            29.85  0.85
9            32.47  0.84
10           34.52  0.83
11           37.15  0.82
12*          39.01  0.82
13           40.43  0.82
14           40.71  0.83
15           40.96  0.83
16           41.11  0.84
17           41.32  0.85
18           41.47  0.85
```

The results show that the optimal model is `model 12` in `lm.loocv$Subsets`, with 12 variables: `long`, `lat`, `bathy`, `dist`, `relief`, `long2`, `lat2`, `lat3`, `bathy2`, `bathy3`, `dist2`, and `dist3`.

7.1.5 Predictive accuracy

We can use the function `glmcv` in the `spm2` package (Li 2021a) to assess the predictive accuracy. The function `glmcv` is a cross-validation function for *GLM* methods based on the function `glm` in the `stats` package. For `glmcv`, we need to provide the following arguments:

(1) `formula`, a formula defining response variable and predictive variables;

(2) `trainxy`, a dataframe contains longitude (long), latitude (lat), predictive variables and the response variable of point samples, with the location information named as 'long' and 'lat';

(3) `y`, a vector of the response variable in the formula, that is, the left part of the formula;

(4) `family`, a description of an error distribution and link function to be used in the model; and

(5) `validation`, validation methods, include 'LOO', and 'CV'.

For other arguments, the defaults could be used, and for details see `?glmcv`.

We can also use the function CVHTF in the bestglm package to assess the predictive accuracy of the models selected. Its arguments are similar to those for the function bestglm.

We will assess the *predictive accuracy* of the models selected in the previous two sections, that is, lm1.2, lm2, lm3, lm.cv, lm.loocv as well as the model selected using regsubsets.

1. Predictive accuracy of lm1.2

Function glmcv

We will apply glmcv with the default family = "gaussian" to assess the predictive accuracy of lm1.2.

```
y <- log(gravel1$gravel +1)
model <- log(gravel + 1) ~  long + lat +  bathy + dist + relief +  I(lat^2) + I(
    lat^3) + I(bathy^2) + I(dist^2)
set.seed(1234)
n <- 100
lmvecv <- NULL

for (i in 1:n) {
  lmcv1 <- glmcv(formula = model, gravel1, y, validation = "CV", predacc = "VEcv")
  lmvecv[i] <- lmcv1
}
```

The median and range of predictive accuracy based on 100 repetitions of 10-fold cross-validation using glmcv are

```
median(lmvecv)
```

```
[1] 27
```

```
range(lmvecv)
```

```
[1] 23.67 28.87
```

Function CVHTF

The function CVHTF performs K-fold cross-validation (McLeod, Xu, and Lai 2020). The predictive accuracy of lm1.2 can also be obtained by CVHTF as follows.

```
X1 <- subset(X, select = c(long, lat, bathy, dist, relief, lat2, lat3, bathy2,
    dist2)) # variables in `lm.cv`
set.seed(1234)
lm1.2.cv.cvhtf <- CVHTF(X1, y, K = 10, REP = 100, family = gaussian)
```

```
lm1.2.cv.cvhtf
```

```
[1] 0.84508 0.08696
```

They are *MSE* and its sd for model lm1.2. The *MSE* can be converted to *VEcv* using the function tovecv in library(spm).

```
lm1.2.cv.VEcv <- tovecv(n = dim(gravel1)[1], mu = mean(y), s = sd(y), m = lm1.2.cv
    .cvhtf[1], measure="mse")
```

The predictive accuracy (lm1.2.cv.VEcv) based on 100 repetitions of CVHTF is

```
lm1.2.cv.VEcv
```

```
[1] 26
```

The predictive accuracy from `glmcv` is slightly higher than that from `CVHTF`, which is probably due to the difference in splitting the training data for cross-validation. In `glmcv`, the function `datasplit` in the `spm2` package is used, and it uses a stratified random sampling technique and resamples the training data based on sample quantiles. Hence, all values in the samples are better presented in each sub-data set than samples derived from random sampling, which may explain why a higher predictive accuracy is produced by `glmcv`. Because of this reason, `CVHTF` will not be applied to the remaining models.

2. Predictive accuracy of the rest models

We can also apply `glmcv` with the default `family = "gaussian"` to assess the predictive accuracy of `lm2`, `lm3`, `lm.cv`, `lm.loocv` as well as the model selected with `regsubsets`, but the R code will not be presented. This is because the R code for model `lm1.2` can be easily modified for these models.

The medians and ranges of predictive accuracy of these models based on 100 repetitions of 10-fold cross-validation using `glmcv` are

Model `lm2`

```
[1]  27.5
```

```
[1]  22.25 30.66
```

Model `lm3`

```
[1]  26.83
```

```
[1]  22.52 28.87
```

Model `lm.cv`

```
[1]  25.56
```

```
[1]  22.08 27.66
```

Model `lm.loocv`

```
[1]  28.59
```

```
[1]  24.50 31.79
```

The model selected with `regsubsets`

```
[1]  25.19
```

```
[1]  21.31 27.96
```

The model `lm.loocv` is of the highest predictive accuracy and should be used to generate spatial predictions.

The fitted values of `lm.loocv$BestModel` are plotted against the observed values in Figure 7.2 by

```
par(font.axis = 2, font.lab = 2)
plot(gravel1$gravel, exp(lm.loocv$BestModel$fitted.values) - 1, xlab = "Observed
    values", ylab = "Fitted values")
lines(gravel1$gravel, gravel1$gravel, col = "blue")
```

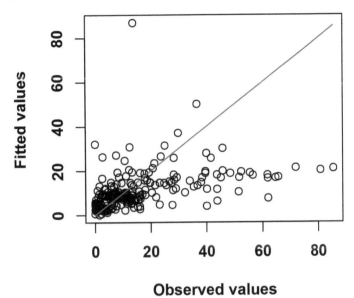

Observed values

FIGURE 7.2: Fitted values vs. observed values.

The optimal model developed, `lm.loocv`, is not fitting the data well. There are a number of tools for model checking (McCullagh and Nelder 1999; Crawley 2007), but this kind of model checking is not helpful for predictive modeling because they are not based on predictive accuracy. When the relationships with the predictors are weak, such as in this case, different predictive methods should be used.

7.1.6 Predictions and standard errors

The predictions and standard errors can be generated based on the model `lm.loocv` by

```
X3 <- subset(X, select = c(long, lat, bathy, dist, relief, long2, lat2, lat3,
    bathy2, bathy3, dist2, dist3))

lm.loocv1 <- lm(y ~ ., X3)

X3.grid <- data.frame(petrel.grid,
long2 = petrel.grid$long^2,
lat2 = petrel.grid$lat^2, lat3= petrel.grid$lat^3,
bathy2 = petrel.grid$bathy^2, bathy3 = petrel.grid$bathy^3,
dist2 = petrel.grid$dist^2, dist3 = petrel.grid$dist^3)

lm.pred <- predict(lm.loocv1, X3.grid, se.fit = TRUE)
names(lm.pred)

lm.pred <- as.data.frame(lm.pred)[, 1:2]
```

The predictions and standard errors are in `log` scale and need to be back-transformed to the original scale (i.e., percentage data for `gravel`).

```
lm.pred$fit <- exp(lm.pred$fit) - 1
lm.pred$se.fit <- exp(lm.pred$se.fit) - 1

range(lm.pred$fit)
```

```
[1]    0.3074 363.7073
```

```
range(lm.pred$se.fit)
```

```
[1] 0.1225 2.4752
```

A small portion of transformed predictions are beyond the range of percentage data (i.e., 0 to 100%) and are corrected by resetting the faulty predictions to the nearest bound of the data range as below.

```
lm.pred1 <- lm.pred
lm.pred1$fit[lm.pred1$fit >= 100] <- 100
```

The predictions of *LM* can also be generated with `glmpred` in the `spm2` package as follows.

```
model.loocv1 <- log(gravel + 1) ~ long + lat + bathy + dist + relief + long2 +
    lat2 + lat3 + bathy2 + bathy3 + dist2 + dist3
```

```
lm.pred2 <- glmpred(formula = model.loocv1, trainxy = X3, longlatpredx = X3.grid[,
    c(1:2)], predx = X3.grid, family = "gaussian")
```

The spatial distribution of predictions and standard errors of the *LM* model are illustrated in Figure 7.3 by

```
lm.pred.df <- cbind(petrel.grid[, 1:2], lm.pred1)
gridded(lm.pred.df) = ~ long + lat
names(lm.pred.df) <- c("predictions", "standard.errors")
```

```
library(raster)
```

```
par(font.axis = 2, font.lab = 2)
plot(brick(lm.pred.df))
```

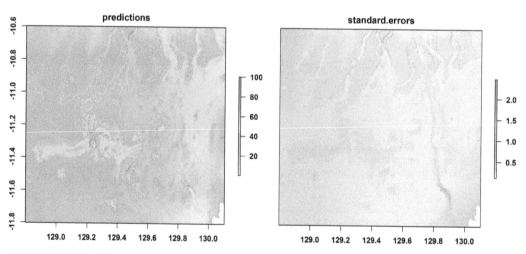

FIGURE 7.3: Spatial patterns of predictions and standard errors using *LM*.

Standard errors are often used to produce prediction intervals that are much wider than confidence intervals of a model fitted (Chambers and Hastie 1992). For *LM* and also for *GLM* that is to be introduced later, prediction intervals are often used as a measure of prediction *uncertainty*. However, such uncertainty does not necessarily reflect the predictive accuracy of the model and could be misleading (Li, Alvarez, et al. 2017; Li 2019a).

7.2 Trend surface analysis

Trend surface analysis (*TSA*) is a special case of *LM*. It only uses geographical coordinates as predictive variables to predict a response variable. It can be extended to include other variables, but in which case it should be classified as *LM*. *TSA* shares the same assumptions as *LM*, and separates the data into regional trends and local variations (Collins and Bolstad 1985). For *TSA*, the degrees (or orders) of geographical coordinates can change, leading to different degrees of trend surfaces (Borcard, Gillet, and Legendre 2011), such as:
(1) first-degree trend surface,
(2) second-degree trend surface, and
(3) third-degree trend surface.

7.2.1 Implementation of *TSA* in `lm`

We will use `lm` and `gravel` in the `gravel1` data set to demonstrate the application of *TSA* with different degrees of trend surfaces below.

1. First-degree trend surface

```
library(MASS)
tsa1 <- lm(log(gravel + 1) ~ long + lat, gravel1)
```

2. Second-degree trend surface

```
tsa2 <- lm(log(gravel + 1) ~ long + lat + I(long^2) + long * lat + I(lat^2) ,
    gravel1)
```

3. Third-degree trend surface

```
tsa3 <- lm(log(gravel + 1) ~ long + lat + I(long^2) + long * lat + I(lat^2) + I(
    long^3) + I(long^2) * lat + long * I(lat^2) + I(lat^3), gravel1)
```

7.2.2 Variable selection

For *TSA*, predictive variables are fixed, so variable selection is not required. However, we can compare and select the most accurate model from models with different degrees of trend surfaces.

7.2.3 Predictive accuracy

The *predictive accuracy* can be assessed with the functions `glmcv` and `CVHTF` in the same way as for *LM* in Section 7.1.5.

7.2.4 Predictions and standard errors

The predictions and standard errors of *TSA* can also be generated and plotted by following the same steps as for *LM* in Section 7.1.6.

7.3   Thin plate splines

Thin plate splines (*TPS*) , formally known as "Laplacian smoothing splines", was developed principally by Wahba and Wendelberger (1980), where the splines consist of polynomials with each polynomial of degree p. For spatial predictive modeling, the polynomials describe pieces of a surface and are fitted together so that they join smoothly (Wahba and Wendelberger 1980; Webster and Oliver 2001). The smoothing parameter for *TPS* is calculated by minimizing the *generalized cross-validation* function (*GCV*). For degree $p = 1, 2$, or 3, a spline is called linear, quadratic or cubic, respectively. *TPS* can be extended to include a multivariate spline function (Wahba 1990; Hutchinson 1995; Mitasova et al. 1995; Burrough and McDonnell 1998).

The predictions of *TPS* can be generated with the function `Tps` in the `fields` package (Nychka et al. 2020). The function `Tps` is for fitting a *TPS* surface to irregularly spaced point data. There are many arguments for `Tps` and the descriptions of relevant arguments in the function are detailed in its help file, which can be accessed by `?Tps`. For *TPS* using `Tps`, the following arguments may need to be specified:
(1) `x`, a matrix of independent variables;
(2) `Y`, a vector of dependent variables;
(3) `m`, `m = NULL` that leads to a default value of 2 for spatial predictive modeling based on `x` containing only location information; but if a number is assigned to `p`, a number must be assigned to `m`;
(4) `p`, polynomial power for Wendland radial basis functions; `p = NULL` that leads to a default value of 2 for spatial predictive modeling based on `x` containing only location information;
(5) `scale.type`, by default `scale.type` is "range", whereby locations are transformed to interval (0,1) by forming `(x-min(x))/range(x)` for each `x`;
(6) `theta`, tapering range; `theta = 3` degrees is a very generous taper range;
(7) `lon.lat`, if `TRUE` locations are interpreted as longitude and latitude and great circle distance is used to find distances among locations; and
(8) `lambda`, smoothing parameter, default is zero which corresponds to interpolation.

If `lambda =0`, it corresponds to no smoothness constraints and the data is interpolated. If `lambda=infinity`, it corresponds to just fitting the polynomial base model by ordinary least squares (i.e., *LM*) (Nychka et al. 2020). Although the smoothing parameter, `lambda`, is with a default value of zero which corresponds to interpolation, a `lambda` estimated may lead to more reliable predictions.

Since *TPS* can use either (1) spatial information, or (2) spatial information and secondary information, we will demonstrate its application to `gravel` in the `petrel` data set for such information sets separately.

7.3.1   Estimation of smoothing parameter `lambda`

Smoothing parameter `lambda` can be estimated using the function `fastTpsMLE` in the `fields` package. For `fastTpsMLE`, the arguments `x`, `y`, and `theta` need to be specified. These arguments are the same as those for `Tps`. We will demonstrate the application of `fastTpsMLE` with `theta = 3` and all other arguments with the default values to `gravel` in the `petrel` data set.

1. Spatial information only

```
library(fields)
library(spm)
```

```
data(petrel)
x1.tps <- petrel[, 1:2]
y.tps <- petrel[, 5]

tps.parameters <- fastTpsMLE(x1.tps, y.tps, theta = 3)
```

The smoothing parameter estimated, `lambda`, is

```
tps.parameters$lambda.best
```

```
[1] 0.9876
```

2. Spatial information and secondary information

```
x2.tps <- petrel[, c(1:2, 6)]

tps.parameters2 <- fastTpsMLE(x2.tps, y.tps, theta = 3)
```

The smoothing parameter estimated, `lambda`, is

```
tps.parameters2$lambda.best
```

```
[1] 0.4869
```

7.3.2 Implementation of *TPS* in Tps

1. Spatial information only

The specification for `lon.lat` has a dramatic impact on the predictions of `Tps` for spatial predictive modeling and `lon.lat = TRUE` should used. Then the `tps.parameters$lambda.best` can be used for `Tps`, with default values for all other arguments in `Tps`.

```
tps1 <- Tps(x1.tps, y.tps, lon.lat = TRUE, lambda = tps.parameters$lambda.best)
#summary(tps1)
```

2. Spatial information and secondary information

Predictive variables can be used in *TPS*. We will use `bathy` in `pb.df` that is sourced from Chapter 1 as a predictive variable for *TPS* as follows.

```
tps2 <- Tps(x2.tps, y.tps, lon.lat = TRUE, lambda = tps.parameters2$lambda.best)
```

The variables fitted and `lambda` estimated are

```
tps2$d
```

```
            [,1]
Intercept 149.04
long      -56.20
lat        87.59
bathy      29.51
```

```
tps2$lambda
```

```
[1] 0.4869
```

The models `tps1` and `tps2` can then be used to generate spatial predictions.

7.3.3 Varible selection and parameter optimization for *TPS*

For *TPS*, the secondary variables need to be selected if two or more predictive variables are available. Relevant arguments for *TPS*, such as m, p, and lambda, may also need to be optimized. Both *variable selection* and *parameter optimization* for *TPS* can be conducted by the function tpscv in the spm2 package based on the predictive accuracy of the *TPS* models resulted.

The arguments of tpscv are similar to those for Tps, with additional arguments or modifications as below:

(1) validation, validation methods, include "LOO" and "CV";

(2) cv.fold, integer; number of folds in the cross-validation; if > 1, then apply n-fold cross-validation; and the default is 10, i.e., 10-fold cross-validation that is recommended; and

(3) predacc, can be either "VEcv" for vecv or "ALL" for all measures in function pred.acc in the spm package.

In the demonstration below, lambda will be estimated to maximize the predictive accuracy of *TPS*. We will also consider all four predictive variables in the petrel data set for *TPS*.

```
tps.lambda <- c(0:100)*0.01
xtps = list(x1.tps, x2.tps,
            x3.tps <- petrel[, c(1:2, 7)],
            x4.tps <- petrel[, c(1:2, 8)],
            x5.tps <- petrel[, c(1:2, 9)],
            x6.tps <- petrel[, c(1:2, 6,7)],
            x7.tps <- petrel[, c(1:2, 6,8)],
            x8.tps <- petrel[, c(1:2, 6,9)],
            x9.tps <- petrel[, c(1:2, 7,8)],
            x10.tps <- petrel[, c(1:2, 7,9)],
            x11.tps <- petrel[, c(1:2, 8,9)],
            x12.tps <- petrel[, c(1:2, 6:8)],
            x13.tps <- petrel[, c(1:2, 6, 7, 9)],
            x14.tps <- petrel[, c(1:2, 6, 8, 9)],
            x15.tps <- petrel[, c(1:2, 7:9)],
            x16.tps <- petrel[, c(1:2, 6:9)])

library(spm2)

tpsopt <- array(0, dim = c(length(tps.lambda), length(xtps)))

for (i in 1:length(tps.lambda)) {
  for (j in 1:length(xtps)) {
    set.seed(1234)
    tpscv1 <- tpscv(xtps[[j]], y.tps, lambda = tps.lambda[i], predacc = "VEcv" )
    tpsopt[i, j] <- tpscv1
  }
}
```

The optimal lambda and predictive variables selected are

```
para.opt <- which (tpsopt == max(tpsopt), arr.ind = T)
tps.lambda[para.opt[, 1]]

[1] 0.45

names(xtps[[para.opt[, 2]]])

[1] "long"  "lat"   "bathy"
```

The variables selected are the same as those in model tps2, but the estimation of lambda is slightly lower than that used in tps2.

The parameter estimated and variables selected should be used to generate spatial predictions.

For *TPS*, the remaining arguments, such as m and p, can also be optimized by following the example for lambda.

7.3.4 Predictive accuracy

To stabilize *predictive accuracy* and get a more reliable accuracy estimation for tps2, we will repeat the cross-validation 100 times. Regarding lambda, we can use either the optimal lambda estimated or the default lambda value (NULL) that will get fastTpsMLE to estimate a value for lambda for each iteration as below.

1. Use the optimal lambda estimated

We will use the optimal lambda estimated, that is 0.45, for all iterations.

```
n <- 100
set.seed(1234)
tpsvecv1 <- NULL

for (i in 1:n) {
  tpscv1 <- tpscv(x2.tps, y.tps, cv.fold = 10, lambda = 0.45, predacc = "VEcv")
  tpsvecv1[i] <- tpscv1
}
```

The median and range of predictive accuracy of *TPS* based on 100 repetitions of 10-fold cross-validation using tpscv are

```
median(tpsvecv1)
```

```
[1] 34.81
```

```
range(tpsvecv1)
```

```
[1] 19.94 41.85
```

2. Estimate the optimal lambda for each iteration

We can also use fastTpsMLE to estimate the optimal lambda for each iteration using the default lambda value, NULL.

```
n <- 100
tpsvecv2 <- NULL

for (i in 1:n) {
  set.seed(1234 + i)
  tpscv1 <- tpscv(x2.tps, y.tps, cv.fold = 10, predacc = "VEcv")
  tpsvecv2[i] <- tpscv1
}
```

It should be noted that a random seed needs to be set for each iteration. If a random seed like set.seed(1234) is set outside the iteration loop, then the results for the second and remaining iterations are the same as if the same random seed were assigned to all iterations except the first one.

The median and range of predictive accuracy of *TPS* based on 100 times of 10-fold cross-validation using tpscv are

```
median(tpsvecv2)
```

```
[1] 35.4
```

```
range(tpsvecv2)
```

```
[1] 27.35 40.91
```

It is clear that *TPS* with the default lambda value (NULL) is more accurate and should be used for generating spatial predictions.

7.3.5 Predictions

We will use *TPS* with the default `lambda` as below to generate predictions for the `pb.df` data set.

```
tps3 <- Tps(x2.tps, y.tps, lon.lat = TRUE)
tps3.pred <- predict(tps3, pb.df)
```

The range of the predictions, `tps3.pred`, is

```
round(range(tps3.pred), 2)
```

```
[1] -6.41 49.06
```

The range of predictions is lower than the expected data range of sample data (see *SK* and *OK* predictions). Some predictions are beyond the range of percentage data and are corrected by resetting the faulty predictions to the nearest bound of the data range.

```
tps3.pred[tps3.pred <= 0] <- 0
```

The predictions of *TPS* are illustrated in Figure 7.4.

Although the default `lambda` is used for `gravel`, it may not always lead to the most accurate predictive *TPS* model. One may try to repeat the above modeling procedure for `mud` data in the `petrel` data set and will see that `lambda` for `Tps` needs to be estimated. Therefore, the value of `lambda` should be determined according to its effect on the predictive accuracy. The other arguments of `Tps` may also need to be optimized to increase its predictive accuracy.

The predictions of *TPS* can also be generated with the function `interpolate` in the `raster` package (Hijmans 2020). *Cubic TPS* is also available in the `bigsplines` package (Helwig 2018).

7.4 Generalized linear models

The predictions of *generalized linear models* (*GLM*) can be presented in the same way as in Equation (7.1) for *LM*.

The *assumptions* of *LM* (i.e., data with normal errors and constant variance) are often not met. *GLM* extend *LM* to accommodate both non-normally distributed response variable and/or with *heterogeneous variance* (Venables and Ripley 2002). *GLM* can deal with data in various distributions, such as:
(1) *Gaussian* (i.e., normal),
(2) *Poisson*,
(3) *Binomial*,

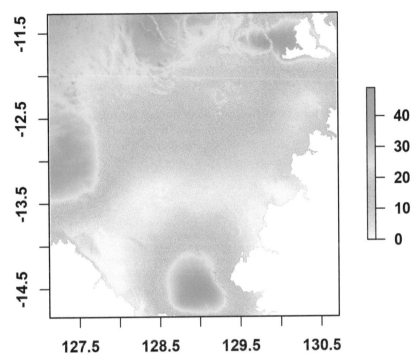

FIGURE 7.4: Spatial patterns of predictions using *TPS* based on location information and bathy

(4) *quasipoisson*, and
(5) *quasibinomial*.

GLM specify a family of *error distribution* and a particular *link function* including relevant variance function (e.g., *poisson* distribution with `log` link for count data or *binomial* with `logit` link for proportion data) (Chambers and Hastie 1992; Venables and Ripley 2002; Crawley 2007).

We will employ the function `glm` in the `stats` package and the function `glmnet` in the `glmnet` package (Friedman et al. 2020) to demonstrate the applications of *GLM* for spatial predictive modeling. Percentage and count data will be used to demonstrate the applications. Relationships of response variable with predictive variables identified for *LM* in Section 7.1 are equally applicable to *GLM*.

7.4.1 Implementation of *GLM* in `glm`

The function `glm` is used to fit *GLM*. For `glm`, the following three arguments are essential.
(1) `formula`, a symbolic description of the model to be fitted;
(2) `family`, a description of the error distribution and link function to be used in the model (see `?glm` for details of family functions); and
(3) `data`, a dataframe, containing response and predictive variables in the model.

1. Percentage data

The `gravel1` point data set and `petrel.grid` grid data set will be used to demonstrate the application of *GLM* to percentage data. We will use `gravel` as response variable and all other variables in the gravel1 data set as predictive variables. Here we consider all available predictive variables and their second- and third-order terms.

```
glm1 <- glm(gravel / 100 ~ . + I(long^2) + I(long^3) + I(lat^2) + I(lat^3) + I(
    bathy^2) + I(bathy^3) + I(dist^2) + I(dist^3) + I(relief^2) + I(relief^3) + I(
    slope^2) + I(slope^3), family = binomial(link=logit), gravel1)
```

In glm1, the interactions among predictive variables are not considered, but such interactions can be easily incorporated. For instance, two-way interactions could be implemented by replacing gravel / 100 ~ . with gravel / 100 ~ . ^2 in glm1. The interactions among relevant predictive variables should be considered if they were believed to be important.

The results of glm1 and anova(glm1) are

```
glm1
```

```
Call:  glm(formula = gravel/100 ~ . + I(long^2) + I(long^3) + I(lat^2) +
    I(lat^3) + I(bathy^2) + I(bathy^3) + I(dist^2) + I(dist^3) +
    I(relief^2) + I(relief^3) + I(slope^2) + I(slope^3), family = binomial(link =
        logit),
    data = gravel3)

Coefficients:
  (Intercept)            long              lat            bathy             dist
 47916.3722519   -1153.9156259    -117.8625230        0.0877277        4.2188839
        relief           slope        I(long^2)        I(long^3)        I(lat^2)
    -0.0487389       1.2280046        9.1700212       -0.0242915       -9.4102024
      I(lat^3)       I(bathy^2)       I(bathy^3)        I(dist^2)        I(dist^3)
    -0.2478311       0.0007853        0.0000024       -4.7372029        1.5373564
    I(relief^2)      I(relief^3)      I(slope^2)       I(slope^3)
     0.0027297      -0.0000199       -1.0040151        0.1536441

Degrees of Freedom: 236 Total (i.e. Null);  218 Residual
Null Deviance:       44.2
Residual Deviance: 26.8      AIC: 143


print(anova(glm1, test="Chi"))

Analysis of Deviance Table

Model: binomial, link: logit

Response: gravel/100

Terms added sequentially (first to last)
```

	Df	Deviance	Resid. Df	Resid. Dev	Pr(>Chi)	
NULL			236	44.2		
long	1	0.12	235	44.1	0.729	
lat	1	1.83	234	42.2	0.176	
bathy	1	2.16	233	40.1	0.141	
dist	1	0.56	232	39.5	0.454	
relief	1	5.86	231	33.7	0.016	*
slope	1	0.01	230	33.7	0.924	
I(long^2)	1	1.30	229	32.4	0.254	
I(long^3)	1	0.03	228	32.3	0.869	
I(lat^2)	1	0.40	227	31.9	0.528	
I(lat^3)	1	2.29	226	29.6	0.130	
I(bathy^2)	1	0.93	225	28.7	0.336	
I(bathy^3)	1	0.46	224	28.3	0.500	
I(dist^2)	1	0.39	223	27.9	0.530	
I(dist^3)	1	0.27	222	27.6	0.606	

```
I(relief^2)   1      0.00      221      27.6      0.988
I(relief^3)   1      0.52      220      27.1      0.471
I(slope^2)    1      0.03      219      27.0      0.852
I(slope^3)    1      0.28      218      26.8      0.599
---
Signif. codes:  0 '***' 0.001 '**' 0.01 '*' 0.05 '.' 0.1 ' ' 1
```

For *GLM*, the default test for `anova` is `Chisq` for the binomial (including negative binomial) and Poisson families, otherwise `F` test should be used (Venables and Ripley 2002).

According to the results of `anova`, only `relief` is significant.

2. Count data

For count data, `glm` can be implemented with either *Poisson distribution* or *negative binomial distribution* as demonstrated below.

Poisson distribution

The `sponge` point data set and `sponge.grid` grid data set in the `spm` package will be used to show the application of *GLM* to count data. The error distribution is assumed to be *Poisson* for the count data as below.

```
data(sponge)
data(sponge.grid)

glm2 <- glm(sponge ~ . + I(easting^2) + I(easting^3) + I(northing^2) + I(northing
     ^3) + I(tpi3^2) + I(tpi3^3) + I(var7^2)+ I(var7^3) + I(entro7^2)+ I(entro7^3)
     + I(bs11^2) + I(bs11^3) + I(bs34^2) + I(bs34^3), family = poisson, sponge)

glm2


Call:  glm(formula = sponge ~ . + I(easting^2) + I(easting^3) + I(northing^2) +
    I(northing^3) + I(tpi3^2) + I(tpi3^3) + I(var7^2) + I(var7^3) +
    I(entro7^2) + I(entro7^3) + I(bs11^2) + I(bs11^3) + I(bs34^2) +
    I(bs34^3), family = poisson, data = sponge)

Coefficients:
  (Intercept)          easting          northing             tpi3             var7
    -3.96e+05         7.31e-05          1.35e-01         2.51e+00        -6.11e-01
       entro7             bs34              bs11      I(easting^2)      I(easting^3)
     1.92e+00         1.14e+00         -2.80e+00         -9.13e-01         8.44e-18
 I(northing^2)    I(northing^3)         I(tpi3^2)         I(tpi3^3)         I(var7^2)
    -1.53e-08         5.80e-16          2.34e-01         -3.07e+00         2.90e-03
    I(var7^3)       I(entro7^2)       I(entro7^3)         I(bs11^2)         I(bs11^3)
    -1.64e-03        -1.35e+00          3.16e-01         -1.04e-01        -1.21e-03
    I(bs34^2)        I(bs34^3)
     3.68e-02         4.62e-04

Degrees of Freedom: 76 Total (i.e. Null);  55 Residual
Null Deviance:      768
Residual Deviance: 232    AIC: 553
```

Negative binomial distribution

For negative binomial distribution, we need to examine the relationships of means and variances of `sponge` data first using the following code.

```
f <- cut(sponge$sponge, c(-Inf, stats::quantile(sponge$sponge, 1:7 / 8), Inf))
mu<-with(sponge, tapply(sponge, list(f), mean))
vars<-with(sponge, tapply(sponge, list(f), var))
```

The results are illustrated in Figure 7.5 by

```
par(mfrow=c(2,2), font.axis = 2, font.lab = 2)
plot(mu, vars)
plot(mu, sqrt(vars))
plot(mu, (vars)^(1/3))
plot(mu, log(vars + 1))
```

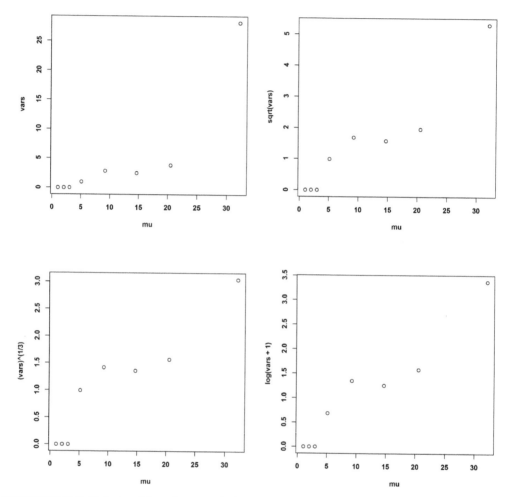

FIGURE 7.5: Relationships of means (mu) and varianes (vars) of sponge data, with variances under four transformations.

The mean-variance relation in Figure 7.5 suggests that a negative binomial model with a square-root or cubic-root transformation may be appropriate for *GLM* (Venables and Ripley 2002).

We will then fit a `negative.binomial` model using all relevant predictive variables in `glm2` by assuming $\theta = 2$ as below.

```
library(MASS)

glm3 <- update(glm2, family = negative.binomial(2))
```

```
summary(glm3)
```

```
Call:
glm(formula = sponge ~ . + I(easting^2) + I(easting^3) + I(northing^2) +
    I(northing^3) + I(tpi3^2) + I(tpi3^3) + I(var7^2) + I(var7^3) +
    I(entro7^2) + I(entro7^3) + I(bs11^2) + I(bs11^3) + I(bs34^2) +
    I(bs34^3), family = negative.binomial(2), data = sponge)
```

```
Deviance Residuals:
   Min      1Q   Median      3Q      Max
-1.899   -0.561   -0.248    0.446    1.378
```

```
Coefficients:
                Estimate Std. Error t value Pr(>|t|)
(Intercept)    -3.64e+05   3.80e+05   -0.96   0.34197
easting        -8.44e-05   1.68e-04   -0.50   0.61821
northing        1.24e-01   1.30e-01    0.95   0.34567
tpi3            2.68e+00   7.63e-01    3.51   0.00091 ***
var7           -1.10e+00   1.14e+00   -0.96   0.34190
entro7         -2.16e-01   2.58e+00   -0.08   0.93360
bs34            8.05e-01   1.33e+00    0.60   0.54780
bs11           -2.03e+00   2.10e+00   -0.96   0.33917
I(easting^2)    2.76e-10   3.95e-10    0.70   0.48821
I(easting^3)   -2.66e-16   2.99e-16   -0.89   0.37750
I(northing^2)  -1.41e-08   1.49e-08   -0.94   0.34939
I(northing^3)   5.32e-16   5.68e-16    0.94   0.35312
I(tpi3^2)       9.61e-01   2.76e+00    0.35   0.72904
I(tpi3^3)      -3.66e+00   2.95e+00   -1.24   0.21998
I(var7^2)       1.94e-01   3.09e-01    0.63   0.53342
I(var7^3)      -1.67e-02   2.30e-02   -0.72   0.47249
I(entro7^2)    -5.90e-02   1.58e+00   -0.04   0.97030
I(entro7^3)     9.50e-02   2.95e-01    0.32   0.74853
I(bs11^2)      -6.22e-02   1.14e-01   -0.54   0.58882
I(bs11^3)      -4.95e-04   2.01e-03   -0.25   0.80670
I(bs34^2)       2.20e-02   6.13e-02    0.36   0.72141
I(bs34^3)       2.72e-04   8.89e-04    0.31   0.76033
---
Signif. codes:  0 '***' 0.001 '**' 0.01 '*' 0.05 '.' 0.1 ' ' 1
```

```
(Dispersion parameter for Negative Binomial(2) family taken to be 0.6881)
```

```
    Null deviance: 146.99  on 76  degrees of freedom
Residual deviance:  41.12  on 55  degrees of freedom
AIC: 479.6
```

```
Number of Fisher Scoring iterations: 14
```

We can also apply the function glm.nb in the MASS package to the data directly, which can estimate the θ parameter automatically as demonstrated below.

```
glm.nb1 <- glm.nb(sponge ~ . + I(easting^2) + I(easting^3) + I(northing^2) + I(
    northing^3) + I(tpi3^2) + I(tpi3^3) + I(var7^2)+ I(var7^3) + I(entro7^2)+ I(
    entro7^3) + I(bs11^2) + I(bs11^3) + I(bs34^2) + I(bs34^3), sponge)
```

The results of glm.nb1 are

```
summary(glm.nb1)
```

```
Call:
glm.nb(formula = sponge ~ . + I(easting^2) + I(easting^3) + I(northing^2) +
    I(northing^3) + I(tpi3^2) + I(tpi3^3) + I(var7^2) + I(var7^3) +
```

```
    I(entro7^2) + I(entro7^3) + I(bs11^2) + I(bs11^3) + I(bs34^2) +
    I(bs34^3), data = sponge, init.theta = 4.280621698, link = log)
```

Deviance Residuals:
```
   Min      1Q   Median      3Q      Max
-2.430  -0.721   -0.313   0.579    1.896
```

Coefficients:

	Estimate	Std. Error	z value	Pr(>|z|)	
(Intercept)	-3.64e+05	3.43e+05	-1.06	0.28907	
easting	-5.34e-05	1.54e-04	-0.35	0.72794	
northing	1.24e-01	1.18e-01	1.05	0.29284	
tpi3	2.62e+00	6.82e-01	3.83	0.00013	***
var7	-1.01e+00	1.06e+00	-0.95	0.34003	
entro7	2.86e-01	2.28e+00	0.13	0.90019	
bs34	9.10e-01	1.24e+00	0.73	0.46283	
bs11	-2.21e+00	1.94e+00	-1.14	0.25434	
I(easting^2)	2.01e-10	3.61e-10	0.56	0.57681	
I(easting^3)	-2.09e-16	2.74e-16	-0.76	0.44484	
I(northing^2)	-1.41e-08	1.35e-08	-1.04	0.29662	
I(northing^3)	5.32e-16	5.14e-16	1.04	0.30042	
I(tpi3^2)	7.81e-01	2.46e+00	0.32	0.75128	
I(tpi3^3)	-3.48e+00	2.64e+00	-1.32	0.18704	
I(var7^2)	1.67e-01	2.83e-01	0.59	0.55490	
I(var7^3)	-1.46e-02	2.11e-02	-0.69	0.48823	
I(entro7^2)	-3.48e-01	1.39e+00	-0.25	0.80287	
I(entro7^3)	1.42e-01	2.61e-01	0.54	0.58604	
I(bs11^2)	-7.21e-02	1.05e-01	-0.68	0.49397	
I(bs11^3)	-6.59e-04	1.85e-03	-0.36	0.72213	
I(bs34^2)	2.65e-02	5.72e-02	0.46	0.64331	
I(bs34^3)	3.28e-04	8.29e-04	0.40	0.69258	

Signif. codes: 0 '***' 0.001 '**' 0.01 '*' 0.05 '.' 0.1 ' ' 1

(Dispersion parameter for Negative Binomial(4.2806) family taken to be 1)

```
    Null deviance: 249.494  on 76  degrees of freedom
Residual deviance:  69.917  on 55  degrees of freedom
AIC: 471.7
```

Number of Fisher Scoring iterations: 1

```
            Theta:  4.28
        Std. Err.:  1.06
```

2 x log-likelihood: -425.74

The θ estimated in glm.nb1 is 4.2806 that is much higher than the value used in glm3.

7.4.2 Implementation of *GLM* in glmnet

GLM can also be implemented with lasso or elasticnet regularization with the glmnet function that fits *GLM* via penalized maximum likelihood (Friedman et al. 2020). For glmnet, besides the arguments described for glm in the previous section, an additional argument alpha is required:

(1) alpha, an elasticnet mixing parameter, with $0 <= alpha <= 1$.

If alpha is 0, then a ridge regression model is fit, and if alpha is 1, then a lasso model is fit (James et al. 2017).

We still use the gravel1 data set to demonstrate its application.

```
library(glmnet)

x <- as.matrix(gravel1[, -7])
y <- gravel1[, 7] / 100

glmnet1 <- glmnet(x, y, family = binomial(link=logit), alpha = 0.5)
```

The `glmnet` can also be applied to other data types such as count data. We will not present the modeling for other data types and leave it to interested readers to test by adopting the example above with an appropriate `family` specification.

7.4.3 Variable selection

Variable selection for *GLM* can be based on either likelihood or predictive accuracy.

1. Variable selection based on likelihood

When variable selection is based on likelihood, it is called model selection. Model selection methods for *LM* can also be used for *GLM* (Venables and Ripley 2002; Ripley 2020). We can use functions, such as `stepAIC`, `dropterm`, `anova`, and `update`, to simplify *GLM* models. However, for `quasi` models (e.g., models developed with `familiy = quasipoisson` or `familiy = quasibinomial`), `stepAIC` cannot be used for model selection, as there is no likelihood, and hence, no AIC (Venables and Ripley 2002).

We will apply `stepAIC` to `glm1` to develop a simplified model as follows.

```
glm1.step1 <- stepAIC(glm1, direction = "backward", k=2, trace = F)
```

The results of `glm1.step1` can be derived with the function `summary`.

```
summary(glm1.step1)

summary(glm1.step1)$coef
```

```
                 Estimate    Std. Error  z value  Pr(>|z|)
(Intercept) -485.83754445 261.61501560   -1.857   0.06330
lat         -123.44230738  64.40791685   -1.917   0.05529
bathy          0.01770756   0.00942539    1.879   0.06028
I(long^3)     -0.00000855   0.00000423   -2.021   0.04324
I(lat^2)      -9.99891533   5.18709019   -1.928   0.05390
I(lat^3)      -0.26754783   0.13825310   -1.935   0.05297
I(relief^2)    0.00025567   0.00016453    1.554   0.12019
```

It is apparent that `I(relief^2)` is insignificant.

```
anova(glm1.step1, test = "Chi")
```

```
Analysis of Deviance Table

Model: binomial, link: logit

Response: gravel/100

Terms added sequentially (first to last)
```

	Df	Deviance	Resid. Df	Resid. Dev	Pr(>Chi)
NULL			236	44.2	
lat	1	1.86	235	42.3	0.17
bathy	1	1.72	234	40.6	0.19
I(long^3)	1	0.58	233	40.0	0.45

```
I(lat^2)     1      0.00      232      40.0      0.97
I(lat^3)     1      6.60      231      33.4      0.01 *
I(relief^2)  1      2.22      230      31.2      0.14
---
Signif. codes:  0 '***' 0.001 '**' 0.01 '*' 0.05 '.' 0.1 ' ' 1
```

The results suggest that only I(lat^3) is significant.

The model glm1.step1 can be further simplified with the function dropterm.

```
dropterm(glm1.step1, test = "Chi")


Single term deletions

Model:
gravel/100 ~ lat + bathy + I(long^3) + I(lat^2) + I(lat^3) +
    I(relief^2)
            Df Deviance AIC  LRT  Pr(Chi)
<none>           31.2 120
lat          1   35.0 122 3.73   0.053 .
bathy        1   35.0 122 3.73   0.053 .
I(long^3)    1   35.2 122 3.94   0.047 *
I(lat^2)     1   35.0 122 3.77   0.052 .
I(lat^3)     1   35.0 122 3.80   0.051 .
I(relief^2)  1   33.4 120 2.22   0.137
---
Signif. codes:  0 '***' 0.001 '**' 0.01 '*' 0.05 '.' 0.1 ' ' 1
```

The results show that no variables should be dropped in terms of *AIC* values, but the variable I(relief^2) is not significant and could be deleted from the model glm1.step1.

This can be conducted with the function update.

```
glm4 <- update(glm1.step1, . ~ . - I(relief^2))
```

The above results show that the *AIC* criterion penalizes terms less severely than Chi test. The *AIC* criterion also penalizes terms less severely than likelihood ratio or Wald's test (Venables and Ripley 2002). A k=log(dim(sponge)[1]) could be used in dropterm to select variables in terms of *BIC* values.

The model glm4 will be used to generate spatial predictions for gravel.

We can also apply stepAIC, anova, dropterm, and update to the *GLM* models for the count data (i.e., glm2 and glm.nb1) by following the steps applied to the *GLM* model for the percentage data (i.e., glm1), so we will not present the applications here.

The function selection in the FWDselect package (Sestelo, Villanueva, and Roca-Pardinas 2015) can also be used to select variables for *LM* and *GLM*, which is also based on *AIC*, *BIC* or other likelihood based criteria.

Again, the models selected based on *AIC* and *BIC* for *GLM* may be the most parsimonious or with maximum likelihood but may not necessarily be the most accurate predictive models (Li, Alvarez, et al. 2017). For predictive modeling, predictive accuracy instead of *AIC* or *BIC* should be used for variable selection (Li, Alvarez, et al. 2017).

2. Variable selection based on predictive accuracy

For *GLM*, the predictive accuracy based *variable selection* will be carried out using function bestglm in the same way as for *LM*.

The sponge data set will be used for demonstration. We need to prepare a data set required for bestglm by considering the second- and third-order terms of each predictive variable. Given that bestglm can only take up to 15 predictive variables, we will consider seven predictive variables in the sponge data set and their second-order terms. We also need to add dist.coast to the data set, as it could be an important predictor for sponge (Li, Alvarez, et al. 2017).

```
library(spm2)
data(sponge2)

X2 <- data.frame(subset(sponge, select = -sponge),
dist.coast = sponge2$dist.coast,
easting2 = sponge$easting^2,
northing2 = sponge$northing^2,
tpi32 = sponge$tpi3^2,
var72 = sponge$var7^2,
entro72 = sponge$entro7^2,
bs342 = sponge$bs34^2,
bs112 = sponge$bs11^2)

y2 <- sponge$sponge
Xy2 <- as.data.frame(cbind(X2, y2))


library(bestglm)

set.seed(1234)

glm.cv <- bestglm(Xy2, IC = "CV", family = poisson)

glm.cv$BestModel


Call:  glm(formula = y ~ ., family = family, data = data.frame(Xy[,
    c(bestset[-1], FALSE), drop = FALSE], y = y))

Coefficients:
    (Intercept)              easting            easting2
 -8.2924185450900    0.0000569015147  -0.0000000000667

Degrees of Freedom: 76 Total (i.e. Null);   74 Residual
Null Deviance:       768
Residual Deviance: 587   AIC: 870


glm.cv$Subsets[, 17:18]

     logLikelihood         CV
0           -522.7  1.163e+02
1           -485.9  1.136e+02
2*          -431.8  1.084e+02
3           -390.4  1.535e+02
4           -356.2  1.132e+02
5           -328.9  3.640e+19
6           -315.7  3.853e+04
7           -303.4  2.248e+73
8           -291.3  1.523e+45
9           -279.7  1.679e+44
10          -273.5  1.746e+94
11          -269.9  1.746e+93
12          -268.2 2.895e+270
13          -267.3 6.929e+129
14          -267.0        Inf
15          -267.0        Inf
```

The results show that the optimal model is with two variables easting and easting2. This is different from the previous findings (Li, Alvarez, et al. 2017) where the second-order terms were not considered. So we will do a further selection by excluding the second-order terms.

```
X2.1 <- data.frame(subset(sponge, select = -sponge),
dist.coast = sponge2$dist.coast)

Xy2.1 <- as.data.frame(cbind(X2.1, y2))

set.seed(1234)

glm.cv2 <- bestglm(Xy2.1, IC = "CV", family = poisson)

glm.cv2$BestModel
```

```
Call:  glm(formula = y ~ ., family = family, data = data.frame(Xy[,
    c(bestset[-1], FALSE), drop = FALSE], y = y))

Coefficients:
(Intercept)    dist.coast
 1.89375964    0.00000413

Degrees of Freedom: 76 Total (i.e. Null);   75 Residual
Null Deviance:        768
Residual Deviance: 695   AIC: 976
```

```
glm.cv2$Subsets[, 10:11]
```

```
   logLikelihood         CV
0         -522.7 1.163e+02
1*        -485.9 1.136e+02
2         -435.0 9.235e+05
3         -400.5 3.282e+02
4         -384.0 1.278e+02
5         -378.1 2.803e+15
6         -373.3 1.140e+13
7         -370.7 1.146e+11
8         -370.0 1.934e+27
```

The results show that the optimal model, `glm.cv2`, is with only one variable `dist.coast`, which is less accurate than `glm.cv`. This suggests that the variable(s) selection of `bestglm` is sensitive to the input variables and `bestglm` should be used with care.

7.4.4   Parameter estimation for `glmnet`

For `glmnet`, the elasticnet mixing parameter, `alpha`, needs to be determined according to its effects on the predictive accuracy. The function `glmnetcv`, a cross-validation function for `glmnet`, in the `spm2` package can be used to estimate an optimal `alpha` that can maximize the predictive accuracy of a `glmnet` model.

For `glmnetcv`, besides the arguments for `glmnet`, the following additional arguments are required:
(1) `validation`, validation methods, include 'LOO' and 'CV';
(2) `cv.fold`, integer; number of folds in the cross-validation; if > 1, then apply n-fold cross-validation; the default is 10; and
(3) `predacc`, can be either "VEcv" for `vecv` or "ALL" for all measures in function `pred.acc`.

We will estimate the optimal `alpha` for `glmnet1` using the `gravel1` data set as an example.

```
x <- as.matrix(gravel1[, -7])
y <- gravel1[, 7] / 100
alpha <- c(0:20) * 0.05
alphaopt <- NULL

for (i in 1:length(alpha)) {
  set.seed(1234)
  glmnetcv1 <- glmnetcv(x, y, family = binomial(link=logit), alpha = alpha[i],
      validation = "CV",  predacc = "VEcv")
  alphaopt[i] <- glmnetcv1
}
```

The predictive accuracy, "VEcv", can be visualized against `alpha` in Figure 7.6 by

```
par(font.axis = 2, font.lab = 2)
plot(alphaopt ~ alpha, xlab = "alpha", ylab = "VEcv (%)", col = "blue")
```

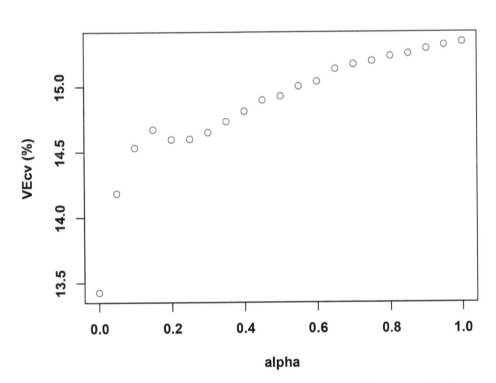

FIGURE 7.6: *VEcv* (%) of *glmnet* for each `alpha` based on 10-fold cross-validation.

The optimal `alpha` that leads to the maximal predictive accuracy is

```
alpha.opt <- which (alphaopt == max(alphaopt))
alpha[alpha.opt]
```

```
[1] 1
```

It shows that `alpha` = 1 should be used.

7.4.5 Predictive accuracy

We can use `glmcv` or `glmnetcv` to assess the *predictive accuracy* of the three *GLM* predictive models selected (i.e., `glm4`, `glm.cv`, and `glmnet1`) as follows.

1. `glm4`

The predictive accuracy of `glm4` can be obtained as follows.

```
y <- gravel1[, 7] / 100
modelglm4 <- gravel / 100 ~ lat + bathy + I(long^3) + I(lat^2) + I(lat^3)
set.seed(1234)
n <- 100
glm4vecv <- NULL

for (i in 1:n) {
  glmcv1 <- glmcv(formula = modelglm4, gravel1, y, family = binomial(link = logit)
      , validation = "CV", predacc = "VEcv")
  glm4vecv[i] <- glmcv1
}
```

The median and range of predictive accuracy of `glm4` based on 100 repetitions of 10-fold cross-validation using `glmcv` are

```
median(glm4vecv)
```

```
[1] 16.78
```

```
range(glm4vecv)
```

```
[1] 10.70 19.89
```

2. `glm.cv`

The predictive accuracy of model `glm.cv` can be assessed by

```
y2 <- sponge$sponge
modelglm.cv <- sponge ~ easting + I(easting^2)

set.seed(1234)
n <- 100
glm.cvvecv <- NULL

for (i in 1:n) {
  glmcv1 <- glmcv(formula = modelglm.cv, sponge, y2, family = poisson, validation
      = "CV", predacc = "VEcv")
  glm.cvvecv[i] <- glmcv1
}
```

The median and range of predictive accuracy of `glm.cv` based on 100 repetitions of 10-fold cross-validation using `glmcv` are

```
median(glm.cvvecv)
```

```
[1] 14.79
```

```
range(glm.cvvecv)
```

```
[1]  8.644 18.111
```

3. `glmnet1`

The predictive accuracy of `glmnet1` can be obtained with `glmnetcv` with the `alpha` estimated as shown below.

```
x <- as.matrix(gravel1[, -7])
y <- gravel1[, 7] / 100
set.seed(1234)
n <- 100
glmnet.vecv <- NULL

for (i in 1:n) {
  glmnetcv1 <- glmnetcv(x, y, family = binomial(link=logit), alpha = 1, validation
      = "CV", predacc = "VEcv")
  glmnet.vecv[i] <- glmnetcv1
}
```

The median and range of predictive accuracy of `glmnet1` based on 100 repetitions of 10-fold cross-validation using `glmnetcv` are

```
median(glmnet.vecv)
```

```
[1] 14.2
```

```
range(glmnet.vecv)
```

```
[1]   8.816 17.287
```

The models developed for `gravel` and `sponge` are of low predictive accuracy. Different modeling methods should be used in these cases. Similar to *LM*, a number of tools for checking models based on *GLM* are available in previous publications (McCullagh and Nelder 1999; Crawley 2007), but they are not predictive accuracy based, so they are not much help for predictive modeling.

7.4.6 Spatial predictions and standard errors

The *spatial predictions* and associated standard errors of *GLM* can be generated based on the most accurate predictive models. This is going to be demonstrated using models `glm4` and `glmnet1`.

1. Model `glm4`

The predictions and standard errors of `glm4` can be generated by

```
glm.pred <- predict(glm4, petrel.grid, type="response", se.fit = TRUE)
glm.pred <- as.data.frame(glm.pred)
```

The predictions and standard errors need to be back-transformed to the original scale (i.e., percentage data) for `gravel` data as below.

```
glm.pred$fit <- glm.pred$fit*100
glm.pred$se.fit <- glm.pred$se.fit*100
```

The properties of predictions and standard errors are

```
range(glm.pred$fit) # predictions
```

```
[1]   5.915 37.490
```

```
range(glm.pred$se.fit) # standard errors estimated
```

```
[1]   3.407 13.219
```

```
summary(glm.pred)
```

```
        fit               se.fit          residual.scale
 Min.   : 5.92     Min.    : 3.41     Min.    :1
 1st Qu.:20.70     1st Qu.: 5.94     1st Qu.:1
 Median :24.40     Median : 7.12     Median :1
 Mean   :24.18     Mean    : 7.39     Mean    :1
 3rd Qu.:28.14     3rd Qu.: 8.73     3rd Qu.:1
 Max.   :37.49     Max.    :13.22     Max.    :1
```

In the results, `residual.scale` is the square root of the dispersion used in computing the standard errors.

The predictions of *GLM* can also be generated with `glmpred`. It is easy to be implemented for *GLM* by following the example in Section 7.1.6 and will not be presented here.

The spatial distribution of predictions and standard errors of the *GLM* model are illustrated in Figure 7.7 by

```
glm.pred.df <- cbind(petrel.grid[, 1:2], glm.pred[, 1:2])
gridded(glm.pred.df) = ~ long + lat
names(glm.pred.df) <- c("predictions", "standard.errors")

library(raster)

par(font.axis = 2, font.lab = 2)
plot(brick(glm.pred.df))
```

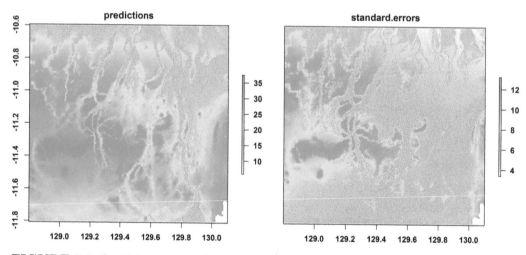

FIGURE 7.7: Spatial patterns of predictions and standard errors using *glm*.

2. Model `glmnet1`

To generate predictions using `glmnet1` with the optimal `alpha`, the default penalty parameter `lambda` can be used (Friedman et al. 2020). The value of `lambda` that gives the minimum mean cross-validated error can also be determined using `cv.glmnet` as below.

```
set.seed(1234)
x <- as.matrix(gravel1[, -7])
y <- gravel1[, 7]/100

enet.cv <- cv.glmnet(x, y, family = binomial(link=logit), alpha = 1, type.measure
    = "mse")

glmnet1.1 <- glmnet(x, y, family = binomial(link=logit), alpha = 1)
```

```
glmnet1.fitted <- predict(glmnet1.1, x, s = enet.cv$lambda.min, type = "response")
    * 100 # back transformed to percentage scale.

range(glmnet1.fitted)

[1]   5.104 37.107
```

We can generate spatial predictions using `glmnet1.1` for `petrel.grid` and then back transform the predictions to percentage scale.

```
x2 <- as.matrix(petrel.grid)

glmnet1.pred <- predict(glmnet1.1, x2, s = enet.cv$lambda.min, type = "response")
    * 100 # back transformed to percentage scale.

range(glmnet1.pred)

[1]   3.51 90.12
```

The spatial distribution of predictions of `glmnet1.1` can be illustrated in Figure 7.8 by

```
glmnet.pred.df <- cbind(petrel.grid[, 1:2], glmnet1.pred)
gridded(glmnet.pred.df) = ~ long + lat
names(glmnet.pred.df) <- "predictions"

library(raster)

par(font.axis = 2, font.lab = 2)
plot(brick(glmnet.pred.df))
```

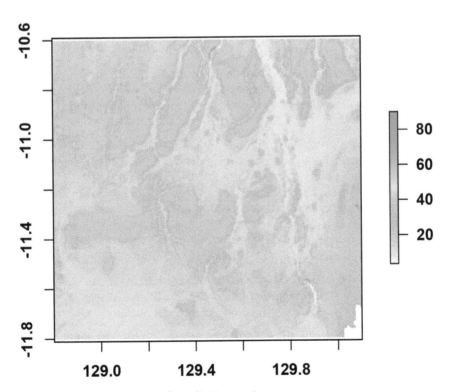

FIGURE 7.8: Spatial patterns of predictions using `glmnet`.

The predictions by `glm4` are much lower than those expected for `gravel` data, while the predictions by `glmnet1` are of the same magnitude as those expected for `garvel` data.

To assess the *reliability of the predictions*, one may need to check the difference between the range of observed values and the range of the predictions. This can be done by plotting the fitted values against the observed values of training samples or comparing the predictions with the observed values that fall into the area predicted. If observed values are unavailable in the area predicted, then the predictions of *IDW* may also serve good references for *assessing* the predictions due to the *exact* and *convex* features of *IDW* (Li and Heap 2014), that is, the predictions of *IDW* are always within the range of training samples, although they are often pulled towards the average of the samples depending on the parameters (i.e., `nmax` and `idp`) used. Moreover, professional knowledge can play a significant role in assessing the predictions by examining their properties such as range and spatial patterns.

7.5   Generalized least squares

Generalized least squares (*GLS*) is an extension of *LM* by allowing the errors to be correlated and/or have unequal variances (Pinheiro and Bates 2000; Pinheiro, Bates, and R-core 2020). Since *GLS* uses the information of spatial correlation, it could be regarded as a multivariate geostatistical method. In this book, because it is an extension of *LM*, it is grouped into modern statistics.

7.5.1   Implementation of *GLS* in `gls`

The function `gls` in the `nlme` package (Pinheiro, Bates, and R-core 2020) can be used to fit a linear model using generalized least squares, with correlated errors and/or unequal variances. We will use `gls` to demonstrate the application of *GLS*. For `gls`, the following three arguments are essential:
(1) `model`, a symbolic description of the model to be fitted;
(2) `data`, a dataframe, containing the variables in the model; and
(3) `correlation`, an optional `corStruct` object describing the within-group correlation structure.

The `correlation` argument is detailed in Pinheiro and Bates (2000) and can be found using the function `corStruct` in `nlme`. For spatial modeling, the following spatial correlation structure classes are available for `gls`:
(1) `corExp`, exponential spatial correlation;
(2) `corGaus`, *Gaussian* spatial correlation;
(3) `corLin`, linear spatial correlation;
(4) `corRatio`, rational quadratics spatial correlation; and
(5) `corSpher`, spherical spatial correlation (Pinheiro and Bates 2000).

We will use `gravel` as response variable and all other variables in the `gravel1` data set as predictive variables. Since *GLS* is an extension of *LM* and `lm.loocv` in Section 7.1.3 is the most accurate model for `gravel` data, here we consider all available predictive variables in `lm.loocv` for the following demonstration.

1. The argument `correlation` with the default value

```
library(nlme)

gls1 <- gls(log(gravel + 1) ~ long + lat +  bathy + dist + relief + I(long^2) + I(
    lat^2)
+ I(lat^3) + I(bathy^2) + I(bathy^3) + I(dist^2) + I(dist^3), gravel1)
```

```
gls1$coefficients
```

(Intercept)	long	lat	bathy	dist
-1534.60062726	17.00459935	-114.27567249	0.06722556	5.38718398
relief	I(long^2)	I(lat^2)	I(lat^3)	I(bathy^2)
0.01985760	-0.06784941	-9.11876420	-0.24036627	0.00061818
I(bathy^3)	I(dist^2)	I(dist^3)		
0.00000202	-5.26871557	1.64947137		

When `correlation` argument in `gls` is not specified, then the default, NULL (corresponding to uncorrelated errors), is assumed and used. Consequently, the model developed, `gls1`, is not different from `lm.loocv` and `gls1$coefficients` are the same as `lm.loocv$BestModel`.

2. The argument `correlation` with a specified value

We will use `corExp` with `range1` and `nugget1` as suggested in Chapter 5 for gravel data.

```
gls2 <- update(gls1, corr = corExp(c(range1, nugget1), form = ~ lat + long, nugget
    = T))
```

The model developed is

```
gls2
```

```
Generalized least squares fit by REML
   Model: log(gravel + 1) ~ long + lat + bathy + dist + relief + I(long^2) +        I
      (lat^2) + I(lat^3) + I(bathy^2) + I(bathy^3) + I(dist^2) +         I(dist^3)
   Data: gravel1
   Log-restricted-likelihood: -324.8
```

```
Coefficients:
```

(Intercept)	long	lat	bathy	dist
-898.7038729210	10.7385366177	-55.5924477332	0.0387215344	9.9857656373
relief	I(long^2)	I(lat^2)	I(lat^3)	I(bathy^2)
0.0171842852	-0.0430258823	-4.4163044129	-0.1157764312	0.0002049351
I(bathy^3)	I(dist^2)	I(dist^3)		
0.0000004658	-11.2462134021	3.8199871740		

```
Correlation Structure: Exponential spatial correlation
 Formula: ~lat + long
 Parameter estimate(s):
 range nugget
0.3478 0.4526
Degrees of freedom: 237 total; 224 residual
Residual standard error: 1.025
```

7.5.2   Variable selection for *GLS*

The *variable selection* can be conducted based on the effects of relevant variables on the predictive accuracy via the function `glscv` in the `spm2` package by following the procedure for *KED* in Section 6.3.2. The function `glscv` is a cross-validation function for *GLS* predictive models and will be introduced below.

7.5.3   Predictive accuracy

We can use `glscv` to assess the accuracy of the *GLS* predictive models. Arguments of `glscv` are similar to those for `gls`, with the following additional arguments:
(1) `trainxy`, a dataframe contains longitude (long), latitude (lat), predictive variables and response variable of point samples; that is, the location information must be names as 'long'

and 'lat';

(2) `y`, a vector of the response variable in the formula, that is, the left part of the formula of the `model` argument;

(3) `validation`, validation methods, including 'LOO' and 'CV';

(4) `cv.fold`, integer; number of folds in the cross-validation; if > 1, then apply n-fold cross-validation; the default is 10;

(5) `predacc`, can be either "VEcv" for `vecv` or "ALL" for all measures in function `pred.acc`.

We will use `glscv` to assess the predictive accuracy of `gls2`.

```
library(spm)
data(petrel)

model <- log(gravel + 1) ~  long + lat +  bathy + dist + relief + I(long^2) + I(
    lat^2) + I(lat^3) + I(bathy^2) + I(bathy^3) + I(dist^2) + I(dist^3)

library(spm2)
n <- 100
gls1vecv <- NULL
set.seed(1234)

for (i in 1:n) {
  glscv1 <- glscv(model = model, gravel1, log(gravel1[, 7] + 1), corr.args =
      corExp(c(range1, nugget1), form = ~ lat + long, nugget = T), validation = "
      CV", predacc = "VEcv")
  gls1vecv[i] <- glscv1
}
```

The median and range of predictive accuracy of `gls2` based on 100 repetitions of 10-fold cross-validation using `glscv` are

```
median(gls1vecv)
```

```
[1]  20.12
```

```
range(gls1vecv)
```

```
[1] 13.73 24.09
```

7.5.4   Predictions

The generation of *GLS* predictions will be demonstrated using `gls2` and back-transformed to percentage data as follows.

```
gls.pred <- exp(predict(gls2, petrel.grid, type = "response")) -1
```

The range of predictions is

```
range(gls.pred)
```

```
[1]    0.4064 154.8139
```

A small number of transformed predictions are beyond the range of percentage data and are corrected by resetting the faulty predictions to the nearest bound of the data range.

```
gls.pred1 <- gls.pred
gls.pred1[gls.pred1 >= 100] <- 100
```

The predictions of *GLS* can also be generated using the function `glspred` in the `spm2` package as below.

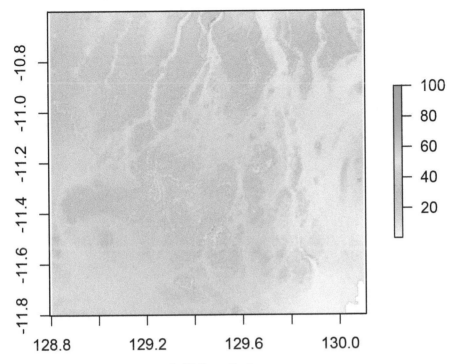

FIGURE 7.9: Spatial patterns of *GLS* predictions.

```
glspred2 <- exp(glspred(model = model, gravel1, longlatpredx = petrel.grid[, c
    (1:2)], predx = petrel.grid, corr.args = corSpher(c(range1, nugget1), form = ~
    lat + long, nugget = T))) -1
```

The spatial distribution of *GLS* predictions can be illustrated in Figure 7.9 by

```
gls.pred1.df <- cbind(petrel.grid[, c(1,2)], gls.pred1)
gridded(gls.pred1.df) = ~ long + lat
names(gls.pred1.df) <- "predictions"

library(raster)

plot(brick(gls.pred1.df))
```

8

Tree-based machine learning methods

This chapter introduces *tree-based machine learning methods*, which are *gradient based/de-trended spatial predictive methods* that use predictive variables.

Three tree-based machine learning methods that are available in R are introduced:
(1) *classification and regression trees (CART)*,
(2) *random forest (RF)*, and
(3) *generalized boosted regression modeling (GBM)*.

Although *CART* is hardly a machine learning method, it forms the foundation of *RF* and *GBM*, so it will be included in this chapter and introduced prior to *RF* and *GBM*.

8.1   Classification and regression trees

Classification and regression trees (CART), also known as *decision trees*, use binary recursive partitioning, whereby the data of response variable are successively split along the gradient of predictive variables into two descendant subsets (or *nodes, leaves*). These splits occur so that at any node the split is selected to maximize the difference between two *split groups* or *branches* (Breiman et al. 1984). Each predictive variable is assessed in turn, and the variable explaining the greatest amount of deviance in the response variable is selected at each node; and splitting continues until nodes are pure (i.e., no further deduction in the deviance) or the data are too sparse to allow further subdivision (Crawley 2007). If the response variable is a categorical variable, then we have a *classification tree*. On the other hand, if the response variable is a numerical variable, then a *regression tree* is resulted (Crawley 2007). The definition of the deviance for a classification tree or a regression tree can be found in Venables and Ripley (2002). The averages of observations for a numerical variable or the (dominate) class for a categorical variable at the terminal nodes of the tree are the predictions of the tree. The observations that fall into the same terminal node will be assigned the same predictive value.

Predictions of *CART* can be represented in Equation (8.1), in a similar way as in Equation (4.1).

$$\widehat{y_{x_i}} = \sum_{m=1}^{M} c_m I\{x_i \in N_m\} \tag{8.1}$$

where $\widehat{y_{x_i}}$ is the predicted value of a response variable y at the location of interest (i), x_i is p predictive variables at location i (i.e., $x_i = (x_{i1}, x_{i2}, ..., x_{ip})$), c_m is a constant value at node m, N_m represents a node of the tree at node m (Hastie, Tibshirani, and Friedman 2009). Here $\widehat{y_{x_i}}$ is equivalent to $\hat{y}(x_0)$ in Equation (4.1).

DOI: 10.1201/9781003091776-8

We use both numerical data and categorical data to demonstrate the applications of *CART* for spatial predictive modeling. *CART* can automatically handle *non-linear relationships* and *interactions*, so we only need to provide the predictive variables in the formula (Equation (8.1)). *CART* can be implemented in the function `rpart` in the `rpart` package (Therneau and Atkinson 2019) or the function `tree` in the `tree` package (Ripley 2019).

8.1.1   Implementation of *CART* in the function `rpart`

The predictions of *CART* can be generated with the function `rpart` in the `rpart` package. The descriptions of relevant arguments of `rpart` are detailed in its help file, which can be accessed by `?rpart`. For *CART* using `rpart`, the following arguments may need to be specified:
(1) `formula`, that defines response variable and predictive variables;
(2) `method`, it is one of "anova", "poisson", "class" or "exp"; if `method` is missing, then the routine tries to make an intelligent guess: if the response variable is a factor, then `method` = "class" is assumed, and if response variable is a numerical variable, method = "anova" is assumed; and
(3) `cp`, complexity parameter; and any split that does not improve the fit by a `cp` is not pursued.

1. Numerical data

The `sponge` point data set and `sponge.grid` grid data set in the `spm` package (Li 2019b) will be used to show the application of *CART* to numerical data.

```
library(spm)
data(sponge)
data(sponge.grid)
```

To make the results reproducible, we need to use the function `set.seed`.

```
library(rpart)
set.seed(1234)
rpart1 <- rpart(sponge ~ . , sponge, method = "anova", cp = 0.001)
```

We can use the function `printcp` in the `rpart` package to display the `cp` table for `rpart1` as below.

```
printcp(rpart1)
```

```
Regression tree:
rpart(formula = sponge ~ ., data = sponge, method = "anova",
    cp = 0.001)

Variables actually used in tree construction:
[1] bs34      easting   northing tpi3

Root node error: 8419/77 = 109

n= 77

       CP nsplit rel error xerror xstd
1 0.3219      0      1.00   1.02 0.17
2 0.0774      1      0.68   0.71 0.13
3 0.0740      2      0.60   0.87 0.18
4 0.0270      3      0.53   0.80 0.16
5 0.0084      4      0.50   0.83 0.16
6 0.0010      5      0.49   0.83 0.16
```

All the errors extracted are proportions of the error for the root tree; and the `xerror` and `xstd` are random and depending on the 10-fold cross-validation computed within the function `rpart` (Venables and Ripley 2002). The `xerror` and `cp` of `rpart1` for each tree size are shown in Figure 8.1 by

```
plotcp(rpart1)
```

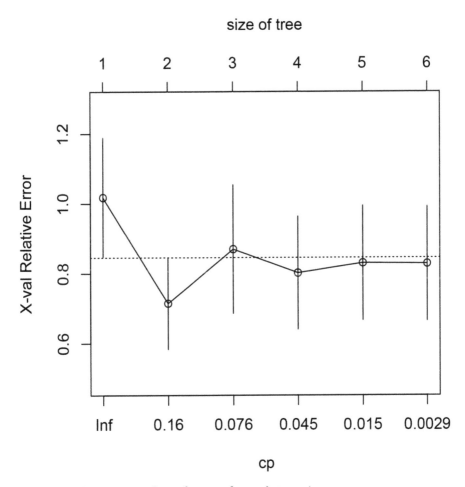

FIGURE 8.1: The `xerror` and `cp` of `rpart1` for each tree size.

The `cp` may be chosen to minimize `xerror` or by following 1SE rule as shown by the horizontal dashed line in Figure 8.1 according to Venables and Ripley (2002). It shows that to minimize `xerror`, `cp` value for pruning the tree `rpart1` should be 0.0774 or any other values below the dashed line.

We use the function `prune` in the `rpart` package to snip off the least important splits in `rpart1` based on `cp` and to produce a trimmed tree model `rpart2`. Given the size of the tree is small, we choose a value of 0.01 to trim `rpart1`, which will lead to a tree with 4 splits and 5 leaves.

```
rpart2 <- prune(rpart1, cp = 0.01)

rpart2

n= 77
```

```
node), split, n, deviance, yval
      * denotes terminal node

  1) root 77 8419.0 10.480
    2) easting>=5.917e+05 27  342.5  2.407 *
    3) easting< 5.917e+05 50 5367.0 14.840
      6) tpi3< 0.003557 24 2458.0 11.080
        12) bs34< -26.23 11  162.7  5.545 *
        13) bs34>=-26.23 13 1672.0 15.770 *
      7) tpi3>=0.003557 26 2258.0 18.310
        14) bs34< -27.91 8  478.9 13.880 *
        15) bs34>=-27.91 18 1552.0 20.280 *
```

```
printcp(rpart2)
```

```
Regression tree:
rpart(formula = sponge ~ ., data = sponge, method = "anova",
    cp = 0.001)

Variables actually used in tree construction:
[1] bs34    easting tpi3

Root node error: 8419/77 = 109

n= 77

      CP nsplit rel error xerror xstd
1 0.322     0      1.00    1.02 0.17
2 0.077     1      0.68    0.71 0.13
3 0.074     2      0.60    0.87 0.18
4 0.027     3      0.53    0.80 0.16
5 0.010     4      0.50    0.83 0.16
```

We can plot the unpruned and pruned trees for *sponge* data in Figure 8.2 by

```
par(mfrow = c(2, 1), font.axis = 2, font.lab = 2)
plot(rpart1)
text(rpart1, digits = 4, cex = 0.55, use.n = T)

plot(rpart2)
text(rpart2, digits = 4, cex = 0.55, use.n = T)
```

The predictions of rpart tree models can be demonstrated with rpart2 as follows.

```
rpart.pred <- predict(rpart2, sponge.grid)
```

The properties of the predictions are

```
range(rpart.pred)
```

```
[1]   5.545 20.278
```

```
range(predict(rpart2, sponge)) # the range of the fitted values
```

```
[1]   2.407 20.278
```

The spatial distribution of predictions of *CART* from rpart2 is illustrated in Figure 8.3 by

```
library(sp)

rpart.pred.df <- cbind(sponge.grid[, c(1:2)], as.data.frame(rpart.pred))
gridded(rpart.pred.df) = ~ easting + northing
names(rpart.pred.df) <- "predictions"
```

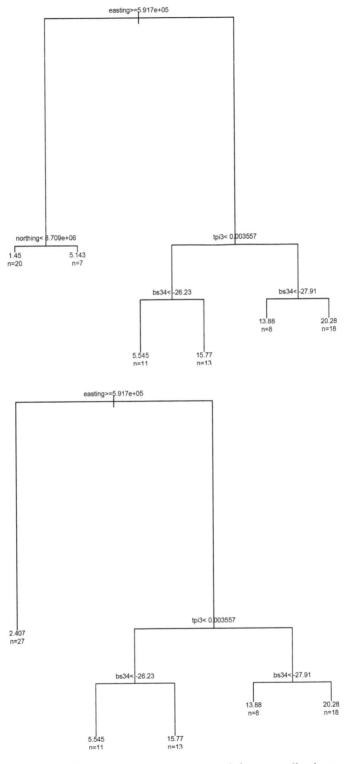

FIGURE 8.2: The trees for *sponge* data. **Top panel** (unpruned): the tree of size 6 produced from rpart1. **Bottom panel** (pruned): the tree of size 5 produced from rpart2.

```
library(raster)

par(font.axis = 2, font.lab = 2)
plot(brick(rpart.pred.df))
```

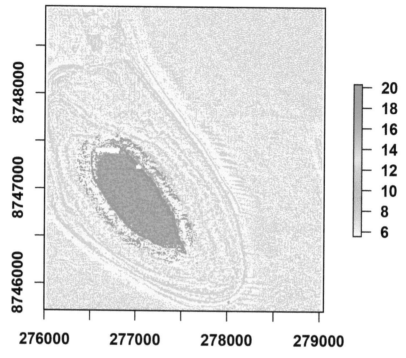

FIGURE 8.3: Spatial patterns of predictions using *CART* from `rpart`.

2. Categorical data

The `hard` point data set in the `spm` package will be used to show the application of *CART* to categorical data.

```
data(hard)
set.seed(1234)

rpart.h1 <- rpart(hardness ~ . , hard[, -1], method = "class", cp = 0.01)

printcp(rpart.h1)


Classification tree:
rpart(formula = hardness ~ ., data = hard[, -1], method = "class",
    cp = 0.01)

Variables actually used in tree construction:
[1] bs     prock

Root node error: 38/137 = 0.28

n= 137

      CP nsplit rel error xerror xstd
1 0.737      0      1.00   1.00 0.14
2 0.026      1      0.26   0.34 0.09
3 0.010      2      0.24   0.34 0.09
```

Given the size of the tree is small, we are not going to trim `rpart.h1`.

The predictions of `rpart.h1` can be generated in the same way as for `rpart2` above.

8.1.2 Implementation of *CART* in the function tree

The predictions of *CART* can also be generated using the function `tree`. The descriptions of relevant arguments of `tree` are detailed in its help file, which can be accessed by `?tree`. For `tree`, the argument, `formula`, that defines response variable and predictive variables, needs to be specified.

1. Numerical data

```
library(tree)

tree1 <- tree(sponge ~ . , sponge)
```

We can use the function `cv.tree` (a cross-validation function for choosing tree complexity, with 10-fold as the default) in the `tree` package to snip off the least important splits in `tree1` based on `cp`.

```
set.seed(1234)
tree1.cv <- cv.tree(tree1, , prune.tree)

for (i in 2:20) tree1.cv$dev <- tree1.cv$dev + cv.tree(tree1, , prune.tree)$dev

tree1.cv$dev <- tree1.cv$dev / 20
```

The function `prune.tree` in the `tree` package, which prunes a `tree` model by recursively "snipping" off the least important splits based on the cost-complexity, is used within `cv.tree`.

```
par(font.axis = 2, font.lab = 2)
plot(tree1.cv)
```

The results suggest that the best size is 8 (Figure 8.4), that is, a tree with 8 nodes or leaves. We can now produce a pruned tree model, `tree2`, based on the best size identified.

```
tree2 <- prune.tree(tree1, best = 8)
```

The trees (i.e., `tree1` and `tree2`) produced by `tree` are shown in Figure 8.5 by

```
par(mfrow = c(2,1), font.axis = 2, font.lab = 2)
plot(tree1)
text(tree1, digits = 4, cex = 0.55)

plot(tree2)
text(tree2, digits = 4, cex = 0.55)
```

The predictions of `tree` models can be demonstrated with `tree2` as follows.

```
tree.pred <- predict(tree2, sponge.grid)
```

The properties of the predictions are

```
range(tree.pred)

[1]  5.545 26.800

range(predict(tree2)) # the range of the fitted values
```

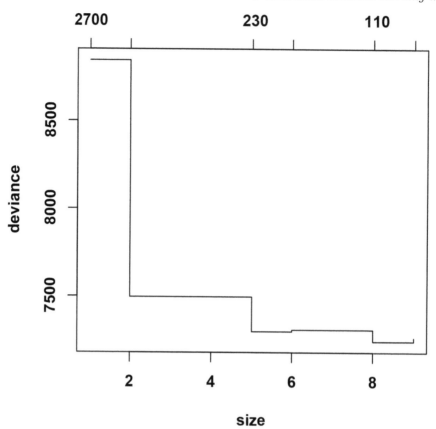

FIGURE 8.4: The 10-fold cross-validation sequence of the deviance and tree size for *sponge* data.

```
[1]   2.407 26.800
```

The spatial distribution of predictions of *CART* from `rpart2` are illustrated in Figure 8.6.

2. Categorical data

Similarly, we can apply the function `tree` to the `hard` data as below.

```
library(tree)

tree.h1 <- tree(hardness ~ . , hard[, -1])
```

We can also use `prune.tree` to snip off the least important splits in `tree.h1` based on `cp`.

```
set.seed(1234)
tree.h1.cv <- cv.tree(tree.h1, , prune.tree)
for (i in 2:20) tree.h1.cv$dev <- tree.h1.cv$dev + cv.tree(tree.h1, , prune.tree)$
    dev
tree.h1.cv$dev <- tree.h1.cv$dev/20

par(font.axis = 2, font.lab = 2)
plot(tree.h1.cv)
```

The results suggest that the best size is 3 (Figure 8.7). We can now produce a pruned tree model, `tree.h2`, based on the best size identified.

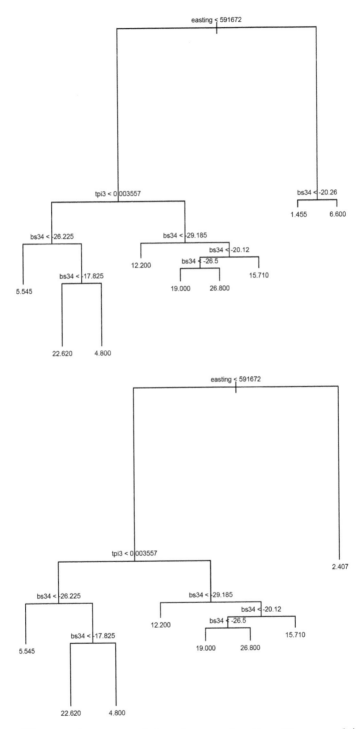

FIGURE 8.5: The trees for *sponge* data using `tree` function. **Top panel** (unpruned): the tree of size 9 produced from `tree1`. **Bottom panel** (pruned): the tree of size 8 produced from `tree2`.

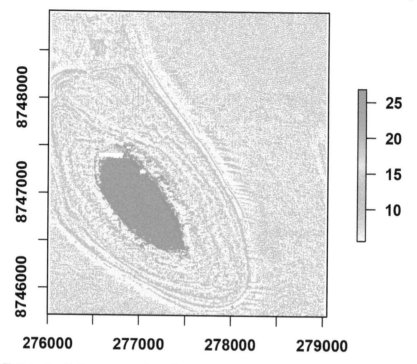

FIGURE 8.6: Spatial patterns of predictions using *CART* from `tree`.

```
tree.h2 <- prune.tree(tree.h1, best = 3)
```

The predictions of `tree.h2` can be generated in a similar way as for `tree2` above.

8.2 Random forest

Random forest (*RF*) is an ensemble method that combines many individual regression or classification trees. For each tree, a bootstrapped sample is drawn from the original sample, and at each split of the tree a portion of predictors is randomly drawn; and then an unpruned regression or classification tree (i.e., an *RF tree*) is fitted to the bootstrapped sample using the sampled predictors for each split. From the complete forest, the status of response variable is usually predicted as the average of the predictions of all *RF* trees for regression or as the classes with majority vote for classification (Breiman 2001). If a *RF* model is built with all predictive variables at each split, then this simply amounts to a *bagging* model (James et al. 2017).

The predictions of *RF* can be represented in Equation (8.2) for regression and in Equation (8.3) for classification, in a similar way as in Equation (4.1).

$$\widehat{y_{x_i}} = \sum_{b=1}^{B} \widehat{T_b(x_i)}/B \tag{8.2}$$

$$\widehat{y_{x_i}} = majority\ vote(\widehat{T_b(x_i)})_1^B \tag{8.3}$$

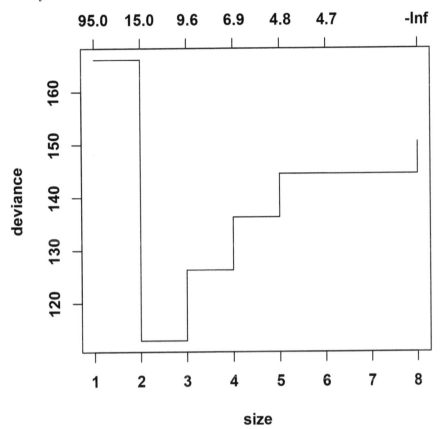

FIGURE 8.7: The 10-fold cross-validation sequence of the deviance and tree size for *hard* data.

where $\widehat{y_{x_i}}$ is the predicted value of a response variable y at the location of interest (i) by RF, x_i are p predictive variables at location i (i.e., $x_i = (x_{i1}, x_{i2}, ..., x_{ip})$), $\widehat{T_b(x_i)}$ is the predicted value at the location (i) by the RF tree T_b based on the bth bootstrapped sample, b represents one of B RF trees in the ensemble (Hastie, Tibshirani, and Friedman 2009).

We use both numerical and categorical data to demonstrate the applications of RF for spatial predictive modeling. Similar to $CART$, RF can automatically handle *non-linear relationships* and *interactions*, so we only need to provide the predictive variables in the formula. RF can be implemented in the `randomForest` package (Liaw and Wiener 2002; Breiman et al. 2018) or the `ranger` package (Wright and Ziegler 2017; Wright, Wager, and Probst 2020). In this chapter, we will use the function `randomForest` in the `randomForest` package to demonstrate the applications of RF.

8.2.1 Application of *RF*

The predictions of RF can be generated with the function `randomForest` that implements Breiman's random forest algorithm (Breiman 2001) for classification and regression. The descriptions of relevant arguments in `randomForest` are detailed in its help file, which can be accessed by `?randomForest`. For `randomForest`, the following arguments may need to be specified:

(1) x, `formula`, a dataframe or a matrix of predictive variables, or a formula defining response variable and predictive variables;

(2) y, a vector of the response variable if x is provided;

(3) `ntree`, the number of trees to grow;

(4) `mtry`, the number of variables randomly sampled as candidates at each split; and

(5) `importance`, importance of predictive variables be assessed or not.

1. Application of the function `randomForest` to numerical data

The `sponge2` point data set in the `spm2` package (Li 2021a) and the `sponge.grid` grid data set in the `spm` package will be used to demonstrate the application of *RF* to numerical data.

```
library(spm2)

data(sponge2)

library(randomForest)

set.seed(1234)

rf1 <- randomForest(sponge2[, -c(3:4)], sponge2[, 3], ntree = 500, importance=TRUE
    )
```

Here we use the default value for `mtry`. If `mrty` is specified as `dim(sponge2[, -c(3:4)])[1]`, that is, all variables are used for `mtry`, then a *bagging* method is used. The importance of each variable is plotted in Figure 8.8 by

```
par(font.axis=2, font.lab=2)
varImpPlot(rf1, cex = 0.8)
```

The fitted values by `rf1` can be generated with `predict` as below.

```
rf1.pred <- predict(rf1, sponge2[, - c(3, 4)])
```

The ranges of fitted values and observed values are

```
range(rf1.pred)
```

```
[1]   1.219 28.395
```

```
range(sponge2$species.richness)
```

```
[1]   1 39
```

The predictions of `rf1` are shrunk considerably in comparison with the observations in terms of their range.

The fitted values are plotted against the observed values in Figure 8.9 by

```
par(font.axis = 2, font.lab = 2)
plot(sponge2$species.richness, rf1.pred, xlab = "Observed values", ylab = "Fitted
    values")
lines(sponge2$species.richness, sponge2$species.richness, col = "blue")
```

2. Application of the function `randomForest` to categorical data

The `hard` point data set will be used to show the application of *RF* to categorical data as follows.

```
data(hard)
```

rf1

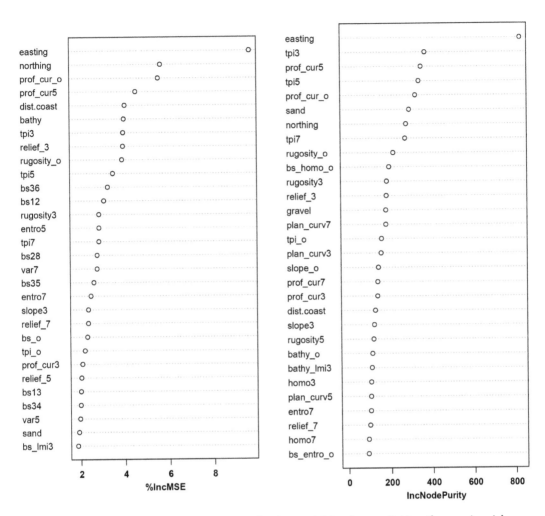

FIGURE 8.8: The importance of the predictive variables for predicting the species richness of sponge using RF.

```
set.seed(1234)

rf.h1 <- randomForest(hard[, -c(1, 17)], hard[, 17], ntree = 500, importance=TRUE)
```

Here we use the default value for `mtry`. If `mrty` is specified as `dim(hard[, -c(1, 17)])[1]`, that is, all variables are used for `mtry`, then a *bagging* method is used. The importance of each variable is plotted in Figure 8.10 by

```
par(font.axis=2, font.lab=2)
varImpPlot(rf.h1, cex = 0.8)
```

The fitted values by `rf.h1` are generated using `predict` and compared with the observed values below.

```
rf.h1.pred <- predict(rf.h1, hard[, -c(1, 17)])
```

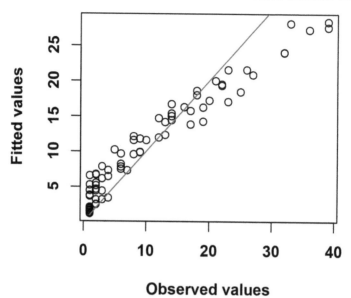

FIGURE 8.9: Observed values vs. fitted values by RF.

```
table(hard$hardness, rf.h1.pred)

      rf.h1.pred
      hard soft
  hard   38    0
  soft    0   99
```

The predictions of rf.h1 are exactly the same as the observations. The predictions of rf.h1 can be generated in the same way as above using predict and replacing hard[, -c(1, 17)] with a relevant data set.

8.2.2 Variable selection for *RF*

We can use *variable selection* techniques to simplify *RF* models such as rf1. Several variable selection techniques have been developed for *RF* (Li 2019a), and relevant functions are available in *R*, including:

(1) function Boruta in the Boruta package (Kursa and Rudnicki 2010, 2020);
(2) function VSURF in the VSURF package (Genuer, Poggi, and Tuleau-Malot 2019);
(3) function rfe in the caret package (Kuhn 2020);
(4) function steprfAVI in the steprf package (Li 2021c);
(5) function steprfAVI1 in the steprf package;
(6) function steprfAVI2 in the steprf package;
(7) function steprf with method = "AVI" in the steprf package;
(8) function steprf with method = "KIAVI" in the steprf package; and
(9) function steprf with method = "KIAVI2" in the steprf package.

The function steprf is based on and is an extension of steprfAVI by including an additional argument method that specifies a variable selection method for *RF*. For method, it can be "AVI", "KIAVI" or "KIAVI2". If "AVI" is used, steprf is steprfAVI.

rf.h1

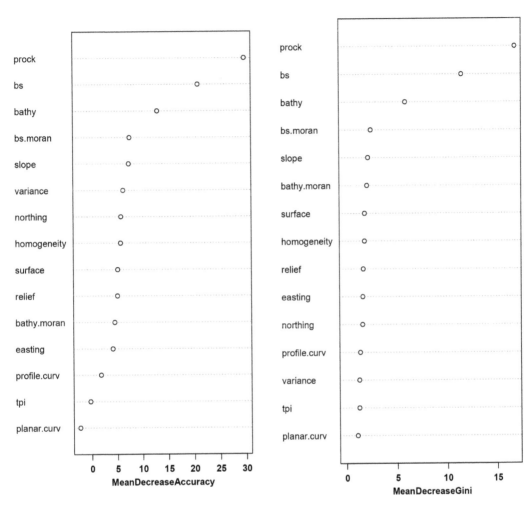

FIGURE 8.10: The importance of the predictors for predicting the species richness of sponge using RF.

We can search and select the most important predictors using these variable selection methods. We will apply these variable selection methods to sponge data as an example for numerical data and to hard data as an example for categorical data.

1. Implementation in Boruta

The function Boruta is a feature selection function with the Boruta algorithm that is an all relevant variable selection wrapper algorithm for any classification and regression methods that output variable importance; and by default, it uses *RF* (Kursa and Rudnicki 2020, 2010). For Boruta, the following arguments may need to be specified:

(1) formula, describing model to be analyzed;
(2) data, dataframe of response and predictive variable;
(3) doTrace, verbosity level; with 0 meaning no tracing; and
(4) maxRuns, maximal number of importance source runs.

Numerical data

```
library(Boruta)

sponge.bor <- Boruta(species.richness ~., data = sponge2[, -4], doTrace = 0,
    maxRuns = 5000)
```

The variables selected are

```
names(sponge2[, getSelectedAttributes(sponge.bor)])
```

```
 [1] "easting"    "northing"   "sand"       "slope_o"    "prof_cur_o"
 [6] "rugosity_o" "dist.coast" "rugosity3"  "rugosity5"  "tpi3"
[11] "tpi5"       "tpi7"       "relief_3"   "var7"
```

The variables selected can now be used to develop a *RF* model as follows.

```
set.seed(1234)

rf.bor1 <- randomForest(sponge2[, getSelectedAttributes(sponge.bor)], sponge2[,
    3], ntree = 500, importance=TRUE)
```

Categorical data

```
hard.bor <- Boruta(hardness ~., data = hard[, -1], doTrace = 0, maxRuns = 5000)
hard.bor
```

The variables selected are

```
names(hard[, -c(1, 17)][, getSelectedAttributes(hard.bor)])
```

```
 [1] "easting"     "northing"   "prock"      "bathy"      "bs"
 [6] "bathy.moran" "relief"     "slope"      "surface"    "homogeneity"
[11] "bs.moran"    "variance"
```

The variables selected can now be used to develop a *RF* model for `hard` data.

2. Implementation in VSURF

The function VSURF is a variable selection function for *RF*. For VSURF, the following two arguments need to be specified:
(1) x, a dataframe of predictive variables; and
(2) y, a vector of response variable.

Numerical data

```
library(VSURF)

rf.vs <- VSURF(sponge2[, -c(3:4)], sponge2[, 3])
```

The variables selected are

```
names(sponge2[, -c(3:4)][, rf.vs$varselect.pred])
```

```
[1] "easting"  "northing" "tpi3"
```

Categorical data

```
library(VSURF)

rf.vs.h1 <- VSURF(hard[, -c(1, 17)], hard[, 17])
```

The variables selected are

```
names(hard[, -c(1, 17)][, rf.vs.h1$varselect.pred])
```

```
[1] "prock" "bs"    "bathy"
```

3. Implementation in `rfe`

The function `rfe` implements backwards selection or recursive feature elimination (*RFE*) algorithm based on the importance ranking of predictive variables, with the less important ones sequentially eliminated. For `rfe`, the following arguments may need to be specified:
(1) `x`, a dataframe of predictive variables;
(2) `y`, a vector of response variable;
(3) `sizes`, the number of features that should be retained;
(4) `metric`, an accuracy metric to be used to select the optimal model; and
(5) `rfeControl`, a list of options, including functions for fitting and prediction; and for details see `?rfe`.

Numerical data

```
library(caret)
```

```
set.seed(1234)
```

```
rfProfile <- rfe(sponge2[, -c(3:4)], sponge2[, 3], sizes = c(4:10, 15), metric = "
    RMSE", rfeControl = rfeControl(functions = rfFuncs, rerank = TRUE, method = "
    repeatedcv", repeats = 20, verbose = FALSE))
```

```
rfProfile
```

```
Recursive feature selection
```

```
Outer resampling method: Cross-Validated (10 fold, repeated 20 times)
```

```
Resampling performance over subset size:
```

Variables	RMSE	Rsquared	MAE	RMSESD	RsquaredSD	MAESD	Selected
4	9.70	0.220	7.35	2.24	0.175	1.73	
5	9.60	0.233	7.28	2.33	0.195	1.78	
6	9.65	0.236	7.28	2.44	0.207	1.89	
7	9.60	0.239	7.25	2.49	0.207	1.91	
8	9.47	0.254	7.18	2.47	0.213	1.89	
9	9.41	0.264	7.12	2.45	0.212	1.89	
10	9.37	0.270	7.09	2.46	0.212	1.86	
15	9.27	0.283	7.02	2.49	0.222	1.90	
79	8.72	0.360	6.72	2.33	0.220	1.81	*

```
The top 5 variables (out of 79):
    easting, northing, tpi3, tpi5, bathy
```

```
rfProfile$bestSubset
```

```
[1] 79
```

```
rfProfile$optsize
```

```
[1] 79
```

```
rfProfile$optVariables
```

```
 [1]  "easting"        "northing"      "tpi3"          "tpi5"          "bathy"
 [6]  "dist.coast"     "prof_cur_o"    "entro7"        "bathy_o"       "rugosity_o"
[11]  "rugosity3"      "relief_3"      "var7"          "bs36"          "sand"
[16]  "prof_cur5"      "tpi7"          "slope_o"       "tpi_o"         "entro5"
[21]  "bs_o"           "rugosity5"     "homo3"         "homo7"         "bs35"
[26]  "relief_5"       "bs27"          "slope3"        "bs26"          "bs10"
[31]  "bs12"           "bs32"          "bs11"          "bathy_lmi5"    "bs31"
[36]  "bs22"           "relief_7"      "bathy_lmi7"    "bathy_lmi3"    "bs30"
[41]  "bathy_lmi_o"    "prof_cur3"     "bs21"          "bs_homo_o"     "var5"
[46]  "plan_curv5"     "prof_cur7"     "bs17"          "bs14"          "bs28"
[51]  "bs33"           "bs23"          "bs16"          "bs34"          "plan_curv3"
[56]  "bs25"           "rugosity7"     "gravel"        "bs18"          "bs20"
[61]  "bs29"           "bs13"          "bs24"          "bs_lmi5"       "homo5"
[66]  "relief_o"       "slope7"        "bs19"          "bs15"          "var3"
[71]  "entro3"         "bs_lmi3"       "bs_lmi7"       "slope5"        "plan_curv7"
[76]  "bs_lmi_o"       "bs_var_o"      "bs_entro_o"    "plan_cur_o"
```

It suggests that all 79 variables are selected by rfe.

Categorical data

```
library(caret)

set.seed(1234)

rfe.h1 <- rfe(hard[, -c(31, 17)], hard[, 17], sizes = c(4:10, 15), rfeControl =
    rfeControl(functions = rfFuncs, rerank = TRUE, method = "repeatedcv", repeats
    = 20, verbose = FALSE))

rfe.h1
```

```
Recursive feature selection

Outer resampling method: Cross-Validated (10 fold, repeated 20 times)

Resampling performance over subset size:
```

Variables	Accuracy	Kappa	AccuracySD	KappaSD	Selected
4	0.912	0.762	0.0673	0.192	
5	0.916	0.772	0.0628	0.185	
6	0.919	0.779	0.0636	0.187	
7	0.921	0.780	0.0640	0.193	
8	0.923	0.784	0.0656	0.200	*
9	0.918	0.773	0.0644	0.190	
10	0.919	0.775	0.0650	0.194	
15	0.916	0.763	0.0684	0.215	
16	0.918	0.775	0.0686	0.202	

```
The top 5 variables (out of 8):
   prock, bs, bathy, bs.moran, homogeneity

rfe.h1$bestSubset

[1] 8

rfe.h1$optsize

[1] 8

rfe.h1$optVariables
```

```
[1] "prock"       "bs"          "bathy"        "bs.moran"      "homogeneity"
[6] "surface"     "variance"    "northing"
```

It suggests that 8 variables are selected by `rfe`.

4. Implementation in `steprfAVI`

The function `steprfAVI` was developed according to previous studies (Liaw and Wiener 2002; Smith, Ellis, and Pitcher 2011; Li 2013b; Li, Siwabessy, Tran, et al. 2014; Li, Alvarez, et al. 2017; Li, Siwabessy, Huang, et al. 2019) and is to select predictive variables for *RF* by their *averaged variable importance* (*AVI*) that is calculated for each model after excluding the least important variable. For more details, see `?steprfAVI`. For `steprfAVI`, the following arguments may need to specified:

(1) `trainx`, a dataframe of predictive variables;
(2) `trainy`, a vector of response variable;
(3) `rpt`, number of iterations of cross-validation;
(4) `predacc`, an accuracy metric to be used to assess the model;
(5) `importance`, importance of predictive variables; and
(6) `nsim`, iteration number for deriving `AVI`.

Numerical data

Application of `steprfAVI` to `spong2` count data that is detailed in Appendix A.

```
library(spm2)
library(steprf)

set.seed(1234)

steprfAVI.1 <- steprfAVI(trainx = sponge2[, -c(3:4)], trainy = sponge2[, 3], rpt =
    20, predacc = "VEcv", importance = TRUE, nsim = 20)
```

The results in `steprfAVI.1` can be retrieved with the function `steprfAVIPredictors` in the `steprf` package.

```
predictors.selected.steprfAVI.1 <- steprfAVIPredictors(steprfAVI.1, trainx =
    sponge2[, -c(3:4)])

predictors.selected.steprfAVI.1$variables.most.accurate
```

```
 [1] "easting"      "northing"    "bathy"        "bs26"        "bs27"
 [6] "bs36"         "prof_cur_o"  "rugosity3"    "tpi3"        "tpi5"
[11] "relief_3"
```

```
predictors.selected.steprfAVI.1$max.predictive.accuracy
```

```
[1] 38.35
```

We can also derive the *predictive accuracy boosting variables* (*PABVs*) and *predictive accuracy reducing variables* (*PARVs*) by

```
predictors.selected.steprfAVI.1$PABV
```

```
 [1] "easting"      "northing"    "sand"         "bathy"       "bs10"
 [6] "bs12"         "bs15"        "bs17"         "bs19"        "bs20"
[11] "bs21"         "bs22"        "bs24"         "bs26"        "bs27"
[16] "bs30"         "bs35"        "bs36"         "bs_homo_o"   "bs_var_o"
[21] "bs_lmi_o"     "bathy_o"     "bathy_lmi_o"  "tpi_o"       "plan_cur_o"
[26] "prof_cur_o"   "relief_o"    "rugosity7"    "tpi3"        "bathy_lmi5"
[31] "plan_curv3"   "plan_curv5"  "relief_3"     "relief_7"    "slope3"
[36] "slope7"       "prof_cur7"   "entro5"       "entro7"      "var7"
```

```
predictors.selected.steprfAVI.1$PARV
```

```
 [1] "bs_entro_o"  "entro3"      "bs_lmi3"     "bs18"        "slope5"
 [6] "bs_lmi7"     "var3"        "bs13"        "gravel"      "homo5"
[11] "plan_curv7"  "bs29"        "bs11"        "bs34"        "bs_lmi5"
[16] "bs28"        "prof_cur3"   "bs14"        "bs16"        "var5"
[21] "bathy_lmi7"  "homo3"       "bathy_lmi3"  "homo7"       "bs25"
[26] "bs33"        "bs_o"        "bs23"        "bs31"        "relief_5"
[31] "rugosity5"   "tpi7"        "slope_o"     "prof_cur5"   "bs32"
[36] "dist.coast"  "rugosity_o"  "rugosity3"   "tpi5"
```

The results show that of the 79 predictive variables, 40 variables were identified as *PABV*, and 39 variables as *PARV*. These identified *PABV*s provide a set of predictive variables to steprf for further variable selection when method = "KIAVI" or method = "KIAVI2" is used.

The predictive accuracy of *RF* model after removing each unimportant predictor is illustrated in Figure 8.11 by

```
library(reshape2)

pa1 <- as.data.frame(steprfAVI.1$predictive.accuracy2)
names(pa1) <- steprfAVI.1$variable.removed
pa2 <- melt(pa1, id = NULL)
names(pa2) <- c("Variable","VEcv")

library(graphics)

par(font.axis = 2, font.lab = 2, cex.lab = 1, cex.axis = 0.5)
with(pa2, boxplot(VEcv ~ Variable, ylab = "VEcv (%)", xlab = NULL, las = 2))
```

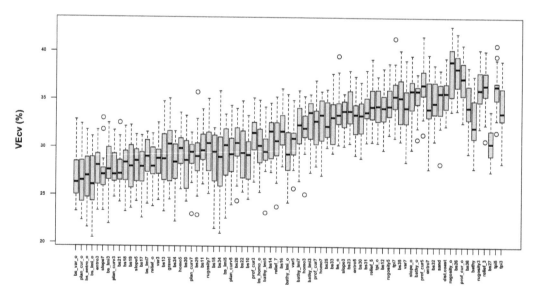

FIGURE 8.11: Predictive accuracy of RF model for sponge data after removing each unimportant predictive variable.

The contribution of each removed predictor to the predictive accuracy of *RF* model is shown in Figure 8.12 by

```
par(font.axis=2, font.lab=2)
barplot(steprfAVI.1$delta.accuracy, col = (1:length(steprfAVI.1$variable.removed))
    ,
```

```
names.arg = steprfAVI.1$variable.removed,
cex.names=0.6, cex.lab = 0.8, cex.axis = 0.6, las = 2, ylab="Increase rate in VEcv
    (%)")
```

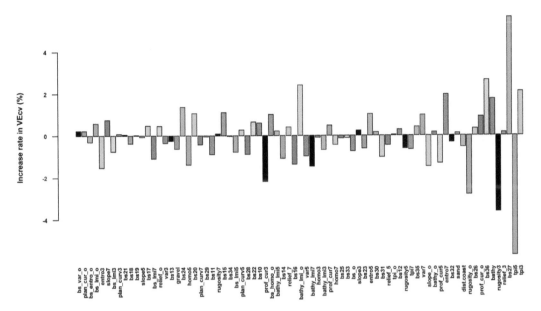

FIGURE 8.12: Contribution of each removed predictive variable to the predictive accuracy
of RF model for sponge data.

Categorical data

Application of steprfAVI to hard data.

```
library(steprf)

set.seed(1234)

steprfAVI.h1 <- steprfAVI(trainx = hard[, -c(1, 17)], trainy = hard[, 17], rpt =
    20, predacc = "ccr", importance = TRUE, nsim = 20)
```

The results in steprfAVI.h1 can be retrieved with steprfAVIPredictors.

```
predictors.selected.steprfAVI.h1 <- steprfAVIPredictors(steprfAVI.h1, trainx =
    hard[, -c(1, 17)])

predictors.selected.steprfAVI.h1$variables.most.accurate

[1] "prock" "bathy" "bs"

predictors.selected.steprfAVI.h1$max.predictive.accuracy

[1] 93.5
```

The predictive accuracy of *RF* model after removing each unimportant predictor is illustrated in Figure 8.13 by

```
library(reshape2)

pa.h1 <- as.data.frame(steprfAVI.h1$predictive.accuracy2)
names(pa.h1) <- steprfAVI.h1$variable.removed
```

```
pa.h2 <- melt(pa.h1, id = NULL)
names(pa.h2) <- c("Variable","CCR")

library(graphics)

par(font.axis = 2, font.lab = 2, cex.lab = 1, cex.axis = 0.8)
with(pa.h2, boxplot(CCR ~ Variable, ylab = "Correct classification rate (%)", xlab
    = NULL, las = 2))
```

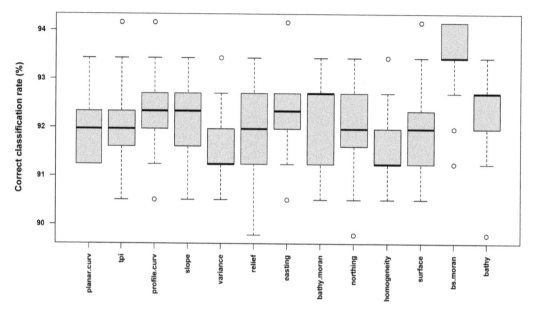

FIGURE 8.13: Predictive accuracy of RF model for hard data after removing each unimportant predictive variable.

The contribution of each removed predictor to the predictive accuracy of *RF* model is shown in Figure 8.14 by

```
par(font.axis=2, font.lab=2)
barplot(steprfAVI.h1$delta.accuracy, col = (1:length(steprfAVI.h1$variable.removed
    )),
names.arg = steprfAVI.h1$variable.removed,
cex.names=0.6, cex.lab = 0.8, cex.axis = 0.6, las = 2, ylab="Increase rate in VEcv
    (%)")
```

5. Implementation in steprfAVI1

The function steprfAVI1 is a simplified version of steprfAVI. This function is to select predictive variables for *RF* based on their *AVI* derived from the full model, by assuming that the order of *AVI* of predictors is not changing with the removal of relevant predictors. That is, the *AVI* is calculated only once from the full model. So steprfAVI1 is faster than steprfAVI.

Numerical data

```
library(steprf)

set.seed(1234)

steprfAVI1.1 <- steprfAVI1(trainx = sponge2[, -c(3:4)], trainy = sponge2[, 3], rpt
    = 20, predacc = "VEcv", importance = TRUE, nsim = 20)
```

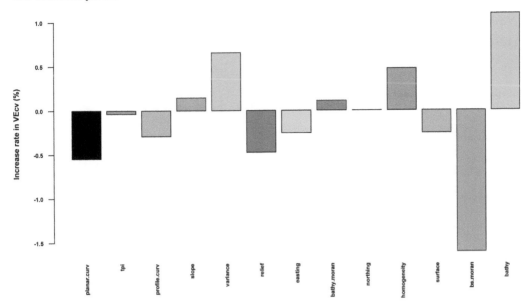

FIGURE 8.14: Contribution of each removed predictive variable to the predictive accuracy of RF model for `hard` data.

The variables selected are retrieved as

```
predictors.selected.steprfAVI1.1 <- steprfAVIPredictors(steprfAVI1.1, trainx =
    sponge2[, -c(3:4)])
```

```
predictors.selected.steprfAVI1.1$variables.most.accurate
```

```
 [1] "easting"    "northing"   "sand"       "bathy"      "bathy_o"
 [6] "tpi_o"      "slope_o"    "prof_cur_o" "rugosity_o" "dist.coast"
[11] "rugosity3"  "rugosity5"  "tpi3"       "tpi5"       "tpi7"
[16] "relief_3"   "prof_cur5"  "entro7"     "var7"
```

Categorical data

```
library(steprf)
```

```
set.seed(1234)
```

```
steprfAVI1.h1 <- steprfAVI1(trainx = hard[, -c(1, 17)], trainy = hard[, 17], rpt =
    20, predacc = "ccr", importance = TRUE, nsim = 20)
```

The variables selected are retrieved as

```
predictors.selected.steprfAVI1.h1 <- steprfAVIPredictors(steprfAVI1.h1, trainx =
    hard[, -c(1, 17)])
```

```
predictors.selected.steprfAVI1.h1$variables.most.accurate
```

```
[1] "northing" "prock"    "bathy"    "bs"       "bs.moran"
```

6. Implementation in `steprfAVI2`

The function `steprfAVI2` is the same as `steprfAVI`, with only one exception, that is, a `set.seed(1234)` being added prior to each chunk of code that involves randomness within the function.

Numerical data

```
library(steprf)

set.seed(1234)

steprfAVI2.1 <- steprfAVI2(trainx = sponge2[, -c(3:4)], trainy = sponge2[, 3], rpt
    = 20, predacc = "VEcv", importance = TRUE, nsim = 20)
```

The variables selected are retrieved as

```
predictors.selected.steprfAVI2.1 <- steprfAVIPredictors(steprfAVI2.1, trainx =
    sponge2[, -c(3:4)])

predictors.selected.steprfAVI2.1$variables.most.accurate
```

```
[1] "easting"    "northing"   "bathy"      "bs27"       "bs36"       "rugosity3"
[7] "tpi3"       "tpi5"
```

Categorical data

```
library(steprf)

set.seed(1234)

steprfAVI2.h1 <- steprfAVI2(trainx = hard[, -c(1, 17)], trainy = hard[, 17], rpt =
    20, predacc = "ccr", importance = TRUE, nsim = 20)
```

The variables selected are retrieved as

```
predictors.selected.steprfAVI2.h1 <- steprfAVIPredictors(steprfAVI2.h1, trainx =
    hard[, -c(1, 17)])

predictors.selected.steprfAVI2.h1 $variables.most.accurate
```

```
[1] "prock" "bathy" "bs"
```

7. Implementation in steprf with method = "AVI"

Since the *RF* model selected by steprf with method = "AVI" is the same as the model selected by steprfAVI, we only include the results of the *RF* models selected by steprfAVI for numerical and categorical data. It should be noted that because steprfAVIPredictors has been implemented within steprf, and the results by steprf are ready for use as demonstrated later.

8. Implementation in steprf with method = "KIAVI"

Numerical data

```
set.seed(1234)

steprf.2 <- steprf(trainx = sponge2[, -c(3:4)], trainy = sponge2[, 3], method = "
    KIAVI", rpt = 20, predacc = "VEcv", importance = TRUE, nsim = 20, delta.
    predacc = 0.005)
```

The results of variable selection can be retrieved as

```
steprf.2$steprfPredictorsFinal$variables.most.accurate
```

```
[1] "easting"    "northing"   "bs27"       "bs36"       "tpi3"
[6] "tpi5"       "bathy"      "prof_cur_o" "bs10"
```

```
steprf.2$max.predictive.accuracy
```

```
[1] 38.35 40.95 40.93
```

```
steprf.2$numberruns
```

```
[1] 3
```

The `steprf.2$steprfPredictorsFinal$variables.most.accurate` shows the variables selected for the last *RF* model and associated results. Whether it is of the highest predictive accuracy needs to be confirmed using `steprf.2$max.predictive.accuracy`. It is apparent that the second run results in a model with the highest accuracy and the variables selected are

```
steprf.2$steprfPredictorsAll[2][[1]]$variables.most.accurate
```

```
[1] "easting"  "northing" "bs27"    "bs36"    "tpi3"    "tpi5"
```

Categorical data

```
set.seed(1234)
```

```
steprf.h2 <- steprf(trainx = hard[, -c(1, 17)], trainy = hard[, 17], method = "
    KIAVI", rpt = 20, predacc = "ccr", importance = TRUE, nsim = 20, delta.predacc
    = 0.005)
```

```
steprf.h2$max.predictive.accuracy
```

```
[1] 93.50 93.28
```

```
steprf.h2$numberruns
```

```
[1] 2
```

Obviously the first run results in a model with higher accuracy. The variables selected are

```
steprf.h2$steprfPredictorsAll[1][[1]]$variables.most.accurate
```

```
[1] "prock" "bathy" "bs"
```

9. Implementation in `steprf` with `method = "KIAVI2"`

Numerical data

```
set.seed(1234)
```

```
steprf.3 <- steprf(trainx = sponge2[, -c(3:4)], trainy = sponge2[, 3], method = "
    KIAVI2", rpt = 20, predacc = "VEcv", importance = TRUE, nsim = 20, delta.
    predacc = 0.005)
```

```
steprf.3$max.predictive.accuracy
```

```
[1] 38.35 43.73 43.45
```

```
steprf.3$numberruns
```

```
[1] 3
```

The second run results in a model with higher accuracy. The variables selected are

```
steprf.3$steprfPredictorsAll[2][[1]]$variables.most.accurate
```

```
[1] "easting"    "northing"    "bathy"      "bs27"       "bs36"
[6] "prof_cur_o" "tpi3"
```

Categorical data

```
set.seed(1234)

steprf.h3 <- steprf(trainx = hard[, -c(1, 17)], trainy = hard[, 17], method = "
    KIAVI2", rpt = 20, predacc = "ccr", importance = TRUE, nsim = 20, delta.
    predacc = 0.005)

steprf.h3$max.predictive.accuracy

[1] 93.50 93.28

steprf.h3$numberruns

[1] 2
```

The model resulted from the first run is of the highest accuracy. The variables selected are

```
steprf.h3$steprfPredictorsAll[1][[1]]$variables.most.accurate

[1] "prock" "bathy" "bs"
```

8.2.3 Predictive accuracy of the *RF* models developed from variable selection methods

Various variable selection methods above have resulted in different sets of predictors. The predictive accuracy of the *RF* models developed from these methods are expected to be different and can be examined using a cross-validation method, such as the function RFcv in the spm package.

For RFcv, the following arguments are essential:
(1) trainx, a dataframe or matrix contains columns of predictor variables;
(2) trainy, a vector of response variable, must have length equal to the number of rows in trainx;
(3) cv.fold, integer; number of folds in the cross-validation; and
(4) predacc, can be either "VEcv" for vecv or "ALL" for all measures in function pred.acc in the spm package.

We can use RFcv to assess the predictive accuracy of the *RF* models and to produce the stabilized predictive accuracy for each model. For the *RF* model by Boruta the results will be presented, while for the rest models, only *R* code will be provided, but the results of all models will be compared and presented in the next section.

1. Accuracy of the *RF* model by Boruta

Numerical data

```
library(Boruta)

set.seed(1234)
n <- 100
VEcv.b <- NULL

for (i in 1:n) {
  rfcv1 <- RFcv(sponge2[, getSelectedAttributes(sponge.bor)], sponge2[, 3],
      predacc = "VEcv")
```

```
  VEcv.b[i] <- rfcv1
}
```

The median and range of the predictive accuracy of the *RF* model by Boruta based on 100 repetitions of 10-fold cross-validation using RFcv are

```
median(VEcv.b)
```

```
[1] 30.26
```

```
range(VEcv.b)
```

```
[1] 18.62 35.10
```

Categorical data

```
library(Boruta)
set.seed(1234)
n <- 100
ccr.b.h <- NULL

for (i in 1:n) {
  rfcv1 <- RFcv2(hard[, -c(1, 17)][, getSelectedAttributes(hard.bor)], hard[, 17],
      predacc = "ccr")
  ccr.b.h[i] <- rfcv1
}
```

The median and range of the predictive accuracy of the *RF* model selected by Boruta based on 100 repetitions of 10-fold cross-validation using RFcv are

```
median(ccr.b.h)
```

```
[1] 91.97
```

```
range(ccr.b.h)
```

```
[1] 89.78 94.16
```

2. Accuracy of the RF model by vsurf

Numerical data

```
set.seed(1234)
n <- 100
VEcv.v <- NULL

for (i in 1:n) {
  rfcv1 <- RFcv(sponge2[, rf.vs$varselect.pred], sponge2[, 3], predacc = "VEcv")
  VEcv.v[i] <- rfcv1
}
```

Categorical data

```
set.seed(1234)
n <- 100
ccr.v <- NULL

for (i in 1:n) {
  rfcv1 <- RFcv2(hard[, -c(1, 17)][, rf.vs.h1$varselect.pred], hard[, 17], predacc
      = "ccr")
  ccr.v[i] <- rfcv1
}
```

3. Accuracy of the RF model by `rfe`

Numerical data

```
set.seed(1234)
n <- 100
VEcv.r <- NULL

for (i in 1:n) {
  rfcv1 <- RFcv(sponge2[, rfProfile$optVariables], sponge2[, 3], predacc = "VEcv")
  VEcv.r[i] <- rfcv1
}
```

Categorical data

```
set.seed(1234)
n <- 100
ccr.r <- NULL

for (i in 1:n) {
  rfcv1 <- RFcv2(hard[, -c(1, 17)][, rfe.h1$optVariables], hard[, 17], predacc = "
      ccr")
  ccr.r[i] <- rfcv1
}
```

4. Accuracy of the RF model by `steprfAVI`

Numerical data

```
set.seed(1234)
n <- 100
VEcv.avi <- NULL

for (i in 1:n) {
  rfcv1 <- RFcv(sponge2[, predictors.selected.steprfAVI.1$variables.most.accurate
      ], sponge2[, 3], predacc = "VEcv")
  VEcv.avi[i] <- rfcv1
}
```

Categorical data

```
set.seed(1234)
n <- 100
ccr.avi <- NULL

for (i in 1:n) {
  rfcv1 <- RFcv2(hard[, -c(1, 17)][, predictors.selected.steprfAVI.h1$variables.
      most.accurate], hard[, 17], predacc = "ccr")
  ccr.avi[i] <- rfcv1
}
```

5. Accuracy of the RF model by `steprfAVI1`

Numerical data

```
set.seed(1234)
n <- 100
VEcv.avi1 <- NULL

for (i in 1:n) {
  rfcv1 <- RFcv(sponge2[, predictors.selected.steprfAVI1.1$variables.most.accurate
      ], sponge2[, 3], predacc = "VEcv")
  VEcv.avi1[i] <- rfcv1
}
```

Categorical data

```
set.seed(1234)
n <- 100
ccr.avi1 <- NULL

for (i in 1:n) {
  rfcv1 <- RFcv2(hard[, -c(1, 17)][, predictors.selected.steprfAVI1.h1$variables.
      most.accurate], hard[, 17], predacc = "ccr")
  ccr.avi1[i] <- rfcv1
}
```

6. Accuracy of the RF model by `steprfAVI2`

Numerical data

```
set.seed(1234)
n <- 100
VEcv.avi2 <- NULL

for (i in 1:n) {
  rfcv1 <- RFcv(sponge2[, predictors.selected.steprfAVI2.1$variables.most.accurate
      ], sponge2[, 3], predacc = "VEcv")
  VEcv.avi2[i] <- rfcv1
}
```

Categorical data

```
set.seed(1234)
n <- 100
ccr.avi2 <- NULL

for (i in 1:n) {
  rfcv1 <- RFcv2(hard[, -c(1, 17)][, predictors.selected.steprfAVI2.h1$variables.
      most.accurate], hard[, 17], predacc = "ccr")
  ccr.avi2[i] <- rfcv1
}
```

7. Accuracy of the RF model by `steprf` with `method = "KIAVI"`

Numerical data

```
set.seed(1234)
n <- 100
VEcv.steprf.kiavi <- NULL

for (i in 1:n) {
  rfcv1 <- RFcv(sponge2[, steprf.2$steprfPredictorsAll[2][[1]]$variables.most.
      accurate], sponge2[, 3], predacc = "VEcv")
  VEcv.steprf.kiavi[i] <- rfcv1
}
```

Categorical data

```
set.seed(1234)
n <- 100
ccr.steprf.kiavi <- NULL

for (i in 1:n) {
  rfcv1 <- RFcv2(hard[, -c(1, 17)][, steprf.h2$steprfPredictorsAll[2][[1]]$
      variables.most.accurate], hard[, 17], predacc = "ccr")
  ccr.steprf.kiavi[i] <- rfcv1
}
```

8. Accuracy of the RF model by `steprf` with `method = "KIAVI2"`

Numerical data

```
set.seed(1234)
n <- 100
VEcv.steprf.kiavi2 <- NULL

for (i in 1:n) {
  rfcv1 <- RFcv(sponge2[, steprf.3$steprfPredictorsAll[2][[1]]$variables.most.
      accurate], sponge2[, 3], predacc = "VEcv")
  VEcv.steprf.kiavi2[i] <- rfcv1
}
```

Categorical data

```
set.seed(1234)
n <- 100
ccr.steprf.kiavi2 <- NULL

for (i in 1:n) {
  rfcv1 <- RFcv2(hard[, -c(1, 17)][, steprf.h3$steprfPredictorsAll[2][[1]]$
      variables.most.accurate], hard[, 17], predacc = "ccr")
  ccr.steprf.kiavi2[i] <- rfcv1
}
```

8.2.4   Comparison of variable selection methods

Numerical data

The predictive accuracy of the *RF* models selected for `sponge2` data (with 79 predictive variables) by eight *variable selection* methods are compared in Figure 8.15.

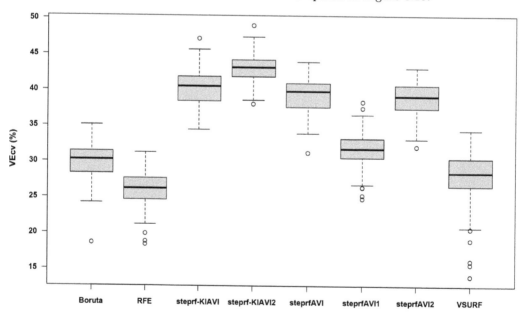

FIGURE 8.15: Predictive accuracy of the RF models developed by eight variable selection methods for `sponge` data.

On the basis of the median predictive accuracy, the variable selection methods from the most accurate to the least accurate are as follows:

```
           Model  VEcv
4 steprf-KIAVI2  43.14
3  steprf-KIAVI  40.52
5      steprfAVI  39.82
7     steprfAVI2  39.16
6     steprfAVI1  31.79
1         Boruta  30.26
8          VSURF  28.48
2            RFE  26.19
```

Categorical data

The predictive accuracy of the *RF* models selected using eight variable selection methods for hard data (with 15 predictive variables) are compared in Figure 8.16.

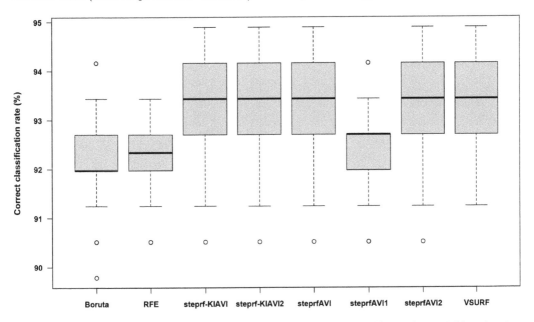

FIGURE 8.16: Predictive accuracy of the RF models developed by eight variable selection methods for hard data.

On the basis of the median predictive accuracy, the variable selection methods from the most accurate to the least accurate are as follows:

```
           Model    CCR
3  steprf-KIAVI  93.43
4 steprf-KIAVI2  93.43
5      steprfAVI  93.43
7     steprfAVI2  93.43
8          VSURF  93.43
6     steprfAVI1  92.70
2            RFE  92.34
1         Boruta  91.97
```

Of these eight variable selection methods, steprf with method = "KIAVI" and method = "KIAVI2" produced the most accurate *RF* models.

8.2.5 Predictions of *RF*

We can use *RF* to generate spatial predictions. The predictions of *RF* can be generated by applying the function `rfpred` in the `spm` package. For `rfpred`, the following arguments need to be specified:

(1) `trainx`, a dataframe or matrix contains columns of predictive variables;

(2) `trainy`, a vector of response variable, must have length equal to the number of rows in trainx;

(3) `longlatpredx`, a dataframe contains longitude and latitude of point locations (i.e., the centers of grids) to be predicted;

(4) `predx`, a dataframe or matrix contains columns of predictive variables for the grids to be predicted; and

(5) `ntree`, number of trees to grow.

We will demonstrate the application of `rfpred` to the `sponge` and `sponge.grid` data sets as follows.

```
library(randomForest)

names(sponge)
names(sponge.grid)
set.seed(1234)

rfpred1 <- rfpred(sponge[, -c(3)], sponge[, 3], sponge.grid[, c(1,2)], sponge.grid
    , ntree = 500)
```

The spatial distribution of predictions of *RF* can be illustrated in Figure 8.17 by

```
library(sp)

rf.pred.df <- rfpred1
gridded(rf.pred.df) = ~ LON + LAT

library(raster)

par(font.axis = 2, font.lab = 2)
plot(brick(rf.pred.df))
```

8.2.6 Notes on *RF*

1. R packages and functions

The function `ranger` in the `ranger` package (Wright and Ziegler 2017; Wright, Wager, and Probst 2020) is a fast implementation of *RF*, particularly for high-dimensional data. The function `rgcv` in the `spm` package is a cross-validation function based on the function `ranger`; and the function `rgpred` in `spm` is also based on the function `ranger` for generating spatial predictions. These `ranger` based functions are equivalent to the functions based on the function `randomForest` (i.e., `RFcv`, `rfpred`). These functions can be easily applied by following the examples provided for the functions `randomForest`, `RFcv`, and `rfpred`.

2. Choice of relevant arguments for *RF*

The argument `mtry` for the function `randomForest` can be estimated based on the result of the function `tuneRF` in the `randomForest` package, whereby the `tuneRF` searches for the optimal value of `mtry`. Given that `NA` is not permitted by `tuneRF`, all `NA`s need to be removed from the data set. For example, `mtry` can be estimated by

```
tuneRF(sponge2[,-c(3)], sponge2[, 3], stepFactor=2, ntreeTry = 100)
```

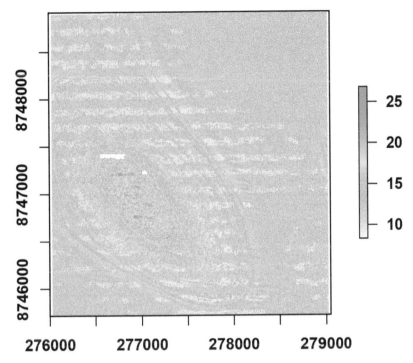

FIGURE 8.17: Spatial patterns of predictions using *RF* for sponge species richness data.

The value estimated is not stable and we need to repeat a number of times to get a reliable value for `mtry`. It can also be estimated using `RFcv` based on predictive accuracy. So do `ntree` and `nodesize`. However, the default values of `mtry`, `ntree`, and `nodesize` are suggested to be good options (Liaw and Wiener 2002; Diaz-Uriarte and Andres 2006), which have also been demonstrated in marine environmental sciences (Li, Potter, Huang, and Heap 2012b; Li, Siwabessy, Tran, et al. 2014; Li, Tran, and Siwabessy 2016), so the default values could be used for these arguments.

3. Function `partialPlot` for *RF*

The relationships of response variable and predictive variables used in a *RF* model may provide useful information to interpret the predictions of the model. Such relationships can be visualized with the function `partialPlot` in the `randomForest` package. We will use `sponge .rf1` to depict the relationships of sponge species richness with seven predictive variables and show the results in Figure 8.18 by

```
library(randomForest)

par(mfrow=c(4, 2), font.axis = 2, font.lab = 2)
partialPlot(sponge.rf1, sponge, easting)
partialPlot(sponge.rf1, sponge, easting)
partialPlot(sponge.rf1, sponge, tpi3)
partialPlot(sponge.rf1, sponge, var7)
partialPlot(sponge.rf1, sponge, entro7)
partialPlot(sponge.rf1, sponge, bs34)
partialPlot(sponge.rf1, sponge, bs11)
```

4. Uncertainty of predictions

For *RF*, several types of *uncertainty* have been produced (Chen, Li, and Wang 2012; Wager,

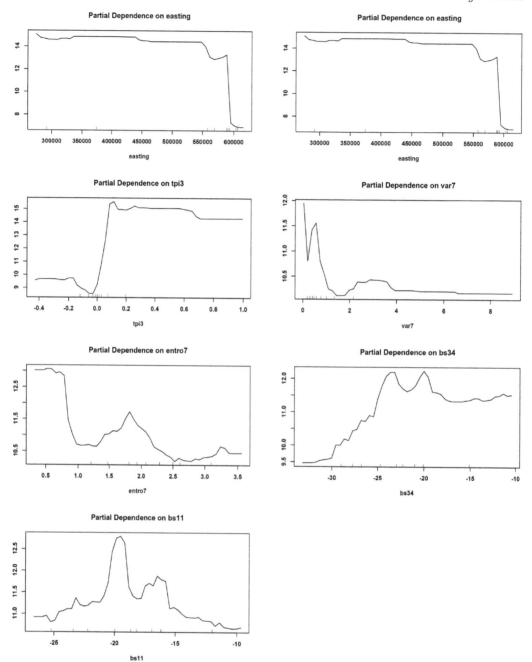

FIGURE 8.18: Partial plots of *RF* model for predicting the spatial distribution of sponge species richness, indicating the relationships of the sponge species richness with the seven predictors in the *RF* model.

Hastie, and Efron 2014; Coulston et al. 2016; Mentch and Hooker 2016; Slaets et al. 2017; Wright and Ziegler 2017), which in fact reflect the differences in sampling strategies (Li 2019a). For example, prediction uncertainty produced for *RF* based on Monte Carlo resampling method in Coulston et al. (2016) was, in fact, measuring the variation in predictions among individual trees instead of the uncertainty of predictions by *RF*. A further example is that the differences among an ensemble of equally probable realizations were used as a measure of prediction uncertainty (Chen, Li, and Wang 2012), which actually only measures the differences among the predictions of various runs of *RF*, that is, measuring the differences resulted from the randomness associated with each run of *RF*. Hence, these values do not relate to predictive accuracy and do not measure prediction uncertainty either (Li 2019a), which is also applicable to all other methods introduced in this book in regards with prediction uncertainty that are produced in a similar way.

8.3 Generalized boosted regression modeling

Generalized boosted regression modeling (*GBM*), like *RF*, is also an ensemble method that combines many individual regression trees, but *GBM trees* are different from *RF* trees. The first *GBM tree* fits training data, each of the remaining *GBM* trees fits the residuals of its previous *GBM tree*. The fitted values of each tree are weighted by a pre-specified weight. The residuals of the first tree are the differences between the training data and the weighted fitted values of the tree; and for each of the remaining trees, the residuals are the differences between the residuals of the previous tree and the weighted fitted values of the tree. The predictions of *GBM* are the sum of the weighted fitted values of all *GBM* trees (James et al. 2017). In addition, *GBM* is for numerical data and cannot be used for categorical data.

Predictions of *GBM* can be represented in Equation (8.4) in a similar way as in Equation (4.1).

$$\widehat{y_{x_i}} = \sum_{b=1}^{B} \lambda \widehat{T_b(x_i)} \tag{8.4}$$

where $\widehat{y_{x_i}}$ is the predicted value of a response variable y at the location of interest (i) by *GBM*, x_i is p predictive variables at location i (i.e., $x_i = (x_{i1}, x_{i2}, ..., x_{ip})$), λ is a shrinkage parameter (or a weight), $\widehat{T_b(x_i)}$ is the predicted value at the location (i) by the *GBM* tree T_b based on the bth iteration, b represents one of B *GBM* trees in the ensemble (James et al. 2017).

Similar to *CART* and *RF*, *GBM* can automatically handle *non-linear relationships* and *interactions*, so we only need to provide predictive variables to *GBM*. *GBM* can be implemented in the `gbm` package (Greenwell et al. 2020) as well as other packages such as the `xgboost` package (Chen et al. 2020). We will use the function `gbm` in the `gbm` package to demonstrate the applications of *GBM* method.

8.3.1 Application of *GBM*

The descriptions of relevant arguments in the function `gbm` are detailed in its help file, which can be accessed by `?gbm`. For gbm, the following arguments may need to be specified:
(1) `formula`, a formula defining response variable and predictive variables;
(2) `distribution`, a character string specifying the name of the distribution to use;
(3) `data`, a dataframe containing the variables in the model;
(4) `weights`, a vector of weights to be used in the fitting process;
(5) `var.monotone`, a vector, the same length as the number of predictors, indicating which variables have a monotone increasing (+1), decreasing (-1), or arbitrary (0) relationship

with the outcome;

(6) `n.trees`, the number of trees to grow;

(7) `interaction.depth`, the maximum depth of each tree (i.e., the highest level of variable interactions allowed);

(8) `n.minobsinnode`, the minimum number of observations in the terminal nodes of the trees;

(9) `shrinkage`, a shrinkage parameter applied to each tree in the expansion; also known as the learning rate or step-size reduction;

(10) `bag.fraction`, the fraction of the training set observations randomly selected to propose the next tree in the expansion;

(11) `train.fraction`, the first `train.fraction * nrows(data)` observations are used to fit the gbm and the remainder are used for computing out-of-sample estimates of the loss function;

(12) `cv.folds`, the number of cross-validation folds to perform; and if `cv.folds > 1` then gbm, in addition to the usual fit, will perform a cross-validation, calculate an estimate of generalization error returned in cv.error; and

(13) `n.cores`, the number of CPU cores to use.

We will demonstrate the application of *GBM* using three data types: (1) count data, (2) percentage data, and (3) binomial data.

1. Application of the function `gbm` to count data

We will use the `sponge` point data set and the `sponge.grid` grid data set in the `spm` package to demonstrate the applications of *GBM* to count data below.

```
require(gbm)

data(sponge); data(sponge.grid)
nc <- ncol(sponge)-1; nr <- nrow(sponge)
set.seed(1234)

sponge.gbm1 <- gbm(sponge ~ easting + northing + tpi3 + var7 + entro7 + bs34 +
    bs11, distribution = "poisson", data = sponge, weights = rep(1, nr), var.
    monotone = rep(0, nc), n.trees = 6000, interaction.depth = 5, n.minobsinnode =
    10, shrinkage = 0.001, bag.fraction = 0.5, cv.fold = 10, keep.data = FALSE,
    verbose = TRUE, n.cores = 6)
```

The *relative influence* of each predictive variable can be retrieved as below.

```
summary(sponge.gbm1, plotit=F)
```

```
                  var  rel.inf
easting       easting   34.211
tpi3             tpi3   22.258
northing     northing   16.411
bs34             bs34   10.242
var7             var7    6.779
bs11             bs11    6.421
entro7         entro7    3.679
```

In fact, `gbm` computes the relative influence of each variable for all trees in `sponge.gbm1`. To generate spatial predictions using a *GBM* model, the optimal number of boosting iterations trees needs to be determined. This can be done with the function `gbm.perf` in the `gbm` package. The function `gbm.step` in the `dismo` package (Hijmans et al. 2017) can also be used to find the optimal number of boosting trees based on k-fold cross-validation. We will use `gbm.perf` to demonstrate how to derive the optimal number. It requires two arguments:

(1) `object`, a gbm.object created from `gbm`; and

(2) `method`, the method used to estimate the optimal number of boosting iterations.

See `?gbm.perf` for details.

The function `gbm.perf` will be used to derive the optimal number of boosting iterations and plot the optimal number and performance measures (i.e., train error and validation error in terms of Poisson deviance versus the iteration number) (Figure 8.19) as below.

```
best.iter  <-  gbm.perf(sponge.gbm1, method = "cv")
```

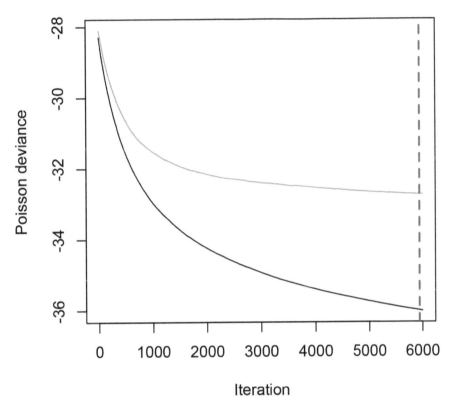

FIGURE 8.19: The optimal number of boosting iterations (blue dashed vertical line), the train error (solid black line), and validation error (solid green line) of *GBM* model versus the iteration number for sponge species richness.

```
best.iter
```

```
[1] 5943
```

The relative influence of predictive variables in the *GBM* model with the optimal number of boosting iterations are illustrated in Figure 8.20 by

```
sponge.gbm1.ri <- summary(sponge.gbm1, n.trees = best.iter, las = 2)
```

```
sponge.gbm1.ri
```

```
                 var rel.inf
easting      easting  34.246
tpi3            tpi3  22.275
northing    northing  16.374
bs34            bs34  10.227
var7            var7   6.775
bs11            bs11   6.418
entro7        entro7   3.685
```

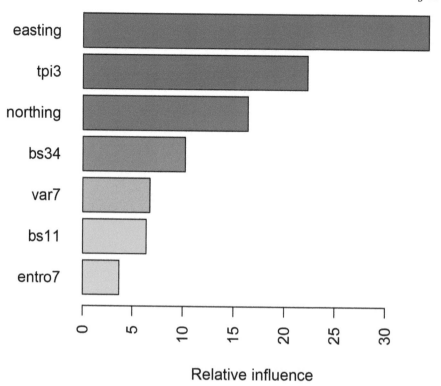

Relative influence

FIGURE 8.20: Relative influence of predictive variables in the *GBM* model with the optimal number of boosting iterations.

2. Application of the function gbm to percentage data

We will use the petrel point data set in the spm package to demonstrate the applications of *GBM* to percentage data as follows.

```
data(petrel)
gravel <- petrel[, c(1, 2, 5:9)]
nc <- ncol(gravel) - 1
nr <- nrow(gravel)

gravel.gbm1 <- gbm(gravel ~ ., data = gravel, var.monotone = rep(0, nc),
    distribution = "gaussian", n.trees = 5000, shrinkage = 0.001, interaction.
    depth = 2, bag.fraction = 0.5, n.minobsinnode = 10, weights = rep(1, nr), cv.
    folds=10, keep.data = TRUE, verbose = TRUE, n.cores = 6)
```

The relative influence of each predictive variable can be retrieved as below.

```
summary(gravel.gbm1, plotit = F)

            var rel.inf
dist       dist  26.855
bathy     bathy  25.644
lat         lat  17.284
long       long  16.160
relief   relief   8.419
slope     slope   5.637
```

3. Application of `gbm` to binomial data

We will use the `hard` point data set in the `spm` package to demonstrate the applications of *GBM* to binomial data as follows.

```
library(gbm)

data(hard)
hard1 <- hard[, -1]
hard1$hardness <- as.numeric(hard1$hardness) - 1 # convert `hardness` from
    categorical into binomial, with soft, 1; hard, 0.
nc <- ncol(hard1) - 1
nr <- nrow(hard1)

hard.gbm1 <- gbm(hardness ~ ., data = hard1, var.monotone = rep(0, nc),
    distribution = "bernoulli", n.trees = 5000, shrinkage = 0.001, interaction.
    depth = 2, bag.fraction = 0.5, n.minobsinnode = 10, weights = rep(1, nr), cv.
    folds=10, keep.data = TRUE, verbose = TRUE, n.cores = 6)
```

The relative influence of each predictive variable is

```
summary(hard.gbm1, plotit=F)
```

```
                       var  rel.inf
prock                prock  58.7496
bs                      bs  17.2963
relief              relief   6.2269
easting            easting   3.7045
surface            surface   3.0402
slope                slope   1.9493
variance          variance   1.7581
homogeneity    homogeneity   1.7057
bs.moran        bs.moran     1.2642
bathy                bathy   0.9516
northing          northing   0.8044
planar.curv    planar.curv   0.7059
tpi                    tpi   0.7020
bathy.moran    bathy.moran   0.6644
profile.curv  profile.curv   0.4772
```

8.3.2 Variable selection for *GBM*

We can use *variable selection* techniques to simplify *GBM* models. Several variable selection techniques can be used to search and select the most important predictors for *GBM*, including two functions in the `stepgbm` package (Li 2021b):

(1) `stepgbmRVI`,
(2) `stepgbm` with `method = "RVI"`,
(3) `stepgbm` with `method = "KIRVI"`, and
(4) `stepgbm` with `method = "KIRVI2"`.

The KIRVI method is from Li, Siwabessy, Huang, et al. (2019); and the KIRVI2 is a novel method and similar to the KIAVI2 method for *RF*, but AVI replaced with RVI.

The function `stepgbm` is based on and is an extension of `stepgbmRVI` by including an additional argument `method` that specifies a variable selection method for *GBM*. For `method`, it can be "RVI", "KIRVI" or "KIRVI2". If "RVI" is used, `stepgbm` is `stepgbmRVI`.

We will apply these variable selection methods to `sponge2` data to demonstrate their applications.

1. Implementation in `stepgbmRVI`

```
library(stepgbm)
data(sponge2)
set.seed(1234)

stepgbmRVI1 <- stepgbmRVI(trainx = sponge2[, -3], trainy = sponge2[, 3],
family = "poisson", cv.fold = 5, n.cores = 8, predacc = "VEcv", min.n.var = 2)
stepgbmRVI1
```

The predictive variables selected by `stepgbmRVI` as retrieved by the function `steprfAVIPredictors` are

```
stepgbmRVI1.res <- steprfAVIPredictors(stepgbmRVI1, trainx = sponge2[, -3])
stepgbmRVI1.res$variables.most.accurate
```

```
[1] "easting"    "sand"       "prof_cur_o" "tpi3"       "relief_5"
[6] "homo3"
```

The predictive accuracy is

```
stepgbmRVI1.res$max.predictive.accuracy
```

```
[1] 39.91
```

We can also derive the *PABVs* and *PARVs* by

```
stepgbmRVI1.res$PABV
```

```
 [1] "easting"    "northing"   "sand"       "bs18"       "bs19"
 [6] "bs21"       "bs22"       "bs27"       "bs30"       "bs31"
[11] "bs32"       "bs34"       "bs_homo_o"  "bathy_o"    "prof_cur_o"
[16] "rugosity_o" "rugosity3"  "tpi3"       "bathy_lmi7" "relief_3"
[21] "relief_5"   "slope3"     "slope7"     "prof_cur5"  "prof_cur7"
[26] "entro7"     "homo3"      "var5"
```

```
stepgbmRVI1.res$PARV
```

```
 [1] "bathy_lmi5" "bs24"       "bs33"       "bs16"       "bs14"
 [6] "bs20"       "bs13"       "bs11"       "bs29"       "bs17"
[11] "bs_lmi7"    "bs_lmi5"    "bs26"       "bs12"       "bs23"
[16] "slope5"     "bs15"       "var3"       "bs28"       "bs10"
[21] "prof_cur3"  "bs_lmi3"    "bs36"       "plan_curv7" "slope_o"
[26] "entro3"     "bs_o"       "plan_cur_o" "rugosity7"  "bs_entro_o"
[31] "bs25"       "bs_lmi_o"   "homo7"      "plan_curv5" "var7"
[36] "bs_var_o"   "homo5"      "mud"        "dist.coast" "tpi7"
[41] "entro5"     "bs35"       "tpi_o"      "rugosity5"  "bathy_lmi3"
[46] "bathy"      "relief_o"   "gravel"     "relief_7"   "plan_curv3"
[51] "tpi5"       "bathy_lmi_o"
```

The results show that of the 80 predictive variables, 28 variables were identified as *PABV*, and the remaining 52 variables are *PARV* for *GBM*. These identified *PABVs* provide a set of predictive variables to `stepgbm` for further variable selection when `method = "KIRVI"` or `method = "KIRVI2"` is used.

The predictive accuracy of each *GBM* model after a certain variable being removed consecutively are illustrated in Figure 8.21 by

```
library(reshape2)

pa1 <- as.data.frame(stepgbmRVI1$predictive.accuracy2)
names(pa1) <- stepgbmRVI1$variable.removed
pa2 <- melt(pa1, id = NULL)
```

```
names(pa2) <- c("Variable", "VEcv")

library(graphics)

par(font.axis = 2, font.lab = 2)
with(pa2, boxplot(VEcv ~ Variable, ylab = "VEcv (%)", xlab = NULL, cex.axis = 0.5,
    las = 2))
```

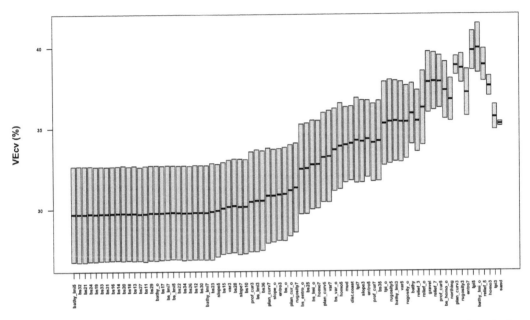

FIGURE 8.21: Predictive accuracy vs. variable removed: each variable removed with the predictive accuracy of its corresponding *GBM* model.

The contribution of each variable removed to the predictive accuracy is depicted in Figure 8.22 by

```
par(font.axis = 2, font.lab = 2)
barplot(stepgbmRVI1$delta.accuracy, col = (1:length(stepgbmRVI1$variable.removed))
    ,
names.arg = stepgbmRVI1$variable.removed,
font.main = 4, cex.names = 0.6, font = 2, las = 2, ylab = "Increase rate in VEcv
    (%)")
```

2. Implementation in `stepgbm` with `method = "RVI"`

```
set.seed(1234)

stepgbm1 <- stepgbm(trainx = sponge2[, -3], trainy = sponge2[, 3],
method = "RVI", family = "poisson", rpt = 2, cv.fold = 5, predacc = "VEcv", min.n.
    var = 2, n.cores = 8, delta.predacc = 0.01)
```

The `stepgbm1` produces the same results as `stepgbmRVI1`, but the results have been processed via `steprfAVIPredictors` within the function `stepgbm` and are ready for use without further application of `steprfAVIPredictors`. We will not repeat it here. It should be noted that `steprfAVIPredictors` was developed for *RF* models, but it can be equally used for *GBM* model.

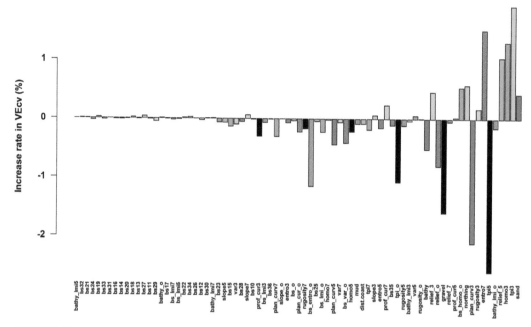

FIGURE 8.22: The contribution of each variable removed to predictive accuracy.

3. Implementation in `stepgbm` with `method = "KIRVI"`

```
data(sponge2)
set.seed(1234)

stepgbm2 <- stepgbm(trainx = sponge2[, -3], trainy = sponge2[, 3],
  method = "KIRVI", family = "poisson", rpt = 2, cv.fold = 5, predacc = "VEcv", min.
    n.var = 2, n.cores = 8, delta.predacc = 0.01)
```

The outputs of `stepgbm2` are

```
names(stepgbm2)
```

```
[1] "stepgbmPredictorsFinal"   "max.predictive.accuracy"
[3] "numberruns"               "laststepRVI"
[5] "stepgbmRVIOutputsAll"     "stepgbmPredictorsAll"
[7] "KIRVIPredictorsAll"
```

The `stepgbm2$stepgbmPredictorsFinal` shows the variables selected for the last *GBM* model and associated results, but the final model may not be necessarily the most accurate model that needs to be confirmed using `stepgbm2$max.predictive.accuracy`. For `stepgbm2`, `stepgbmRVI` was run three times (see `stepgbm2$numberruns`) and three accuracy values are produced as below.

```
stepgbm2$max.predictive.accuracy
```

```
[1] 39.91 42.68 42.68
```

As expected, the first value is the same as that for `stepgbmRVI1` above. The last two values are the same and higher than the first one. This suggests that the last two *GBM* models are the same and are the most accurate *GBM* models. They should be used for further modeling. The predictive variables selected are

```
stepgbm2$stepgbmPredictorsAll[[2]]$variables.most.accurate
```

```
[1] "easting"    "sand"       "prof_cur_o" "tpi3"       "relief_5"
[6] "homo3"      "bathy_lmi7" "relief_3"   "entro7"
```

Since the last two models are the same, `stepgbm2$stepgbmPredictorsAll[[3]]$variables.most.accurate` would produce the same list of predictive variables as above.

4. Implementation in stepgbm with method = "KIRVI2"

```
data(sponge2)
set.seed(1234)
```

```
stepgbm3 <- stepgbm(trainx = sponge2[, -3], trainy = sponge2[, 3],
method = "KIRVI2", family = "poisson", rpt = 2, cv.fold = 5, predacc = "VEcv", min
    .n.var = 2, n.cores = 8, delta.predacc = 0.01)
```

The outputs of `stepgbm3` are

```
names(stepgbm3)
```

```
[1] "stepgbmPredictorsFinal"  "max.predictive.accuracy"
[3] "numberruns"              "laststepRVI"
[5] "stepgbmRVIOutputsAll"    "stepgbmPredictorsAll"
[7] "KIRVIPredictorsAll"
```

```
stepgbm3$max.predictive.accuracy
```

```
[1] 39.91 42.65 42.01
```

The results show that the second *GBM* model selected via KIRVI2 is of the highest accuracy. It should be used for further modeling. The predictive variables in the second *GBM* model are

```
stepgbm3$stepgbmPredictorsAll[[2]]$variables.most.accurate
```

```
[1] "easting"    "sand"       "prof_cur_o" "tpi3"       "bathy_lmi7"
[6] "relief_3"   "relief_5"   "entro7"     "homo3"
```

However, the model selected with KIRVI is slightly more accurate than this one and will be used for further modeling instead, although the predictive variables selected are the same by both KIRVI and KIRVI2. The difference in their predictive accuracy may be due to the difference in the order of the predictive variables in the *GBM* models resulted.

8.3.3 Parameter optimization for *GBM* models

For the function gbm, values of relevant arguments, such as, `interaction.depth`, `train.fraction` and `n.minobsinnode`, need to be optimized. We will use the function gbmcv in the spm package to demonstrate the optimization. Prior to parameter optimization, we will introduce gbmcv.

For gbmcv, the following arguments may need to be specified:
(1) `trainx`, a dataframe or matrix contains the columns of predictive variables;
(2) `trainy`, a vector of response variable;
(3) `var.monotone`, family (i.e., distribution in gbm), n.trees, learning.rate (i.e., shrinkage in gbm), `interaction.depth`, `bag.fraction`, `train.fraction`, `n.minobsinnode`, `cv.fold`, and `n.cores`, which are the same as those for gbm;
(4) `predacc`, can be either "VEcv" for vecv or "ALL" for all measures in function pred.acc.

1. Optimization of `interaction.depth` and `train.fraction`

We can estimate the optimal `interaction.depth` and `train.fraction` for a *GBM* model, which is demonstrated based on the above `stepgbm2` model for `sponge2` data set below.

```
library(spm)
```

```
interact.depth <- c(1:10)
train.fract <- c(5:10) * 0.1
gbmopt <- matrix(0, length(interact.depth), length(train.fract))

for (i in 1:length(interact.depth)) {
  for (j in 1:length(train.fract)) {
    set.seed(1234)
    gbmcv1 <- gbmcv(trainx = sponge2[, stepgbm2$stepgbmPredictorsAll[[2]]$
        variables.most.accurate], trainy = sponge2[, 3], family = "poisson",
        learning.rate = 0.001, interaction.depth = interact.depth[i], train.
        fraction = train.fract[j], n.minobsinnode = 5, cv.fold = 5, predacc = "
        VEcv", n.cores = 8)
    gbmopt[i, j] <- gbmcv1
  }
}
```

Given the sample size of the `sponge2` data set, we reduced `n.minobsinnode` from 10, the default value, to 5 in order to run a small `train.fraction` without errors.

The optimal values for `interact.depth` and `train.fract` are

```
gbm.opt <- which (gbmopt == max(gbmopt), arr.ind = T)
interact.depth[gbm.opt[, 1]]
```

```
[1] 3
```

```
train.fract[gbm.opt[, 2]]
```

```
[1] 1
```

2. Optimization of `n.minobsinnode` and `interaction.depth`

Given that `train.fraction = 1` is estimated, we can use it to estimate the optimal `n.minobsinnode` together with `interaction.depth` based on `stepgbm2` for the `sponge2` data set by

```
interaction.depth <- c(1:5)
n.minobsinnode <- c(1:10)
gbmopt2 <- matrix(0, length(interaction.depth), length(n.minobsinnode))

for (i in 1:length(interaction.depth)) {
  for (j in 1:length(n.minobsinnode)) {
    set.seed(1234)
    gbmcv1 <- gbmcv(trainx = sponge2[, stepgbm2$stepgbmPredictorsAll[[2]]$
        variables.most.accurate], trainy = sponge2[, 3], family = "poisson",
        learning.rate = 0.001, interaction.depth = interaction.depth[i], train.
        fraction = 1, n.minobsinnode = n.minobsinnode[j], cv.fold = 5, predacc = "
        VEcv", n.cores = 8)
    gbmopt2[i, j] <- gbmcv1
  }
}
```

The optimal values for `interaction.depth` and `n.minobsinnode` are

```
gbm.opt2 <- which (gbmopt2 == max(gbmopt2), arr.ind = T)
interaction.depth[gbm.opt2[, 1]]
```

```
[1] 2 3 4 5
```

```
n.minobsinnode[gbm.opt2[, 2]]
```

```
[1] 8 8 8 8
```

The results show that `interaction.depth` = c(2:5) and `n.minobsinnode` = 8 were estimated for *GBM*.

8.3.4 Predictive accuracy of *GBM*

We can use `gbmcv` to produce a stabilized *predictive accuracy* of the *GBM* models based on the predictive variables selected and optimal parameters. We will use `stepgbm2` and the parameters estimated to demonstrate the stabilization of the predictive accuracy of a *GBM* model below.

```
set.seed(1234)
n <- 100
gbmvecv1 <- NULL

for (i in 1:n) {
  gbmcv1 <- gbmcv(trainx = sponge2[, stepgbm2$stepgbmPredictorsAll[[2]]$variables.
      most.accurate], trainy = sponge2[, 3], family = "poisson", learning.rate =
      0.001, interaction.depth = 5, train.fraction = 1, n.minobsinnode = 8, cv.
      fold = 5, predacc = "VEcv", n.cores = 8)
  gbmvecv1[i] <- gbmcv1
}
```

The median and range of predictive accuracy of the optimal *GBM* model based on 100 repetitions of 10-fold cross-validation using `gbmcv` are

```
median(gbmvecv1)
```

```
[1] 37.77
```

```
range(gbmvecv1)
```

```
[1] 30.87 45.43
```

8.3.5 Partial dependence plots for *GBM*

We can use `stepgbm2` above to depict the relationships of sponge species richness with nine predictive variables.

```
sponge2x <- sponge2[, stepgbm2$stepgbmPredictorsAll[[2]]$variables.most.accurate]
sponge2y <- sponge2[, 3]
set.seed(1234)

sponge.stepgbm2 <- gbm(sponge2y ~ ., distribution = "poisson", data = sponge2x, n.
    trees = 3000, interaction.depth = 5, n.minobsinnode = 8, shrinkage = 0.001,
    bag.fraction = 0.5, cv.fold = 10, keep.data = FALSE, verbose = TRUE, n.cores =
    8)

best.iter.stepgbm2 <- gbm.perf(sponge.stepgbm2, method = "cv", plot.it = FALSE)
```

The relationships of response variable and each of nine predictive variables used in the optimal *GBM* model can now be visualized by

```
par(mfrow = c(2, 5), font.axis = 2, font.lab = 2)
plot(sponge.stepgbm2, i.var = 1, n.trees = best.iter.stepgbm2)
```

```
plot(sponge.stepgbm2, i.var = 2, n.trees = best.iter.stepgbm2)
plot(sponge.stepgbm2, i.var = 3, n.trees = best.iter.stepgbm2)
plot(sponge.stepgbm2, i.var = 4, n.trees = best.iter.stepgbm2)
plot(sponge.stepgbm2, i.var = 5, n.trees = best.iter.stepgbm2)
plot(sponge.stepgbm2, i.var = 6, n.trees = best.iter.stepgbm2)
plot(sponge.stepgbm2, i.var = 7, n.trees = best.iter.stepgbm2)
plot(sponge.stepgbm2, i.var = 8, n.trees = best.iter.stepgbm2)
plot(sponge.stepgbm2, i.var = 9, n.trees = best.iter.stepgbm2)
```

The resulted plot is similar to those presented in Figure 8.18 and will not be displayed.

The relationships of response variable and two most influencing predictive variables used in the optimal *GBM* model can also be visualized in Figure (8.23) by

```
par(font.axis = 2, font.lab = 2)
plot(sponge.stepgbm2, i.var = 1:2, n.trees = best.iter.stepgbm2)
```

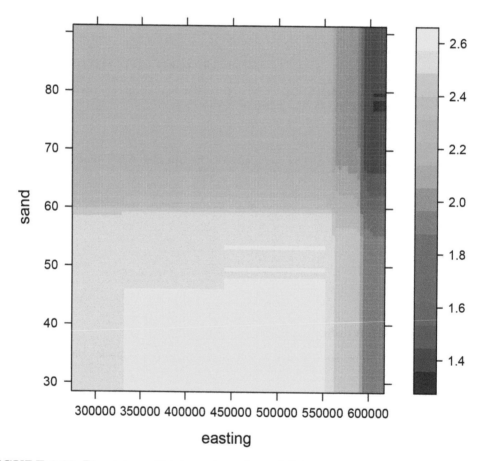

FIGURE 8.23: Bivariate partial dependent plots of *GBM* model for predicting the spatial distribution of sponge species richness, illustrating the relationships of the sponge species richness with two predictors (i.e., easting and sand) in the *GBM* model.

8.3.6   Predictions of *GBM*

For demonstration purposes, we will use `sponge.gbm1` with the above identified optimal number of boosting iterations `best.iter` to generate spatial predictions for areas based on `sponge`

.grid data set. We are unable to use sponge.stepgbm2 in this demonstration because the sponge.grid data set does not contain relevant variables used in sponge.stepgbm2.

```
pred.gbm1 <- predict.gbm(sponge.gbm1, sponge.grid, n.trees = best.iter, type = "
    response")
```

```
range(pred.gbm1)
```

```
[1]  9.947 52.613
```

The predictions of *GBM* can also be generated with the function gbmpred in the spm package based on the predictive variables in sponge.gbm1 as below.

```
pred.gbm2 <- gbmpred(sponge[, -c(3)], sponge[, 3], sponge.grid[, c(1:2)], sponge.
    grid, family = "poisson", n.cores = 8)
```

The spatial distribution of predictions of *GBM* is illustrated in Figure 8.24 by

```
library(sp)
```

```
sponge.pred.gbm <- cbind(sponge.grid[, c(1,2)], pred.gbm1)
gridded(sponge.pred.gbm) = ~ easting + northing
names(sponge.pred.gbm) <- "predictions"
```

```
library(raster)
```

```
par(font.axis = 2, font.lab = 2)
plot(brick(sponge.pred.gbm))
```

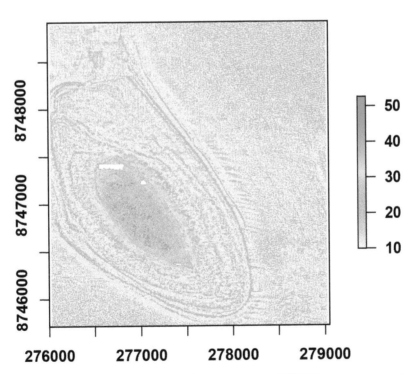

FIGURE 8.24: Spatial patterns of predictions using *GBM* for sponge species richness data.

9

Support vector machines

This chapter introduces *support vector machines* (*SVM*). It is a kind of machine learning method and also is one of *gradient based/detrended spatial predictive methods* that can use predictive variables. It is available in *R*.

SVM constructs a $(p-1)-$dimensional *hyperplane* or a set of $(p-1)-$dimensional *hyperplanes* in a *p*- or higher-dimensional space for a response variable with p predictive variables (i.e., x_1, x_2, ..., x_p, in a $p-$dimensional space) and n observations, which can be used for classification (Cortes and Vapnik 1995) and regression (Drucker et al. 1996). A binary (or two classes) response variable with two predictive variables (x_1 and x_2) can be used as an example to introduce *SVM* (Figure 9.1) (James et al. 2017). In the two-dimensional space of x_1 and x_2, a hyperplane is a flat one-dimensional subspace, i.e., a line (Figure 9.1(a)).

In a $p-$dimensional space, a *hyperplane* is defined in Equation (9.1).

$$f(x) = \beta_0 + \sum_{i=1}^{p} \beta_i x_i \tag{9.1}$$

where x is an observation with p predictive variables; and β_i are coefficients.

For a hyperplane, $f(x) = 0$, x is on the hyperplane. If $f(x) > 0$, then x lies to one side of the hyperplane. On the other hand, if $f(x) < 0$, then x lies on the other side of the hyperplane. So the hyperplane is also called a *separating hyperplane*. Since this $f(x)$ is a linear combination of x_i, the hyperplane is a linear hyperplane for a linear boundary between observations (Figure 9.1(a)).

To construct an SVM based upon a *separating hyperplane*, we need to select an *optimal separating hyperplane* (Figure 9.1(b)). The optimal hyperplane is chosen to maximize its distance to the nearest observations on each side, that is, the Euclidean distance from it to the nearest observations on each side is largest. The minimal distance from the observations to the hyperplane is known as the *margin*. This *optimal separating hyperplane* is known as the *maximum margin hyperplane*. The observations that lie directly on the margin are known as *support vectors*. In short, an *SVM* finds the hyperplane that is oriented so that the margin between the support vectors is maximized.

If non-separable observations exist, a non-negative *tuning parameter* λ is introduced to consider non-separable observations (Figure 9.1(c)). If $\lambda = 0$, then no observation violates the margin; and if $\lambda > 0$, no more than λ observations can be on the wrong side of the hyperplane. In practice, λ is generally chosen via cross-validation. These observations are also known as *support vectors*.

For possibly non-linear boundaries between observations, $f(x)$ can consider quadratic, cubic, higher-order polynomial functions, even interaction terms of the predictors, or certain *kernels* to accommodate the non-linear boundaries (Figure 9.1(d)).

DOI: 10.1201/9781003091776-9

SVM was initially developed for classification of categorical data (Cortes and Vapnik 1995) and then extended for regression of numerical data (Drucker et al. 1996).

The predictions of *SVM* can be represented in Equation (9.2).

$$\widehat{y_{x_0}} = \widehat{f(x_0)}$$

$$= \beta_0 + \sum_{i=1}^{s} \alpha_i K(x_0, x_i) \tag{9.2}$$

where $\widehat{y_{x_0}}$ is the predicted value of a response variable y at the location of interest (0), x_0 is p predictive variables at location 0 (i.e., $x_0 = (x_{01}, x_{02}, ..., x_{0p})$), s is the number of support vectors derived from n observations, x_i is p predictive variables at location i (i.e., $x_i = (x_{i1}, x_{i2}, ..., x_{ip})$), $K(x_0, x_i)$ is a *kernel* at 0. Here $\widehat{y_{x_0}}$ is equivalent to $\hat{y}(x_0)$ in Equation (4.1).

If $K(x_0, x_i) = \sum_{j=1}^{p} x_{0j} x_{ij}$, it is a *linear kernel*, and the prediction resulted is equivalent to $\widehat{f(x_0)} = \widehat{\beta_0} + \widehat{\beta_1} x_1 + \widehat{\beta_2} x_2 + ... + \widehat{\beta_p} x_p$ (James et al. 2017). $K(x_0, x_i)$ can also be other *kernels*, such as *polynomial kernel*, *radial kernel*, and *sigmoid* (or *neural network*) (Hastie, Tibshirani, and Friedman 2009; James et al. 2017; Meyer et al. 2019).

For multiclass classification with k classes ($k > 2$), $k(k-1)/2$ *SVMs* are trained for 'one-versus-one' or all-pairs approach, with each *SVM* comparing a pair of classes, and the prediction is the most frequently assigned class by these *SVMs* (James et al. 2017; Meyer et al. 2019). For 'one-versus-all' approach, k *SVMs* are trained, each time comparing one class to the remaining $k-1$ classes, and the prediction is the class with the largest $\widehat{y_{x_0}}$ (James et al. 2017).

For *support vector regression (SVR)* (i.e., *SVM* for regression), only observations with residuals larger in absolute value than some positive constant (i.e., ϵ, insensitive loss) that is equivalent to the margin of *SVM* for classification are considered as *support vectors* (Drucker et al. 1996).

The functions `svm` and `tune.svm` in the `e1071` package (Meyer et al. 2019) in R can be used for *SVM* modeling. The `LiblineaR` package (Helleputte, Gramme, and Paul 2017) is an alternative option, which is useful for very large linear problems (James et al. 2017); and a further option is the `kernlab` package (Karatzoglou, Smola, and Hornik 2019). The function `svmcv` in the `spm2` package (Li 2021a) is a cross-validation function based on the function `svm` regression and can be used to demonstrate the parameter estimation and accuracy assessment of *SVM* predictive models. In this chapter, we will use `svm`, `tune.svm`, and `svmcv` to demonstrate the implementation of *SVM*.

9.1   Application of *SVM*

The predictions of *SVM* can be generated with the function `svm`. The descriptions of relevant arguments of `svm` are detailed in its help files. For `svm`, the following arguments may need to be specified:

(1) `formula`, a formula defining response variable and predictive variables;
(2) `data`, a dataframe containing the variables in the formula;
(3) `scale`, a logical vector indicating the variables to be scaled;

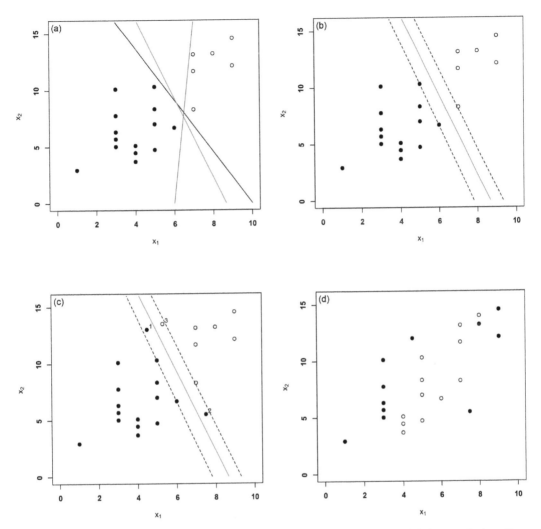

FIGURE 9.1: For observations in two classes (one class in black solid dots and the other in black circles) along two predictive variables (x_1 and x_2): (a) three separating hyperplanes, out of numerous possibles, are shown in black, blue and red; (b) maximal margin hyperplane in a solid red line, with the margin that is the distance from the solid line to either of the dashed black lines; and the two black dots and one black circle that lie on the dashed lines are the support vectors; (c) support vector classifier with a hyperplane (solid red line) and the margins (dashed black lines); and all observations in black dots, besides two support vectors, are on the correct side of the margin, except observation 1 that is on the wrong side of the margin and observation 2 that is on the wrong side of the hyperplane and the wrong side of the margin; and all observations in black circles, besides one support vector, are on the correct side of the margin, except for the observation 3 that is on the wrong side of the margin; and (d) observations in two classes, with a non-linear boundary between them; they cannot be classified using the support vector classifier with a linear boundary but can be classfied using SVM with a certain kernel for a non-linear boundary (modified from (James et al. 2017)).

(4) `type`, the default setting for classification is `C-classification` and for regression is `eps-regression`;

(5) `kernel`, the `kernel` used in training and predicting, including `linear`, `polynomial`, `radial basis` and `sigmoid`;

(6) `degree`, parameter needed for kernel `polynomial` (default is 3);

(7) `gamma`, parameter needed for all `kernels` except `linear` (default is 1/(data dimension));

(8) `cost`, cost of constraints violation (default is 1); and the higher the cost, the narrower the margins, and vice versa;

(9) `nu`, a parameter needed for `nu-classification`, `nu-regression`, and `one-classification` (default is 0.5);

(10) `tolerance`, tolerance of termination criterion (default is 0.001);

(11) `epsilon`, for the insensitive-loss function (default is 0.1);

(12) `cross`, if a integer value k > 0 is specified, a k-fold cross-validation on the training data is performed to assess the quality of the model, with accuracy rate for classification and *MSE* for regression.

The `hard` data set in the `spm` package will be used to show the application of *SVM* to categorical data. The `sponge2` point data set in the `spm2` package and the `sponge.grid` grid data set in `spm` will be used to show the application of *SVM* to numerical data.

1. Application of `svm` to categorical data

To apply `svm`, relevant arguments (e.g., `cost` and `gamma`) need to be specified. For this demonstration, the default values are used for the arguments.

```
svm.h1 <- svm(hardness ~ ., data = hard[, -1])
```

The details of the *SVM* model `svm.h1` are

```
svm.h1
```

```
Call:
svm(formula = hardness ~ ., data = hard[, -1])


Parameters:
   SVM-Type:  C-classification
 SVM-Kernel:  radial
       cost:  1

Number of Support Vectors:  66
```

In this application, the default setting, "C-classification", was used for `type`; `radial` used for `Kernel` and 1 used for `cost`. The optimal `type`, `Kernel`, and `cost` can be determined according to their influences on the predictive accuracy as shown in Section 9.3.

The fitted values by `svm.h1` can be retrieved using `svm.h1$fitted` and can also be produced via the function `predict`.

```
svm.h1.pred <- predict(svm.h1, hard[, -c(1, 17)])
```

```
identical(svm.h1$fitted, svm.h1.pred)
```

```
[1] TRUE
```

The fitted values by `svm.h1` are compared with the observed values below.

```
table(hard$hardness, svm.h1$fitted)
```

```
          hard soft
   hard   27   11
   soft    0   99
```

The fitted values of `svm.h1` are exactly the same as the observations for `soft` class, but 11 out of 38 `hard` class are incorrectly classified.

2. Application of `svm` to numerical data

```
svm.sponge1 <- svm(species.richness ~ ., data = sponge2[, -4])
```

The details of the *SVM* model `svm.sponge1` are

```
svm.sponge1
```

```
Call:
svm(formula = species.richness ~ ., data = sponge2[, -4])


Parameters:
   SVM-Type:  eps-regression
 SVM-Kernel:  radial
       cost:  1
      gamma:  0.01266
    epsilon:  0.1


Number of Support Vectors:  66
```

In this application, the default setting, "eps-regression", for `type` was used; `radial` used for `Kernel`, 1 used for `cost`; 0.01266 used for `gamma` and 0.1 used for `epsilon`. These arguments can be optimized based on their influences on the predictive accuracy as shown in Section 9.3.

The fitted values by `svm.sponge1` can be retrieved using `svm.sponge1$fitted` and can also be produced via `predict` as below.

```
svm.sponge1.pred <- predict(svm.sponge1, sponge2[, -4])

identical(svm.sponge1$fitted, svm.sponge1.pred)
```

```
[1] TRUE
```

The relationship between fitted values of `svm.sponge1` and observed values is depicted in Figure 9.2 by

```
par(font.axis = 2, font.lab = 2)
plot(sponge2$species.richness, svm.sponge1.pred, xlab = "Observed values", ylab =
    "Fitted values")
lines(sponge2$species.richness, sponge2$species.richness, col = "blue")
```

The ranges of fitted values and observations are

```
range(svm.sponge1.pred)
```

```
[1] -0.05481 22.26907
```

```
range(sponge2$species.richness)
```

```
[1]  1 39
```

The fitted values of `svm.sponge1` are shrunk considerably in comparison with the observations in terms of their ranges, and negative fitted values are produced for the count data.

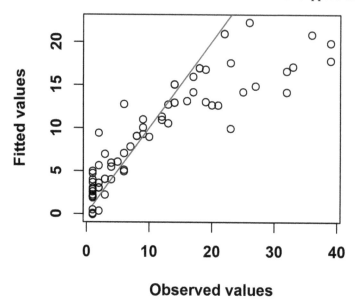

FIGURE 9.2: Observed values vs. fitted values by SVM.

9.2 Variable selection for *SVM*

A few *variable selection* methods are available for *SVM*:
(1) function rfe in the caret package,
(2) function sigFeature in the sigFeature package (Das 2020), and
(3) function svmfs in the penalizedSVM package (Becker, Werft, and Benner 2018).

Both svmfs and sigFeature are for *SVM* classification. Although rfe seems also for *SVM* classification, its authors have applied it to numerical data.

We will use rfe to search and select the most important predictive variables for *SVM* classification and regression models.

1. Implementation in rfe for categorical data

```
library(caret)
set.seed(1234)

svmrfe.hard <- rfe(hard[, -c(1, 17)], hard[, 17], sizes = c(4:10, 15, 20, 30),
    metric = "Kappa", rfeControl = rfeControl(functions = caretFuncs, method = "
    repeatedcv", repeats = 5), method = "svmRadial", tuneLength = 12, trControl =
    trainControl(method = "cv"), verbose = FALSE)

svmrfe.hard


Recursive feature selection

Outer resampling method: Cross-Validated (10 fold, repeated 5 times)

Resampling performance over subset size:

 Variables Accuracy Kappa AccuracySD KappaSD Selected
```

```
         4   0.892 0.694      0.0763   0.234
         5   0.907 0.734      0.0670   0.211        *
         6   0.895 0.700      0.0614   0.218
         7   0.893 0.691      0.0763   0.247
         8   0.893 0.685      0.0805   0.260
         9   0.891 0.677      0.0672   0.223
        10   0.873 0.630      0.0738   0.246
        15   0.862 0.602      0.0637   0.208
```

```
The top 5 variables (out of 5):
   bs, prock, bathy, bs.moran, bathy.moran
```

```
svmrfe.hard$bestSubset
```

```
[1] 5
```

```
svmrfe.hard$optsize
```

```
[1] 5
```

```
svmrfe.hard$optVariables
```

```
[1] "bs"           "prock"       "bathy"        "bs.moran"    "bathy.moran"
```

The results show that five variables are selected by rfe: bs, prock, bathy, bs.moran, and bathy
.moran.

2. Implementation in rfe for numerical data

The rfe seems able to deal with numerical data via svmRadial method.

```
library(caret)
set.seed(1234)
```

```
svmrfe.sponge <- rfe(sponge2[, -c(3:4)], sponge2[, 3], sizes = c(4:10, 15, 20, 30)
    , metric = "RMSE", rfeControl = rfeControl(functions = caretFuncs, method = "
    repeatedcv", repeats = 5), method = "svmRadial", tuneLength = 12, trControl =
    trainControl(method = "cv"), verbose = FALSE)
```

```
svmrfe.sponge
```

```
Recursive feature selection
```

```
Outer resampling method: Cross-Validated (10 fold, repeated 5 times)
```

```
Resampling performance over subset size:
```

Variables	RMSE	Rsquared	MAE	RMSESD	RsquaredSD	MAESD	Selected
4	9.80	0.235	6.98	2.69	0.239	1.88	
5	9.83	0.242	7.03	2.72	0.259	1.97	
6	9.96	0.233	7.16	2.73	0.238	1.91	
7	9.82	0.236	7.21	2.76	0.250	1.95	
8	9.93	0.237	7.19	2.56	0.224	1.85	
9	9.91	0.264	7.23	2.60	0.244	1.86	
10	9.67	0.278	7.05	2.86	0.260	2.15	
15	9.65	0.266	7.01	2.69	0.241	1.87	
20	9.31	0.289	6.83	2.58	0.227	1.85	
30	8.67	0.381	6.39	2.54	0.236	1.90	
79	7.88	0.511	6.18	2.18	0.254	1.81	*

```
The top 5 variables (out of 79):
   easting, tpi5, prof_cur5, sand, tpi3
```

```
svmrfe.sponge$bestSubset

[1] 79

svmrfe.sponge$optsize

[1] 79

svmrfe.sponge$optVariables
```

```
 [1] "easting"     "tpi5"        "prof_cur5"   "sand"        "tpi3"
 [6] "relief_7"    "relief_5"    "slope3"      "northing"    "relief_3"
[11] "tpi_o"       "rugosity3"   "slope_o"     "slope5"      "tpi7"
[16] "bathy_lmi7"  "gravel"      "bathy_lmi5"  "bathy_lmi3"  "slope7"
[21] "rugosity5"   "bathy_lmi_o" "relief_o"    "prof_cur7"   "bs_var_o"
[26] "rugosity7"   "prof_cur_o"  "dist.coast"  "bathy"       "bathy_o"
[31] "prof_cur3"   "homo3"       "plan_curv3"  "bs_lmi3"     "bs_lmi_o"
[36] "bs_lmi5"     "bs_lmi7"     "bs_homo_o"   "plan_curv5"  "entro3"
[41] "homo5"       "rugosity_o"  "entro5"      "homo7"       "bs22"
[46] "bs21"        "plan_cur_o"  "entro7"      "bs19"        "bs20"
[51] "bs_entro_o"  "bs23"        "bs26"        "bs18"        "bs25"
[56] "bs27"        "bs10"        "bs24"        "bs17"        "bs_o"
[61] "var5"        "bs16"        "plan_curv7"  "bs15"        "bs35"
[66] "bs28"        "bs34"        "var7"        "bs14"        "bs13"
[71] "bs11"        "bs36"        "bs29"        "bs33"        "bs32"
[76] "bs12"        "bs31"        "bs30"        "var3"
```

The results suggest that all 79 variables are selected by rfe, with the following top five predictive variables: easting, tpi5, prof_cur5, sand, and tpi3.

9.3 Parameter optimization for *SVM* models

For svm, several arguments may need to be optimized. We will use three important arguments, type, cost, and gamma to demonstrate the *parameter optimization* for *SVM*. The type argument can be selected by either svm or svmcv, and the remaining two arguments can be determined by either tune.svm or svmcv, based on their effects on predictive accuracy. The function tune .svm will be applied to categorical data and svmcv will be used for numerical data.

The arguments of tune.svm are largely the same as those for svm as shown in the examples below.

For svmcv, the following arguments may need to be specified:
(1) formula, scale, type, kernel, degree, gamma, cost, nu, tolerance, and epsilon, which are the same as those for svm;
(2) trainxy, a dataframe contains longitude (long), latitude (lat), predictive variables and response variable of point samples;
(3) y, a vector of response variable in the formula, that is, the left part of the formula;
(4) validation, validation methods, including "LOO" and "CV";
(5) cv.fold, number of folds in the cross-validation; and
(6) predacc, can be either "VEcv" for vecv or "ALL" for all measures; and for classification, always use "All".

1. Selection of type

We will use functions svm and svmcv to demonstrate the parameter selection for type.

Categorical data

For categorical data, two options, "C-classification" and "nu-classification", are available for the argument `type` and for selection as below.

```
set.seed(1234)

svm.hard.type1 <- svm(hardness ~ ., data = hard[, -1], type = "C-classification",
    cross = 10)

svm.hard.type1$tot.accuracy

[1] 89.05

set.seed(1234)

svm.hard.type2 <- svm(hardness ~ ., data = hard[, -1], type = "nu-classification",
    cross = 10)

svm.hard.type2$tot.accuracy

[1] 81.75
```

It shows that the default setting "C-classification" results in a more accurate *SVM* model.

Numerical data

For numerical data, two options, "eps-regression" and "nu-regression", can be selected from for the argument `type` as below.

```
type <- c("eps-regression", "nu-regression")
typeopt <- NULL

for (i in 1:length(type)) {
  set.seed(1234)
  svmcv1 <- svmcv(species.richness ~ ., trainxy =  sponge2[, -4], y = sponge2[,
     3], type = type[i], validation = "CV", predacc = "VEcv")
  typeopt[i] <- svmcv1
}

typeopt

[1] 35.50 34.02
```

It shows that the `type` with "eps-regression" resulted in a slightly more accurate *SVM* regression model.

2. Estimation of `cost` and `gamma`

We will use `tune.svm` for categorical data, and `tune.svm` and `svmcv` for numerical data.

Categorical data

For the `hard` point data set, we assume the optimal `cost` lies within `c(1:10)*0.25` and the optimal `gamma` is within `c(1:10)*0.1`. We use 10-fold cross-validation to estimate the optimal parameters.

```
library(e1071)

svm.tune.hard <- tune.svm(hardness ~ ., data = hard[, c(svmrfe.hard$optVariables,
    "hardness")], cross = 10, cost = c(1:10)*0.25, gamma = c(1:10)*0.1)
```

```
svm.tune.hard


Parameter tuning of 'svm':

- sampling method: 10-fold cross validation

- best parameters:
 gamma cost
   0.9    1

- best performance: 0.08187

names(svm.tune.hard)

[1] "best.parameters"    "best.performance"  "method"           "nparcomb"
[5] "train.ind"          "sampling"          "performances"     "best.model"
```

The parameters are plotted in Figure 9.3 by

```
par(font.axis=2, font.lab=2)
plot(svm.tune.hard, cex = 0.5, main = "Classification error")
```

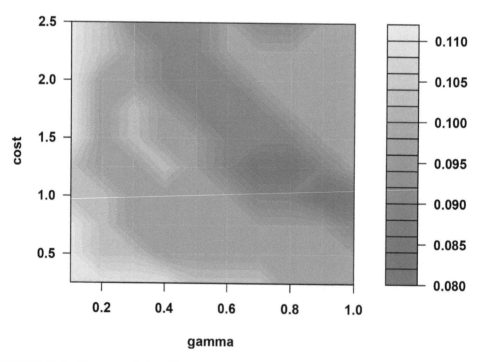

FIGURE 9.3: Changes of classification error with 'cost' and 'gamma' for hardness using SVM.

The results show that the optimal value of gamma is 0.9, and the optimal value of cost is 1. These parameters should be used for svm.

```
svm.tune.h1 <- svm(hardness ~ ., data = hard[, c(svmrfe.hard$optVariables, "
    hardness")], cost = svm.tune.hard$best.parameters$cost, gamma = svm.tune.hard$
    best.parameters$gamma)
```

The fitted values by `svm.tune.h1` are compared with the observed values below.

```
table(hard$hardness, svm.tune.h1$fitted)
```

```
      hard soft
hard   31    7
soft    0   99
```

The results show that the fitted values of `svm.tune.h1` are exactly the same as the observations for `soft` class, and 7 out of 38 `hard` class are incorrectly classified.

Numerical data

Function `tune.svm`

In a similar way as for `hard` data, we will estimate optimal values of `cost` and `gamma` using `tune.svm` with 10-fold cross-validation for `species.richness` in `sponge2` data.

```
svm.tune.sponge <- tune.svm(species.richness ~ ., data = sponge2[, -4], cross =
    10, cost = c(1:10)*0.5, gamma = c(1:20)*0.001)
```

```
svm.tune.sponge
```

```
Parameter tuning of 'svm':

- sampling method: 10-fold cross validation

- best parameters:
 gamma cost
  0.01  3.5

- best performance: 63.46
```

These parameters are plotted in Figure 9.4 by

```
par(font.axis=2, font.lab=2)
plot(svm.tune.sponge, cex = 0.5, main = "MSE")
```

The results show that the optimal value of `gamma` is 0.01, and the optimal value of `cost` is 3.5. These parameters should be used for `svm`.

```
svm.tune.sponge1 <- svm(species.richness ~ ., data = sponge2[, -4], cost = svm.
    tune.sponge$best.parameters$cost, gamma = svm.tune.sponge$best.parameters$
    gamma)
```

The fitted values by `svm.tune.h1` are compared with the observed values by

```
par(font.axis = 2, font.lab = 2)
plot(sponge2$species.richness, svm.tune.sponge1$fitted, xlab = "Observed values",
    ylab = "Fitted values")
lines(sponge2$species.richness, sponge2$species.richness, col = "blue")
```

The ranges of fitted values and observed values are

```
range(svm.tune.sponge1$fitted)
```

MSE

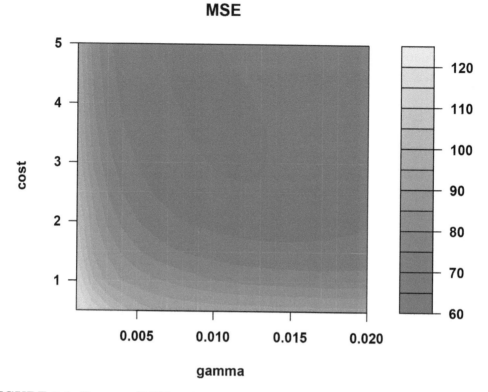

FIGURE 9.4: Changes of MSE with 'cost' and 'gamma' for the species richness of sponge using SVM.

```
[1]  -0.05678 36.42252
```

```
range(sponge2$species.richness)
```

```
[1]   1 39
```

The fitted values of `svm.tune.sponge1` are also shrunk in comparison with the observations in terms of their range (Figure 9.5), but much closer to the observations than that of `svm.sponge1` above (Figure 9.2), and also negative fitted values are produced for the count data.

Function svmcv

The `gamma` and `cost` can also be estimated via `svmcv` as shown in the following example.

```
library(spm2)
```

```
cost = c(1:10)*0.5
gamma = c(1:20)*0.001
svmopt <- matrix(0, length(cost), length(gamma))

for (i in 1:length(cost)) {
  for (j in 1:length(gamma)) {
    set.seed(1234)
    svmcv1 <- svmcv(species.richness ~ ., trainxy =  sponge2[, -4], y = sponge2[,
        3], cost = cost[i], gamma = gamma[j], validation = "CV", predacc = "VEcv")
    svmopt[i, j] <- svmcv1
```

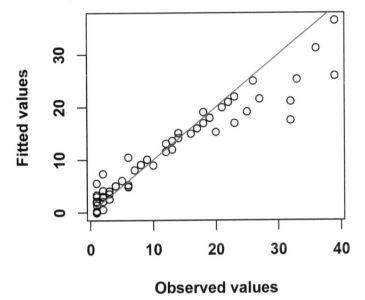

FIGURE 9.5: Observed values vs. fitted values by SVM.

```
  }
}
```

The optimal values of `cost` and `gamma` are

```
svm.opt <- which (svmopt == max(svmopt), arr.ind = T)
cost[svm.opt[, 1]]
```

```
[1] 4
```

```
gamma[svm.opt[, 2]]
```

```
[1] 0.009
```

Besides `type`, `cost`, and `gamma`, other arguments, such as `nu`, `kernel`, and `degree`, for `svm` can also be optimized by `svmcv`. These R code for optimizing `type`, `cost`, and `gamma` can be easily modified for the optimization of the other arguments, so their optimization will not be presented.

9.4   Predictive accuracy of *SVM* models

1. Categorical data

We will use `svm` with the parameters estimated and variables selected to determine the predictive accuracy of the *SVM* classification model developed for the `hard` data set.

```
set.seed(1234)
n <- 100
ccr.svm.h <- NULL

for (i in 1:n) {
```

```
svm.hard.acc <- svm(hardness ~ ., data = hard[, c(svmrfe.hard$optVariables, "
    hardness")], type = "C-classification", cost = svm.tune.hard$best.parameters
    $cost, gamma = svm.tune.hard$best.parameters$gamma, cross = 10)
  ccr.svm.h[i] <- svm.hard.acc$tot.accuracy
}
```

The median and range of predictive accuracy of the *SVM* model based on 100 repetitions of 10-fold cross-validation using `svm` are

```
median(ccr.svm.h)
```

```
[1] 91.24
```

```
range(ccr.svm.h)
```

```
[1] 88.32 91.97
```

2. Numerical data

Predictive accuracy of *SVM* regression models for numerical data can be assessed by `svm` and `svmcv`.

We will use `svm` and `svmcv` with the parameters estimated and predictive variables selected (i.e., all 79 variables) to assess the predictive accuracy of the *SVM* regression model developed for the `sponge2` data set.

Function `svm`

```
set.seed(1234)
```

```
svm.sponge.acc <- svm(species.richness ~ ., data = sponge2[, -4], type = "nu-
    regression", cost = svm.tune.sponge$best.parameters$cost, gamma = svm.tune.
    sponge$best.parameters$gamma, cross = 10)
```

```
svm.sponge.acc$tot.MSE
```

```
        [,1]
[1,]  62.83
```

which is *MSE* for model `svm.sponge.acc`. We can convert this *MSE* to *VEcv* using the `tovecv` function in the `spm` package by

```
svm.VEcv1 <- tovecv(n = dim(sponge2)[1], mu = mean(sponge2$species.richness), s =
    sd(sponge2$species.richness), m = svm.sponge.acc$tot.MSE, measure="mse")
```

```
svm.VEcv1
```

```
        [,1]
[1,]  42.54
```

It shows that the predictive accuracy, `svm.VEcv1`, of `svm.sponge` is 42.54%. This accuracy changes with the random seed, and a stabilized accuracy needs to be derived in the same way as in previous examples and will not be presented.

Function `svmcv`

```
set.seed(1234)
svmvecv1 <- svmcv(species.richness ~ ., trainxy = sponge2[, -4], y = sponge2[, 3],
    cost = 4, gamma = 0.009, validation = "CV", predacc = "VEcv")
```

The predictive accuracy, *VEcv*, of `svmvecv1` is

```
svmvecv1
```

```
[1] 47.12
```

This is higher than `svm.VEcv1` for `svm.sponge.acc`. Again, this accuracy changes with the random seed used, and a stabilized accuracy needs to be derived in the same way as in previous examples but will not be presented here either.

9.5 Predictions of *SVM*

We can use *SVM* to generate the spatial predictions, which can be demonstrated using the `sponge` and `sponge.grid` data sets as below.

We use the default `type` that is optimal for `sponge` data and the parameters optimized for `cost` and `gamma` for `sponge` data to generate spatial predictions.

```
svm.sponge2 <- svm(sponge ~ ., data = sponge, cost = 6.5, gamma = 0.039)
```

```
svm.sponge2.pred <- predict(svm.sponge2, sponge.grid)
```

The ranges of predictions and observations are

```
range(svm.sponge2.pred)
```

```
[1] -9.486 20.399
```

```
range(sponge$sponge)
```

```
[1]  1 39
```

The results show that the range of the predictions is shifted to negative domain by -9.49, which needs to be corrected. They will be corrected by resetting the negative predictions to 0. The spatial distribution of predictions of *SVM* are illustrated in Figure 9.6 by

```
library(sp)
svm.sponge2.pred2 <- svm.sponge2.pred
svm.sponge2.pred2[svm.sponge2.pred2 <= 0] <- 0

svm.pred.df <- cbind(sponge.grid[, c(1:2)], as.data.frame(svm.sponge2.pred2))
gridded(svm.pred.df) = ~ easting + northing
names(svm.pred.df) <- "predictions"

library(raster)

par(font.axis = 2, font.lab = 2)
plot(brick(svm.pred.df))
```

The predictions of *SVM* can also be generated using the function `svmpred` in the spm2 package as below.

```
model <- sponge ~ easting + northing + tpi3 + var7 + entro7 + bs34 + bs11
```

```
svmpred1 <- svmpred(formula = model, trainxy = sponge, longlatpredx = sponge.grid
    [, c(1:2)], predx = sponge.grid, cost = 4, gamma = 0.009)
```

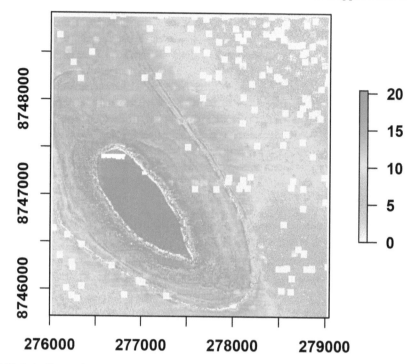

FIGURE 9.6: Spatial patterns of predictions using *SVM* for sponge species count data.

A note: since the whole range of the predictions by the *SVM* regression model is shifted down by roughly 10 in comparison with that of the observations, one might think it is reasonable to shift the predictions up by making the minimum of the range of the predictions to 0 by adding 9.4855, which is equivalent to the shift, to the predictions, thus the range of the predictions could match with the range of the observations. However, this may lead to wrong predictions because the predicted area is only a small portion of the sampling area and such shifting up of the range of predictions could produce falsely large predictions in the predicted area. In practice, a reasonable way is to use alternative modeling methods or improve the data for modeling.

9.6   Further modeling methods

There are also some other machine learning methods that have been or could be used for spatial predictive modeling (Li 2019a). These methods may include: (1) general regression neural network (*GRNN*), a version of probabilistic neural network (Specht 1990; Li, Potter, Huang, Daniell, et al. 2010; Li, Heap, Potter, Huang, et al. 2011); (2) hybrid methods of *GRNN* with geostatistical methods (*GRNNOK, GRNNIDW*) (Li, Potter, Huang, and Heap 2012b); (3) neural network (*NNET*) (Appelhans et al. 2015); (4) neural network residual kriging (*NNRK*) and hybrid of regression kriging and *NNRK* (*RKNNRK*) (Seo, Kim, and Singh 2015); (5) *cubist* and hybrid of *cubist* and *OK* (*cubistOK*) (Appelhans et al. 2015); (6) multivariate adaptive regression splines (MARS) (Leathwick, Elith, and Hastie 2006); and (7) Bayesian methods (Stephens and Diesing 2014). Deep learning methods that are based on *NNET* could also be useful tools for spatial predictive modeling, but their applications to spatial predictive modeling are still rare. These methods may need to be considered in future.

10

Hybrids of modern statistical methods with geostatistical methods

This chapter introduces the hybrids of *modern statistical methods* with geostatistical methods, where the geostatistical methods refer to *mathematical methods, univariate geostatistical methods*, or both. They are *gradient based/detrended predictive methods*. The *modern statistical methods* are the methods introduced in Chapter 7, while the mathematical and univariate geostatistical methods are the methods introduced in Chapters 4 and 5, respectively. A range of *hybrid methods* could be developed for spatial predictive modeling (Li and Heap 2008). The *hybrid methods* are also called *combined methods* (Li and Heap 2014; Li, Siwabessy, Huang, et al. 2019; Li 2019a). In this chapter, we will focus on the hybrid methods of modern statistical methods with *IDW* and/or *kriging*. *IDW* also covers *NN* and *KNN*, as they can be regarded as special cases of *IDW*, while *kriging* methods include *OK, SK, BOK*, and *BSK*. They are sourced from previous studies (Li, Potter, Huang, Daniell, et al. 2010; Li, Alvarez, et al. 2017) as well as from Li and Heap (2008) and are listed below:
(1) hybrid method of *LM* and *IDW* (*LMIDW*),
(2) hybrid method of *LM* and *OK* (*LMOK*),
(3) hybrid methods of *LM, OK*, and *IDW* (*LMOKIDW*),
(4) hybrid method of *GLM* and *IDW* (*GLMIDW*),
(5) hybrid method of *GLM* and *OK* (*GLMOK*),
(6) hybrid methods of *GLM, OK*, and *IDW* (*GLMOKIDW*),
(7) hybrid method of *GLS* and *IDW* (*GLSIDW*),
(8) hybrid method of *GLS* and *OK* (*GLSOK*), and
(9) hybrid methods of *GLS, OK*, and *IDW* (*GLSOKIDW*).

We will also discuss various extensions of these hybrid methods (i.e., the hybrid methods of modern statistical methods with *NN, KNN, SK, BOK*, and *BSK*). These methods are contained in the:
1) hybrid methods of *LM, kriging*, and *IDW* (*LMkrigingIDW*) (see Table 10.1),
2) hybrid methods of *GLM, kriging*, and *IDW* (*GLMkrigingIDW*) (see Table 10.2), and
3) hybrid methods of *GLS, kriging*, and *IDW* (*GLSkrigingIDW*) (see Table 10.3).

We firstly provide a general prediction formula for all hybrid methods to be introduced in this and the next chapters, and then demonstrate how to implement these methods in *R*.

General prediction formula

Prior to defining the formula, we need to make a couple of clarifications. As stated in Chapter 3, the *observed values* are referring to the values of validation samples or new samples. We also need to clarify the difference between *residuals* and *predictive errors*. The *residuals* are the differences between the *fitted values* for and the observed values of training samples, which are going to be used in the hybrid methods, while the *predictive errors* are the differences between the predicted values for and the observed values of validation samples or new samples.

DOI: 10.1201/9781003091776-10

All hybrid methods share the same general prediction formula. The predictions of the hybrid methods can be generated according to Equations (10.1) to (10.7).

Firstly, we need to obtain the *residuals* (ϵ) that are the differences between the fitted values ($\hat{y}$) for and the observed values (y) of training samples using Equation (10.1).

$$\epsilon_{x_i} = y_{x_i} - \widehat{y_{x_i}} \tag{10.1}$$

where ϵ_{x_i} is the residual value of an attribute at the location of interest x_i; y_{x_i} is the observed value of an attribute at the location of interest x_i; $\widehat{y_{x_i}}$ is the fitted value of an attribute at the location of interest x_i by a modern statistical method introduced in Chapter 7 or a machine learning method introduced in Chapters 8 and 9.

The next step is to generate the *predictions based on the residuals* using a mathematical or univariate geostatistical method according to Equation (10.2).

$$\hat{y}(\epsilon_{x_0}) = \sum_{i=1}^{n} \lambda_i \cdot y(\epsilon_{x_i}) \tag{10.2}$$

where $\hat{y}(\epsilon_{x_0})$ is the predicted value of an attribute at the location of interest x_0 by a mathematical or univariate geostatistical method, n represents the number of samples used to generate the predictions, λ_i is the weight assigned to the sample at x_i, $y(\epsilon_{x_i})$ is the residual value at the sampled location x_i, and finally the *predictions of a hybrid method* can be generated according to Equation (10.3).

$$\hat{y}(x_0) = \widehat{y_{x_0}} + \hat{y}(\epsilon_{x_0}) \tag{10.3}$$

where $\hat{y}(x_0)$ is the predicted value of an attribute at the location of interest x_0 by a hybrid method, and $\widehat{y_{x_0}}$ is the predicted value of an attribute at the location of interest x_0 by a modern statistical method introduced in Chapter 7 or a machine learning method introduced in Chapters 8 and 9.

In Equation (10.3), the hybrid methods are assumed to be the hybrid methods of modern statistical methods or a machine learning method with a mathematical method or a univariate geostatistical method, respectively. For the hybrid methods of modern statistical methods or a machine learning method with both mathematical and univariate geostatistical methods together, $\hat{y}(\epsilon_{x_0})$ can be extended to Equation (10.4).

$$\hat{y}(\epsilon_{x_0}) = \lambda_m \widehat{y_m}(\epsilon_{x_0}) + \lambda_k \widehat{y_k}(\epsilon_{x_0}) \tag{10.4}$$

where $\widehat{y_m}(\epsilon_{x_0})$ is the predicted value of an attribute at the location of interest x_0 by a mathematical interpolation method (e.g., *IDW*) based on the residuals, $\widehat{y_k}(\epsilon_{x_0})$ is the predicted value of an attribute at the location of interest x_0 by a univariate geostatistical method (e.g., *OK*) based on the residuals, and λ_m and λ_k are weight, each with value ranging from 0 to 1, and their sum is either 1 or 2/3.

Finally, the *predictions of the hybrid methods* of modern statistical methods or a machine learning method with a mathematical method and a univariate geostatistical method (e.g., *IDW* and *OK*) together can be generated according to Equation (10.5).

$$\hat{y}(x_0) = \widehat{y_{x_0}} + \lambda_m \widehat{y_m}(\epsilon_{x_0}) + \lambda_k \widehat{y_k}(\epsilon_{x_0}) \tag{10.5}$$

Changes in the value of the sum, λ_m and λ_k will result in various hybrid methods. Taking the hybrid methods of *GLM*, *OK*, and *IDW* as an example, in Equation (10.5):

(1) if the sum is 1 and $\lambda_m = \lambda_k$, then the hybrid method resulted is the average of *GLMOK* and *GLMIDW* (*GLMOKGLMIDW*) or *GLMOKIDW* with $\lambda_m = 1/2$ and $\lambda_k = 1/2$;

(2) if the sum is 1 and $\lambda_m \neq \lambda_k$, then the hybrid method resulted is the sum of *GLMOK* weighted by λ_k and *GLMIDW* by λ_m;

(3) if the sum is 1 and $\lambda_m = 1$ and $\lambda_k = 0$, then the hybrid method resulted is *GLMIDW*;

(4) if the sum is 1 and $\lambda_m = 0$ and $\lambda_k = 1$, then the hybrid method resulted is *GLMOK*;

(5) if the sum is 2/3 and $\lambda_m = \lambda_k$, then the hybrid method resulted is the average of *GLM*, *GLMOK* and *GLMIDW* (*GLMGLMOKGLMIDW*) or *GLMOKIDW* with $\lambda_m = 1/3$ and $\lambda_k = 1/3$; and

(6) if the sum is 2/3 and $\lambda_m \neq \lambda_k$, then the hybrid method resulted is sum of *GLM*, *GLMOK* weighted by λ_k and *GLMIDW* by λ_m.

When the sum is 2/3 and λ_k is 0, the predictions of the hybrid methods of modern statistical or machine learning methods with a mathematical method can be generated according to Equation (10.6).

$$\hat{y}(x_0) = \widehat{y_{x_0}} + \lambda_m \widehat{y_m}(\epsilon_{x_0})2/3 \qquad (10.6)$$

Taking the hybrid methods of *GLM* and *IDW* as an example, according to Equation (10.6), the hybrid method resulted is *GLMIDW* with $\lambda_m = 2/3$.

When the sum is 2/3 and λ_m is 0, the predictions of the hybrid methods of modern statistical or machine learning methods with a univariate geostatistical method can be generated according to Equation (10.7).

$$\hat{y}(x_0) = \widehat{y_{x_0}} + \lambda_k \widehat{y_k}(\epsilon_{x_0})2/3 \qquad (10.7)$$

Taking the hybrid methods of *GLM* and *OK* as an example, according to Equation (10.7), the hybrid method resulted is *GLMOK* with $\lambda_k = 2/3$.

10.1 Hybrid method of *LM* and *IDW*

We will use the `petrel` data set to demonstrate the application of the hybrid method of *LM* and *IDW* (*LMIDW*). For the `petrel` data, an optimal model based on `lm.loocv` was developed in Chapter 7. We will use the predictive variables in `lm.loocv` to construct the optimal *LM* model for *LMIDW* first.

```
library(spm)
data(petrel)

lm.opt <- data.frame(petrel[, c(1, 2, 6:8, 5)], long2 = petrel$long^2, lat2 =
    petrel$lat^2, lat3= petrel$lat^3, bathy2 = petrel$bathy^2, bathy3 = petrel$
    bathy^3, dist2 = petrel$dist^2, dist3 = petrel$dist^3)

lm1 <- lm(log(gravel + 1) ~ ., lm.opt)
```

Then we will produce the residuals of `lm1` for *IDW* and develop an *IDW* model based on the residuals.

```
library(gstat)
lm.opt$res1 <- log(petrel$gravel + 1) - lm1$fitted.values
```

```
idw1 <- gstat(id = "res1", formula = res1 ~ 1, locations = ~ long + lat, data = lm
    .opt, set = list(idp = 2), nmax = 12)
```

Finally, we will produce the predictions of *LMIDW* using `lm1` and `idw1`. We will still use the training data `lm.opt` to make the predictions. Since the training data are used, the predictions are actually the fitted values of *LMIDW*.

```
lm1.pred <- predict(lm1, lm.opt)
idw.pred <- predict(idw1, lm.opt)$res1.pred
lmidw.pred1 <- lm1.pred + idw.pred
lmidw.pred <- exp(lmidw.pred1) - 1
```

Now we can check the fitted values of *LMIDW*, `lmidw.pred`, against the observed values.

```
sum(abs(lmidw.pred - petrel$gravel))
```

```
[1] 0.000000003167
```

Since *IDW* is an *exact* method (Li and Heap 2014), the hybrid method, *LMIDW*, is also an exact method and should fit the data perfectly. As expected, *LMIDW* indeed produces a perfect fit to the data. In fact, all hybrid methods of modern statistical methods with the mathematical methods are *exact* method. However, the *goodness of fit* is not necessarily related to the predictive accuracy (Li, Alvarez, et al. 2017) and should not be used to assess the performance of predictive models.

In addition, if `idp` is specified as 0 in model `idw1`, then *KNN* is used, and the hybrid method resulted would be *LMKNN*. If `nmax` is specified as 1, then *NN* is used, and the hybrid method resulted would be *LMNN*.

10.1.1 Variable selection and parameter optimization based on predictive accuracy

For *LMIDW*, *variable selection* can be conducted with the methods and procedures introduced in Chapter 7 for *LM*. We will only demonstrate how to estimate the optimal parameters for *LMIDW* based on the residuals of the optimal *LM* predictive model. This optimization can be done using the function `glmidwcv` with `family = "gaussian"` in the spm2 package (Li 2021a). We will introduce `glmidwcv` first and then use it to conduct parameter optimization.

1. Function `glmidwcv`

For `glmidwcv`, the following arguments are essential:
(1) `formula`, a formula defining response variable and predictive variables;
(2) `longlat`, a dataframe contains the longitude and latitude of point samples;
(3) `trainxy`, a dataframe contains the longitude (long), latitude (lat), predictive variables and response variable of point samples;
(4) `y`, a vector of the response variable in the formula, that is, the left part of the formula;
(5) `family`, a description of error distribution and link function;
(6) `idp`, a numeric number specifying the inverse distance weighting power;
(7) `nmaxidw`, number of nearest observations that should be used for a prediction;
(8) `validation`, validation methods, including "LOO" and "CV";
(9) `cv.fold`, number of folds in the cross-validation; and
(10) `predacc`, can be either "VEcv" for `vecv` or "ALL" for all measures.

2. Parameter optimization based on predictive accuracy

We can estimate the optimal `idp` and `nmaxidw` for *LMIDW* based on the residuals of the optimal *LM* predictive model. We will use `glmidwcv` with the same formula as that for `lm1` to demonstrate the *parameter optimization* below.

```
model <- log(gravel + 1) ~ long + lat + bathy + dist + relief + I(long^2) + I(lat
    ^2) + I(lat^3) + I(bathy^2) + I(bathy^3) + I(dist^2) + I(dist^3)
longlat <- petrel[, c(1, 2)]
y <- log(petrel$gravel +1)
idp <- (1:20) * 0.2
nmax <- c(5:40)
idwopt <- array(0, dim = c(length(idp), length(nmax)))

for (i in 1:length(idp)) {
  for (j in 1:length(nmax)) {
    set.seed(1234)
    lmidwcv1 <- glmidwcv(formula = model, longlat = longlat, trainxy = petrel, y =
        y, family = "gaussian", idp = idp[i], nmaxidw = nmax[j], validation = "CV
        ", predacc = "VEcv")
    idwopt[i, j] <- lmidwcv1
  }
}
```

The optimal values of `idp` and `nmax` that lead to the maximal predictive accuracy are

```
para.opt <- which (idwopt == max(idwopt), arr.ind = T)
idp[para.opt[, 1]]
```

```
[1] 0.6
```

```
nmax[para.opt[, 2]]
```

```
[1] 10
```

These parameters should be used for the *LMIDW* predictive model.

In addition, `glmidwcv` can also be used to assess the predictive accuracy of *LMKNN* and *LMNN* if `idp` is specified as 0 and `nmaxidw` as 1 in `glmidwcv`, respectively.

10.1.2   Predictive accuracy

The *predictive accuracy* of *LMIDW* changes with the `set.seed()` used. To stabilize it and get a reliable accuracy estimation for *LMIDW*, we need to repeat the cross-validation. We will use `glmidwcv` with the above `model`, `y`, and the optimal `idp` and `nmax` to demonstrate the *accuracy stabilization* below.

```
set.seed(1234)
n <- 100
lmidwvecv <- NULL
for (i in 1:n) {
  lmidwcv1 <- glmidwcv(formula = model, longlat = longlat, trainxy = petrel, y = y
      , family = "gaussian", idp = 0.6, nmax = 10, validation = "CV", predacc = "
      VEcv")
  lmidwvecv[i] <- lmidwcv1
}
```

The median and range of predictive accuracy of *LMIDW* based on 100 repetitions of 10-fold cross-validation using `glmidwcv` are

```
median(lmidwvecv)
```

```
[1] 34.61
```

```
range(lmidwvecv)
```

```
[1] 29.42 38.42
```

10.1.3 Predictions

The predictions of *LMIDW* can be generated using the function `glmidwpred` with `family = "gaussian"` in the `spm2` package.

1. Function `glmidwpred`

For function `glmidwpred`, the following arguments are required:

(1) `formula`, `longlat`, `trainxy`, `y`, `family`, `idp`, and `nmaxidw`, which are the same as those for `glmidwcv`;

(2) `longlatpredx`, a dataframe contains the longitude and latitude of point locations (i.e., the centers of grids) to be predicted; and

(3) `predx`, a dataframe of predictive variables for the grids to be predicted.

2. Generation of spatial predictions using `glmidwpred`

We will use the `model` and `y` in Section 10.1.1 and the optimal `idp` and `nmaxidw` to produce spatial predictions for the area based on the `petrel.grid` data set.

```
lmidwpred1 <- glmidwpred(formula = model, longlat = longlat, trainxy = petrel,
y = y, longlatpredx = petrel.grid[, c(1:2)], predx = petrel.grid, family = "
    gaussian", idp = 0.6, nmaxidw = 10)
```

The *LM* predictions need to be back-transformed to percentage data.

```
lmidwpred1$predictions <- exp(lmidwpred1$predictions) - 1
```

The range of the predictions are

```
range(lmidwpred1$predictions)
```

```
[1]    0.7896 302.0587
```

A portion of the transformed predictions are beyond the range of percentage data and need to be corrected by resetting the faulty predictions to the nearest bound of the data range.

```
lmidwpred <- lmidwpred1
lmidwpred$predictions[lmidwpred$predictions >= 100] <- 100
```

The spatial distribution of *LMIDW* predictions is illustrated in Figure 10.1.

10.2   Hybrid method of *LM* and *OK*

For the hybrid method of *LM* and *OK* (*LMOK*), we will also use the `petrel` data set to demonstrate its application. To develop an optimal predictive model and generate spatial predictions, we will apply *OK* to the residuals of `lm1` that are stored in `lm.opt` in Section 10.1. We will firstly conduct a variogram modeling for *LMOK*.

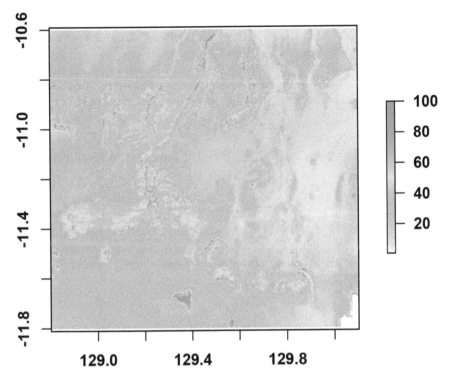

FIGURE 10.1: The spatial patterns of predictions using *LMIDW*.

```
library(sp)
data.train <- lm.opt
coordinates(data.train) = ~ long + lat
vgm1 <- variogram(object = res1 ~ 1, data.train)

model.1 <- fit.variogram(vgm1, vgm(mean(vgm1$gamma), "Sph", mean(vgm1$dist), min(
    vgm1$gamma) / 10))
```

Then use `model.1` to make *OK* predictions based on `res1` in the training data set, and finally make the predictions of *LMOK*.

```
pred.krige1 <- krige(res1 ~ 1, data.train, data.train, model = model.1, nmax = 12)

names(as.data.frame(pred.krige1))
lmok.pred1 <- as.data.frame(pred.krige1)[, 3] + lm1.pred
lmok.pred <- exp(lmok.pred1) - 1
```

Since the training data set are used, the predictions are actually the fitted values of *LMOK*. Similar to *LMIDW*, the fitted values of *LMOK* are exactly the same as the observed values. This is because *OK* is an *exact* method (Li and Heap 2014), and all hybrid methods of modern statistical methods with the univariate geostatistical methods are *exact* methods. Again, the *goodness of fit* is not a measure of the predictive accuracy, which also applies to *LMOK*. The predictive accuracy of *LMOK* will be assessed in Section 10.2.2.

In addition, if `beta` is specified for `pred.krige1`, then *SK* is used, and the hybrid method resulted would be *LMSK*. If `block` is specified, then *BK* is used, and the hybrid method resulted would be *LMBOK* or *LMBSK* depending on the specification of `beta`. For ways to specify these arguments, see relevant sections in Chapter 5.

If the `res1 ~ 1` in `vgm1` is extended to include external variable(s) (e.g., `res1 ~ bathy`), then the hybrid method resulted would be *LMKED*; but this extension should not be used because any possible external trends should have already been taken care of in *LM* model. This note is also applicable to all other hybrid methods that are to be introduced in this and the next chapters. Despite this note, relevant cross-validation functions (e.g., `glmkrigecv`) introduced below and in the next chapter could be used to assess the predictive accuracy of such hybrid methods.

10.2.1   Variable selection and parameter optimization based on predictive accuracy

Similar to *LMIDW*, *variable selection* for *LMOK* can be conducted using the methods and procedures introduced in Chapter 7 for *LM*. We only need to demonstrate how to estimate the optimal parameters for *LMOK* based on the residuals of the optimal *LM* predictive model. This optimization can be carried out using the function `glmkrigecv` with `family = "gaussian"` in the `spm2` package. We will introduce `glmkrigecv` first and then use it to conduct parameter optimization.

1. Function `glmkrigecv`

For `glmkrigecv`, the following arguments are essential:
(1) `formula.glm`, the same as the `formula` for `glmidwcv`;
(2) `longlat`, `trainxy`, `y`, `family`, `validation`, `cv.fold`, and `predacc`, the same as those for `glmidwcv`;
(3) `formula.krige`, the same as the `formula` for `krigecv`;
(4) `transformation`, `delta`, `vgm.args`, `anis`, `alpha`, `block`, and `beta`, the same as those for `krigecv`; and
(5) `nmaxkrige`, the same as `nmax` for `krigecv`.

2. Parameter optimization based on predictive accuracy

We can select the optimal `vgm.args` and estimate the optimal `nmaxkrige` for *LMOK*, which will be based on the residuals of the optimal *LM* predictive model. We will demonstrate the *parameter optimization* using `glmkrigecv` with the same formula as that for `lm1`.

```
nmax.ok <- c(5:40)
vgm.args <- c("Exp", "Gau", "Sph", "Exc", "Mat", "Ste", "Lin")
okopt <- matrix(0, length(nmax.ok), length(vgm.args))

for (i in 1:length(nmax.ok)) {
  for (j in 1:length(vgm.args)) {
    set.seed(1234)
    lmkrigecv1 <- glmkrigecv(formula.glm = model, longlat = petrel[, c(1, 2)],
        trainxy = petrel, y = log(petrel$gravel +1), family = "gaussian",
        nmaxkrige = nmax.ok[i], formula.krige = res1 ~ 1, vgm.args = vgm.args[j],
        validation = "CV", predacc = "VEcv")
    okopt[i, j] <- lmkrigecv1
  }
}
```

The optimal `nmaxkrige` and `vgm.args` that result in the maximal predictive accuracy are

```
okopt[is.na(okopt)] <- -9999 # NA's were produced for "Exp" and need to be
    replaced to run the following *R* code.
ok.opt <- which(okopt == max(okopt), arr.ind = T)
nmax.ok[ok.opt[, 1]]
```

[1] 11

```
vgm.args[ok.opt[, 2]]
```

```
[1] "Lin"
```

These parameters should be used for the *LMOK* predictive model.

All other parameters (e.g, `transformation`, `alpha`) can be optimized by following the procedures introduced in Chapter 5 and their optimization will not be presented here.

Furthermore, `glmkrigewcv` can also be used to assess the predictive accuracy of other hybrid methods such as *LMSK* and *LMBSK* if `beta` or both `beta` and `block` are specified, respectively.

10.2.2 Predictive accuracy

We will use the the above `model`, `y`, and the optimal `nmaxkrige` and `vgm.args` to stabilize the *predictive accuracy* of *LMOK*.

```
set.seed(1234)
n <- 100
lmokvecv <- NULL
for (i in 1:n) {
  lmokcv1 <- glmkrigecv(formula.glm = model, longlat = petrel[, c(1, 2)], trainxy
      = petrel, y = log(petrel$gravel +1), family = "gaussian", nmaxkrige = 11,
      formula.krige = res1 ~ 1, vgm.args = "Lin", validation = "CV", predacc = "
      VEcv")
  lmokvecv[i] <- lmokcv1
}
```

The median and range of predictive accuracy of *LMOK* based on 100 repetitions of 10-fold cross-validation using `glmkrigecv` are

```
median(lmokvecv)
```

```
[1] 33.89
```

```
range(lmokvecv)
```

```
[1] 26.61 38.34
```

10.2.3 Predictions

The predictions can be generated using the function `glmkrigepred` with `family = "gaussian"` in the `spm2` package for *LMOK*.

1. Function `glmkrigepred`

For `glmkrigepred`, the following arguments may need to be specified;
(1) `formula.glm`, `longlat`, `trainxy`, `y`, `family`, `transformation`, `delta`, `formula.krige`, `vgm.args`, `anis`, `alpha`, `block`, `beta`, and `nmaxkrige`, which are the same as those for `glmkrigecv`;
(2) `longlatpredx`, a dataframe contains the longitude and latitude of point locations (i.e., the centers of grids) to be predicted; and
(3) `predx`, a dataframe of predictive variables for the grids to be predicted.

2. Generation of spatial predictions using `glmkrigepred`

We will use the `model` in Section 10.1.1, and the optimal `vgm.args` and `nmaxkrige` to produce spatial predictions for the area based on the `petrel.grid` data set.

```
lmkrigepred1 <- glmkrigepred(formula = model, longlat = petrel[, c(1, 2)], trainxy
    = petrel, y = log(petrel$gravel +1), longlatpredx = petrel.grid[, c(1:2)],
    predx = petrel.grid, family = "gaussian", transformation = "none", formula.
    krige = res1 ~ 1, vgm.args = "Lin", nmaxkrige = 11)
```

The *LM* predictions need to be back-transformed to percentage data.

```
lmkrigepred1$predictions <- exp(lmkrigepred1$predictions) - 1

range(lmkrigepred1$predictions)
```

```
[1]    0.9968 316.3382
```

A portion of the transformed predictions that are beyond the range of percentage data need to be corrected by resetting the faulty predictions to the nearest bound of the data range by following the example provided in Section 10.1.3.

The spatial distribution of *LMOK* predictions is illustrated in Figure 10.2.

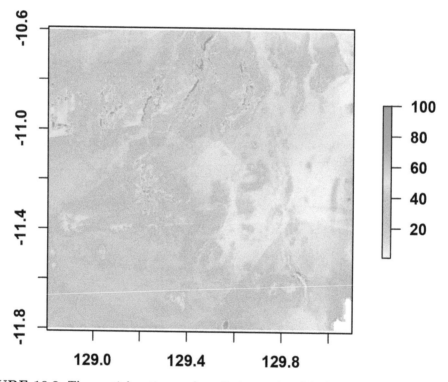

FIGURE 10.2: The spatial patterns of predictions using *LMOK*.

10.3   Hybrid methods of *LM*, *OK*, and *IDW*

For hybrid methods of *LM*, *OK*, and *IDW* (*LMOKIDW*), we will also use the petrel data set to demonstrate their applications. To develop an optimal predictive model and generate spatial predictions for *LMOKIDW*, firstly, we need to develop an optimal *LM* predictive model, and then we apply *IDW* and *OK* to the residuals of the model. These have been done and stored in `lm1.pred`, `idw.pred`, and `pred.krige1` in Sections 10.1 and 10.2 for *LMIDW* and *LMOK*. We will use them to demonstrate the generation of the predictions of *LMOKIDW* methods.

1. Average of *LMOK* and *LMIDW*

The average of *LMOK* and *LMIDW* (*LMOKLMIDW*) is a *LMOKIDW* method with $\lambda_m = \lambda_k = 1/2$ according to Equation (10.5). We will produce the predictions of *LMOKLMIDW* using `lm1.pred`, `idw.pred` and `pred.krige1` as below.

```
lambdai <- 1 / 2
lambdak <- 1 / 2
lmoklmidw.pred1 <- lm1.pred + lambdak * as.data.frame(pred.krige1)[, 3] + lambdai
    * idw.pred

lmoklmidw.pred <- exp(lmoklmidw.pred1) - 1
```

Since the training data were used for `lm1.pred`, `idw.pred`, and `pred.krige1`, the predictions are actually the fitted values of *LMOKLMIDW*.

Because both *IDW* and *OK* are exact methods, the hybrid method, *LMOKLMIDW*, is also an *exact* method and fits the data perfectly.

2. Average of *LM*, *LMOK*, and *LMIDW*

The average of *LM*, *LMOK*, and *LMIDW* (*LMLMOKLMIDW*) is a *LMOKIDW* method with $\lambda_m = \lambda_k = 1/3$ according to Equation (10.5). We will produce the predictions of *LMLMOKLMIDW* using `lm1.pred`, `idw.pred`, and `pred.krige1`.

```
lambdai <- 1 / 3
lambdak <- 1 / 3
lmlmoklmidw.pred1 <- lm1.pred + lambdak * as.data.frame(pred.krige1)[, 3] +
    lambdai * idw.pred

lmlmoklmidw.pred <- exp(lmlmoklmidw.pred1) - 1
```

Since the training data were used for `lm1.pred`, `idw.pred`, and `pred.krige1`, the predictions are also the fitted values of *LMLMOKLMIDW*. Due to $\lambda_m + \lambda_k = 2/3$, the fitted values are no longer the same as the observed values, thus *LMLMOKLMIDW* is not an exact method. This can be checked against the observed values.

```
sum(abs(lmlmoklmidw.pred - petrel$gravel))
```

```
[1] 844.4
```

In fact, λ_m and λ_k for *LMOKIDW* can take any values within their range (i.e., 0 to 1) as long as their sum is 1 or 2/3. Their values can be determined according to the predictive accuracy of the hybrid method resulted as shown below.

10.3.1 Variable selection and parameter optimization based on predictive accuracy

Similar to *LMIDW* and *LMOK*, for *LMOKIDW* *variable selection* can be conducted with the methods and procedures introduced in Chapter 7 for *LM*. We only need to demonstrate how to estimate the *optimal parameters* for *IDW* and *OK* components of the hybrid methods based on the residuals of the optimal *LM* predictive model. This optimization can be done using the function `glmkrigeidwcv` with `family = "gaussian"` in the `spm2` package. We will introduce `glmkrigeidwcv` first and then use it to conduct parameter optimization.

1. Function `glmkrigeidwcv`

For `glmkrigeidwcv`, the following arguments may need to be specified:

(1) `formula.glm`, `longlat`, `trainxy`, `y`, `family`, `transformation`, `delta`, `formula.krige`, `vgm.args`, `anis`, `alpha`, `block`, `beta`, `nmaxkrige`, `idp`, `nmaxidw`, `validation`, `cv.fold`, and `predacc`, which are the same as those for `glmkrigecv` and `glmidwcv`;

(2) `hybrid.parameter`, either 2 or 3, with the default value of 2; and

(3) `lambda`, ranging from 0 to 2, with the default value of 1.

If `lambda` is < 1, more weight is placed on *kriging*; and if it is > 1, more weight is placed on *IDW*.

When the default `hybrid.parameter` is used, if `lambda` is 0, *IDW* is not considered and the method resulted is *LMkriging*, and if `lambda` is 2, then the method resulted is *LMIDW*.

When `hybrid.parameter` is 3, if `lambda` is 0, then the method resulted is equivalent to the average of *LM*, *LMkriging*, and *LMkriging* (*LMLMkrigingLMkriging*), and if `lambda` is 2, then the method resulted is equivalent to the average of *LM*, *LMIDW*, and *LMIDW* (*LMLMID-WLMIDW*).

The relationships of λ_m and λ_k in Equation (10.5) with `lambda` (λ) and `hybrid.parameter` (ϕ) are presented in Equations (10.8) and (10.9), respectively.

$$\lambda_m = \lambda/\phi \tag{10.8}$$

$$\lambda_k = (2 - \lambda)/\phi \tag{10.9}$$

Thus *LMLMkrigingLMkriging* can be expressed as *LMkriging* with $\lambda_k = 2/3$, and *LMLMID-WLMIDW* as *LMIDW* with $\lambda_m = 2/3$.

In addition, if `idp` is specified as 0, then *KNN* is used; and if `nmaxidw` is specified as 1, then *NN* is used. If `beta` is specified, then *SK* is used; if `block` is specified, then *BOK* is used; and if both `beta` and `block` are specified, then *BSK* is used. Consequently, with the changes in the specifications of `idp`, `nmaxidw`, `beta`, `block`, `lambda`, and `hybrid.parameter`, many hybrid methods of *LM*, *kriging*, and *IDW* (*LMkrigingIDW*) can be resulted (Table 10.1).

When `0 < lambda < 2`, then numerous hybrid methods of *LM*, *kriging*, and *IDW* could be possibly derived according to Equation (10.5) and implemented via `glmkrigeidwcv`, such as *LMOKIDW* with $\lambda_m = 1.26/3$ and $\lambda_k = 0.72/3$, and *LMSKKNN* with $\lambda_m = 1.35/2$ and $\lambda_k = 0.65/2$.

The hybrid methods from No. 1 to No. 31 in Table 10.1 are exact methods and the remaining methods are not exact methods.

All these methods can be implemented and assessed using `glmkrigeidwcv` with `family = "guassian"`. Furthermore, the most accurate method can be selected from all these hybrid methods based on predictive accuracy with `glmkrigeidwcv` as demonstrated below.

2. Parameter optimization based on predictive accuracy

We can use the optimal `idp`, `nmaxidw`, `vgm.args`, and `nmaxkrige` in Sections 10.1 and 10.2 to conduct further *parameter optimization*.

TABLE 10.1: *LMkrigingIDW* methods can be derived and implemented via `glmkrigeidwcv` in relation to `idp`, `nmaxidw`, `beta`, `block`, `lambda`, and `hybrid.parameter` (ϕ); and λ_m and λ_k are from Equations (10.5), (10.6) and (10.7).

No.	idp	nmaxidw	beta	block	lambda	ϕ	Hybrid methods
1	!= 0	> 0	NULL	0	1	2	*LMOKLMIDW*
2	0	> 1	NULL	0	1	2	*LMOKLMKNN*
3	0	1	NULL	0	1	2	*LMOKLMNN*
4	!= 0	> 0	not NULL	0	1	2	*LMSKLMIDW*
5	0	> 1	not NULL	0	1	2	*LMSKLMKNN*
6	0	1	not NULL	0	1	2	*LMSKLMNN*
7	!= 0	> 0	NULL	> 0	1	2	*LMBOKLMIDW*
8	0	> 1	NULL	> 0	1	2	*LMBOKLMKNN*
9	0	1	NULL	> 0	1	2	*LMBOKLMNN*
10	!= 0	> 0	not NULL	> 0	1	2	*LMBSKLMIDW*
11	0	> 1	not NULL	> 0	1	2	*LMBSKLMKNN*
12	0	1	not NULL	> 0	1	2	*LMBSKLMNN*
13	!= 0	> 0	NULL	0	0	2	*LMOK*
14	!= 0	> 0	not NULL	0	0	2	*LMSK*
15	!= 0	> 0	NULL	> 0	0	2	*LMBOK*
16	!= 0	> 0	not NULL	> 0	0	2	*LMBSK*
17	!= 0	> 0	NULL	0	2	2	*LMIDW*
18	0	> 1	NULL	0	2	2	*LMKNN*
19	0	1	NULL	0	2	2	*LMNN*
20	!= 0	> 0	NULL	0	1	3	*LMLMOKLMIDW*
21	0	> 1	NULL	0	1	3	*LMLMOKLMKNN*
22	0	1	NULL	0	1	3	*LMLMOKLMNN*
23	!= 0	> 0	not NULL	0	1	3	*LMLMSKLMIDW*
24	0	> 1	not NULL	0	1	3	*LMLMSKLMKNN*
25	0	1	not NULL	0	1	3	*LMLMSKLMNN*
26	!= 0	> 0	NULL	> 0	1	3	*LMLMBOKLMIDW*
27	0	> 1	NULL	> 0	1	3	*LMLMBOKLMKNN*
28	0	1	NULL	> 0	1	3	*LMLMBOKLMNN*
29	!= 0	> 0	not NULL	> 0	1	3	*LMLMBSKLMIDW*
30	0	> 1	not NULL	> 0	1	3	*LMLMBSKLMKNN*
31	0	1	not NULL	> 0	1	3	*LMLMBSKLMNN*
32	!= 0	> 0	NULL	0	0	3	*LMOK* with $\lambda_k = 2/3$
33	!= 0	> 0	not NULL	0	0	3	*LMSK* with $\lambda_k = 2/3$
34	!= 0	> 0	NULL	> 0	0	3	*LMBOK* with $\lambda_k = 2/3$
35	!= 0	> 0	not NULL	> 0	0	3	*LMBSK* with $\lambda_k = 2/3$
36	!= 0	> 0	NULL	0	2	3	*LMIDW* with $\lambda_m = 2/3$
37	0	> 1	NULL	0	2	3	*LMKNN* with $\lambda_m = 2/3$
38	0	1	NULL	0	2	3	*LMNN* with $\lambda_m = 2/3$

These optimal parameters were estimated or selected separately for *LMIDW* and *LMOK*. If it is believed that the parameters are not optimal for *LMOKIDW*, then we need to re-do the optimization. This can be done with `glmkrigeidwcv`, but we will not present it here as it can be easily implemented by following the examples provided for parameter optimization for *LMIDW* and *LMOK* in Sections 10.1 and 10.2 as well as the example below for estimating the optimal `lambda`.

Below is an example to estimate the optimal value of `lambda` for *LMOKIDW* with `hybrid.`
`parameter = 2`.

```
lambda <- c(0:100)*0.02
lambdaopt <- NULL
for (i in 1:length(lambda)) {
  set.seed(1234)
  lmkrigelmidwcv1 <- glmkrigeidwcv(formula.glm = model, longlat = longlat, trainxy
      = petrel, y = log(petrel$gravel +1), family = "gaussian", transformation =
      "none", formula.krige = res1 ~ 1, vgm.args = "Lin", nmaxkrige = 11, idp =
      0.6, nmaxidw = 10, lambda = lambda[i], hybrid.parameter = 2, validation = "
      CV", predacc = "VEcv")
  lambdaopt[i] <- lmkrigelmidwcv1
}
```

The predictive accuracy of the hybrid method, "VEcv", can be visualized against `lambda` in
Figure 10.3 by

```
par(font.axis = 2, font.lab = 2)
plot(lambdaopt ~ lambda, xlab = "lambda", ylab = "VEcv (%)", col = "blue")
```

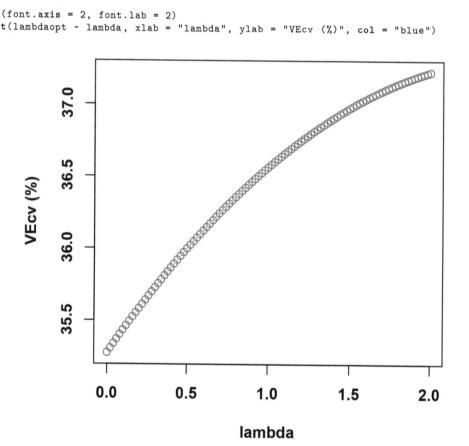

FIGURE 10.3: *VEcv* (%) of the hybrid method of *LM*, *IDW*, and *OK* with `hybrid.parameter`
`= 2` for each `lambda` based on 10-fold cross-validation.

The optimal `lambda` that leads to the maximal predictive accuracy is

```
lambda.opt <- which (lambdaopt == max(lambdaopt))
lambda[lambda.opt]
```

```
[1] 2
```

The maximal predictive accuracy is

```
max(lambdaopt)
```

```
[1] 37.22
```

Since the value estimated for `lambda` is 2, the hybrid method resulted is, in fact, *LMIDW* according to Equation (10.5).

Similar to *LMOKIDW* with `hybrid.parameter = 2`, the parameters selected can also be used to optimize `lambda` for *LMOKIDW* with `hybrid.parameter = 3`. The results are in `lambda2opt` and the optimal `lambda` is

```
lambda2.opt <- which (lambda2opt == max(lambda2opt))
lambda2[lambda2.opt]
```

```
[1] 2
```

The maximal predictive accuracy is

```
max(lambda2opt)
```

```
[1] 37.44
```

The value estimated for `lambda` is again 2, which suggests that the hybrid method resulted is *LMIDW* with $\lambda_m = 2/3$ according to Equation (10.6).

The predictive accuracy of the hybrid model resulted is higher than that of *LMOKIDW* with `hybrid.parameter = 2`. Hence, the *LMOKIDW* with `hybrid.parameter = 3` and a `lambda = 2`, that is, *LMIDW* with $\lambda_m = 2/3$, will be used.

10.3.2 Predictive accuracy

We will use the `model` and `y` in Section 10.1.1 and the optimal `nmaxkrige`, `vgm.args`, `idp`, and `nmaxidw` to demonstrate how to derive the *stabilized predictive accuracy* of *LMIDW* with $\lambda_m = 2/3$ (see Equation (10.6)).

```
set.seed(1234)
n <- 100
lmkrigelmidwvecv <- NULL
for (i in 1:n) {
  lmkrigelmidwcv1 <- glmkrigeidwcv(formula.glm = model, longlat = petrel[, c(1, 2)
      ], trainxy = petrel, y = log(petrel$gravel +1), family = "gaussian",
      transformation = "none", formula.krige = res1 ~ 1, vgm.args = "Lin",
      nmaxkrige = 11, idp = 0.6, nmaxidw = 10, lambda = 2, hybrid.parameter = 3,
      validation = "CV", predacc = "VEcv")
  lmkrigelmidwvecv[i] <- lmkrigelmidwcv1
}
```

The median and range of predictive accuracy of *LMIDW* with $\lambda_m = 2/3$ based on 100 repetitions of 10-fold cross-validation using `glmkrigeidwcv` are

```
median(lmkrigelmidwvecv)
```

```
[1] 35.04
```

```
range(lmkrigelmidwvecv)
```

```
[1] 30.72 38.83
```

10.3.3   Predictions

The predictions can be generated using the function `glmkrigeidwpred` with `family = "gaussian"` in the `spm2` package for the hybrid methods.

1. Function `glmkrigeidwpred`

For `glmkrigeidwpred`, the following arguments may need to be specified:

(1) `formula.glm`, `longlat`, `trainxy`, `predx`, `y`, `longlatpredx`, `family`, `transformation`, `delta`, `formula.krige`, `vgm.args`, `anis`, `alpha`, `block`, `beta`, `nmaxkrige`, `idp`, and `nmaxidw`, which are the same as the arguments for `glmidwcv`, `glmkrigecv`, and `glmkrigepred`; and

(2) `hybrid.parameter` and `lambda`, which are the same as the arguments for `glmkrigeidwcv`.

2. Generation of spatial predictions using `glmkrigeidwpred`

We will use the most accurate hybrid method, *LMIDW* with $\lambda_m = 2/3$, with the `model` in Section 10.1.1 and the optimal `vgm.args`, `nmaxkrige`, `idp`, and `nmaxidw` to produce spatial predictions for the area based on the `petrel.grid` data set.

```
lmkrigelmidwpred1 <- glmkrigeidwpred(formula = model, longlat = petrel[, c(1, 2)],
    trainxy = petrel, y = log(petrel$gravel +1), longlatpredx = petrel.grid[, c
    (1:2)], predx = petrel.grid, family = "gaussian", transformation = "none",
    formula.krige = res1 ~ 1, vgm.args = "Lin", nmaxkrige = 11, idp = 0.6, nmaxidw
    = 11, lambda = 2, hybrid.parameter = 3)
```

The predictions need to be back-transformed to percentage data.

```
lmkrigelmidwpred1$predictions <- exp(lmkrigelmidwpred1$predictions) - 1
```

```
range(lmkrigelmidwpred1$predictions)
```

```
[1]    0.6796 296.6838
```

A portion of the transformed predictions that are beyond the range of percentage data need to be corrected by resetting the faulty predictions to the nearest bound of the data range by following the example provided in Section 10.1.3.

The spatial distribution of predictions by *LMIDW* with $\lambda_m = 2/3$ is illustrated in Figure 10.4.

10.4   Hybrid method of *GLM* and *IDW*

We will use the `sponge` data set to demonstrate the application of the hybrid method of *GLM* and *IDW* (*GLMIDW*). For the `sponge` data, an optimal model based on `glm.cv` was developed in Chapter 7. We will use the predictive variables in `glm.cv` to construct the optimal *GLM* model for *GLMIDW* first.

```
library(spm)
data(sponge)
glm.opt <- data.frame(sponge[, c(1:3)], easting2 = sponge$easting^2)
```

```
glm1 <- glm(sponge ~ easting + easting2, glm.opt, family = poisson)
```

Then produce the residuals of `glm1` for *IDW* and develop an *IDW* model based on the residuals.

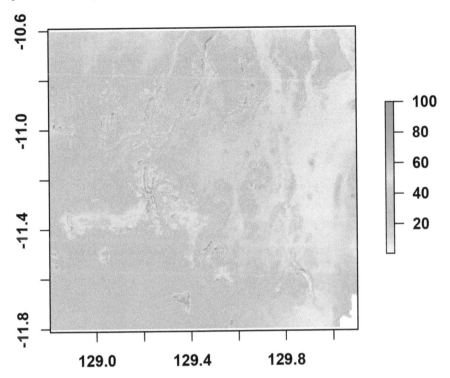

FIGURE 10.4: The spatial patterns of predictions using *LMIDW* with `lambda_m` = 2/3.

```
library(gstat)
glm.opt$res1 <- sponge$sponge - glm1$fitted.values
idw1 <- gstat(id = "res1", formula = res1 ~ 1, locations = ~ easting + northing,
    data = glm.opt, set = list(idp = 2), nmax = 12)
```

Finally, produce the predictions of *GLMIDW* using `glm1` and `idw1` for the training data `glm.opt`.

```
glm1.pred <- predict(glm1, glm.opt, type = "response")
idw.pred <- predict(idw1, glm.opt)$res1.pred
glmidw.pred <- glm1.pred + idw.pred
```

Similar to *LMIDW*, the hybrid method, *GLMIDW*, is also an *exact* method. Again, the *goodness of fit* cannot be used as a measure of predictive accuracy of *GLMIDW*.

If `idp` is specified as 0 in model `idw1`, then *KNN* is used, and the hybrid method resulted would be *GLMKNN*. If `nmax` is specified as 1, then *NN* is used, and the hybrid method resulted would be *GLMNN*.

10.4.1 Variable selection and parameter optimization based on predictive accuracy

For *GLMIDW*, *variable selection* can be conducted with the methods and procedures introduced in Chapter 7 for *GLM*. We will only demonstrate how to estimate the optimal parameters for *GLMIDW* based on the residuals of the optimal *GLM* predictive model. This optimization can be done using `glmidwcv` that has been introduced in Section 10.1.

We will demonstrate the *parameter optimization* of `idp` and `nmaxidw` for *GLMIDW* using `glmidwcv` with the same formula as that for `glm1`.

```
glm.opt2 <- data.frame(sponge[, c(1:3)], easting2 = sponge$easting^2)
names(glm.opt2) <- c("long", "lat", "sponge", "long2")
model <- sponge ~ long + long2

longlat <- sponge[, c(1, 2)]
names(longlat) <- c("long", "lat")


y <- sponge$sponge
idp <- (1:20) * 0.2
nmax <- c(5:40)
glmidwopt <- array(0, dim = c(length(idp), length(nmax)))

for (i in 1:length(idp)) {
  for (j in 1:length(nmax)) {
    set.seed(1234)
    glmidwcv1 <- glmidwcv(formula = model, longlat = longlat, trainxy = glm.opt2, y
        = y, family = "poisson", idp = idp[i], nmaxidw = nmax[j], validation = "CV",
        predacc = "VEcv")
    glmidwopt[i, j] <- glmidwcv1
  }
}
```

The optimal `idp` and `nmax` that lead to the maximal predictive accuracy are

```
glmidw.para.opt <- which (glmidwopt == max(glmidwopt), arr.ind = T)
idp[glmidw.para.opt[, 1]]
```

```
[1] 1
```

```
nmax[glmidw.para.opt[, 2]]
```

```
[1] 38
```

These parameters should be used for the *GLMIDW* predictive model.

In addition, `glmidwcv` can also be used to assess the predictive accuracy of *GLMKNN* and *GLMNN* if `idp` is specified as 0 and `nmaxidw` as 1 in `glmidwcv`, respectively.

10.4.2   Predictive accuracy

We will use the `model` and `y` in the previous section and the optimal `idp` and `nmaxidw` to produce the *stabilized predictive accuracy* of *GLMIDW*.

```
set.seed(1234)
n <- 100
glmidwvecv <- NULL
for (i in 1:n) {
  glmidwcv1 <- glmidwcv(formula = model, longlat = longlat, trainxy = glm.opt2, y
      = y, family = "poisson", idp = 1, nmaxidw = 38, validation = "CV", predacc =
      "VEcv")
  glmidwvecv[i] <- glmidwcv1
}
```

The median and range of predictive accuracy of *GLMIDW* based on 100 repetitions of 10-fold cross-validation using `glmidwcv` are

```
median(glmidwvecv)
```

```
[1] 32.27

range(glmidwvecv)

[1] 14.58 40.43
```

10.4.3 Predictions

The predictions can be generated with the function `glmidwpred` in the `spm2` package.

We will use the `model` and the optimal `idp` and `nmaxidw` to produce spatial predictions for the area based on the `sponge.grid` data set.

```
glm.opt.grid <- data.frame(sponge.grid[, c(1:2)], easting2 = sponge.grid$easting
    ^2)
names(glm.opt.grid) <- c("long", "lat", "long2")

glmidwpred1 <- glmidwpred(formula = model, longlat = longlat, trainxy = glm.opt2,
y = glm.opt2$sponge, longlatpredx = glm.opt.grid[, c(1:2)], predx = glm.opt.grid,
    family = "poisson", idp = 1, nmaxidw = 38)
```

The range of the predictions changes from 13.7 to 14.6, which is pretty narrow.

The spatial distribution of *GLMIDW* predictions is illustrated in Figure 10.5. The patterns of the predictions largely represent the changes in longitude and are not as expected in comparison with the predictions of other methods in Chapter 8 and in Li, Alvarez, et al. (2017). This suggests that the predictions are largely artifact and the method used does not suit the data, so different modeling methods should be used instead. It should be noted that the abrupt changes in the predictions are due to the choice of `nmaxidw`. The guidelines on how to assess the reliability of the predictions provided in Section 7.4.6 are also applicable to *GLMIDW*.

10.5 Hybrid method of *GLM* and *OK*

We will also use the `sponge` data set and `glm.cv` to demonstrate the application of the hybrid method of *GLM* and *OK* (*GLMOK*). To develop an optimal predictive model and generate spatial predictions, we need to apply *OK* to the residuals of `glm1` that are stored in `glm.opt`. We will firstly conduct a variogram modeling for *GLMOK*.

```
data.train2 <- glm.opt
coordinates(data.train2) = ~ easting + northing

vgm2 <- variogram(object = res1 ~ 1, data.train2)

model.2 <- fit.variogram(vgm2, vgm(mean(vgm2$gamma), "Sph", mean(vgm2$dist), min(
    vgm2$gamma) / 10))
```

Then use `model.2` to make *OK* predictions based on `res1` in the training data set `data.train2`, and finally make the predictions of *GLMOK*.

```
pred.krige2 <- krige(res1 ~ 1, data.train2, data.train2, model = model.2, nmax =
    12)

names(as.data.frame(pred.krige2))
glmok.pred <- as.data.frame(pred.krige2)[, 3] + glm1.pred
```

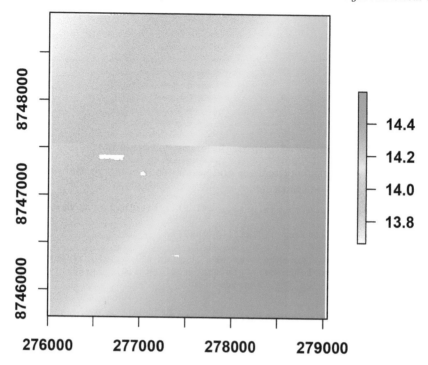

FIGURE 10.5: The spatial patterns of predictions using *GLMIDW*.

Similar to *LMOK*, the hybrid method, *GLMOK*, is also an *exact* method. Again, the *goodness of fit* cannot be used as a measure of predictive accuracy of *GLMOK*.

If beta is specified for pred.krige2, then *SK* is used, and the hybrid method resulted would be *GLMSK*. If block is specified, then *BK* is used, and the hybrid method resulted would be *GLMBOK* or *GLMBSK* depending on the specification of beta. For ways to specify these arguments, see relevant sections in Chapter 5.

If the res1 ~ 1 in vgm2 above is extended to include external variable(s) (e.g., res1 ~ tpi3), then the hybrid method resulted would be *GLMKED*; but this extension should not be used because any possible external trends should have already been taken care of in *GLM* model, and regarding such extension further information has been provided in Section 10.2.

10.5.1 Variable selection and parameter optimization based on predictive accuracy

Similar to *GLMIDW*, for *GLMOK variable selection* can be conducted using the methods and procedures introduced in Chapter 7 for *GLM*. We only need to demonstrate how to optimize parameters for *GLMOK* based on the residuals of the optimal *GLM* predictive model. This optimization can be done with the function glmkrigecv in the spm2 package. The details of glmkrigecv are introduced in Section 10.2.1.

We will use glmkrigecv with the model, longlat, y, and glm.opt2 in Section 10.4.1 to perform the *parameter optimization* of vgm.args and nmaxkrige for *GLMOK*.

```
nmax.ok <- c(5:40)
vgm.args <- c("Exp", "Gau", "Sph", "Exc", "Mat", "Ste", "Lin")
glmokopt <- matrix(0, length(nmax.ok), length(vgm.args))
```

```
for (i in 1:length(nmax.ok)) {
  for (j in 1:length(vgm.args)) {
    set.seed(1234)
    glmkrigecv1 <- glmkrigecv(formula.glm = model, longlat = longlat, trainxy =
        glm.opt2, y = y, family = "poisson", nmaxkrige = nmax.ok[i], formula.krige
        = res1 ~ 1, vgm.args = vgm.args[j], validation = "CV", predacc = "VEcv")
    glmokopt[i, j] <- glmkrigecv1
  }
}
```

The optimal `nmaxkrige` and `vgm.args` that lead to the maximal predictive accuracy are

```
glmok.opt <- which(glmokopt == max(glmokopt), arr.ind = T)
nmax.ok[glmok.opt[, 1]]
```

```
[1] 14
```

```
vgm.args[glmok.opt[, 2]]
```

```
[1] "Exc"
```

These parameters should be used for the *GLMOK* predictive model.

As shown in Chapter 5, all other parameters (e.g, `transformation`, `alpha`) can also be optimized and their optimization will not be presented here.

Furthermore, `glmkrigewcv` can also be used to assess the predictive accuracy of other hybrid methods such as *GLMSK* and *GLMBSK* if `beta` or both `beta` and `block` are specified, respectively.

10.5.2 Predictive accuracy

We will use the `model`, `longlat`, `y`, and `glm.opt2` in Section 10.4.1 and the above optimal `nmaxkrige` and `vgm.args` to produce a *stabilized predictive accuracy*.

```
set.seed(1234)
n <- 100
glmokvecv <- NULL
for (i in 1:n) {
  glmokcv1 <- glmkrigecv(formula.glm = model, longlat = longlat, trainxy = glm.
      opt2, y = y, family = "poisson", nmaxkrige = 14, formula.krige = res1 ~ 1,
      vgm.args = "Exc", validation = "CV", predacc = "VEcv")
  glmokvecv[i] <- glmokcv1
}
```

The median and range of predictive accuracy of *GLMOK* based on 100 repetitions of 10-fold cross-validation using `glmkrigecv` are

```
median(glmokvecv)
```

```
[1] 27.71
```

```
range(glmokvecv)
```

```
[1]  8.847 36.369
```

10.5.3   Predictions

The predictions of *GLMOK* can be generated with the function `glmkrigepred` in the `spm2` package.

We will use the `model`, `longlat`, `y`, and `glm.opt2` in Section 10.4.1 and the above optimal `vgm.args` and `nmaxkrige` to produce spatial predictions of `sponge` for the area based on the `glm.opt.grid` data set in Section 10.4.3.

```
glmkrigepred1 <- glmkrigepred(formula.glm = model, longlat = longlat, trainxy =
    glm.opt2, y = glm.opt2$sponge, longlatpredx = glm.opt.grid[, c(1:2)], predx =
    glm.opt.grid, family = "poisson", transformation = "none", formula.krige =
    res1 ~ 1, vgm.args = "Exc", nmaxkrige = 14)
```

The range of the predictions changes from 14.27 to 15.1, which is pretty narrow.

The spatial distribution of *GLMOK* predictions is illustrated in Figure 10.6. The patterns of the predictions are largely similar to that of *GLMIDW*. As discussed in the previous section for *GLMIDW*, such predictions are largely artifact and different modeling methods should be used.

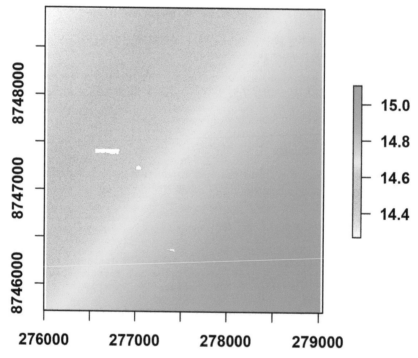

FIGURE 10.6: The spatial patterns of predictions using *GLMOK*.

10.6   Hybrid methods of *GLM*, *OK*, and *IDW*

For the hybrid methods of *GLM*, *OK*, and *IDW* (*GLMOKIDW*), we will also use the `sponge` data set to demonstrate their applications. To develop an optimal predictive model for and generate spatial predictions of *GLMOKIDW*, firstly, we need to develop an optimal *GLM* predictive model, and then we apply *IDW* and *OK* to the residuals of the model, which have been done and stored in `glm1.pred` and `idw.pred` in Section 10.4 and `pred.krige2` in Section 10.5 for *GLMIDW* and *GLMOK*. We will use them to demonstrate the generation of the predictions of *GLMOKIDW* methods.

1. Average of *GLMOK* and *GLMIDW*

The average of *GLMOK* and *GLMIDW* (*GLMOKGLMIDW*) is a *GLMOKIDW* method with $\lambda_m = \lambda_k = 1/2$ according to Equation (10.5). We can produce the predictions of *GLMOKGLMIDW* as follows:

```
lambdai <- 1 / 2
lambdak <- 1 / 2
glmokglmidw.pred <- glm1.pred + lambdak * as.data.frame(pred.krige2)[, 3] +
    lambdai * idw.pred
```

Similar to *LMOKLMIDW*, the hybrid method, *GLMOKGLMIDW*, is also an *exact* method.

2. Average of *GLM*, *GLMOK*, and *GLMIDW*

The average of *GLM*, *GLMOK*, and *GLMIDW* (*GLMGLMOKGLMIDW*) is a *GLMOKIDW* method with $\lambda_m = \lambda_k = 1/3$ according to Equation (10.5). Its predictions can be produced by

```
lambdai <- 1 / 3
lambdak <- 1 / 3
glmglmokglmidw.pred <- glm1.pred + lambdak * as.data.frame(pred.krige2)[, 3] +
    lambdai * idw.pred
```

Since the training data were used for `glm1.pred`, `idw.pred`, and `pred.krige2`, the predictions are also the fitted values of *GLMGLMOKGLMIDW*. Due to $\lambda_m + \lambda_k = 2/3$, the fitted values are no longer the same as the observed values, thus *GLMGLMOKGLMIDW* is not an exact method.

Similar to *LMOKIDW*, λ_m and λ_k for *GLMOKIDW* can also take any values within their range (i.e., 0 to 1) as long as their sum is 1 or 2/3. Their values can be determined according to the predictive accuracy of the hybrid method resulted as shown below.

10.6.1 Variable selection and parameter optimization based on predictive accuracy

Similar to *LMkrigingIDW* methods, with various specifications of `idp`, `nmaxidw`, `beta`, `block`, `lambda`, and `hybrid.parameter` in `glmkrigeidwcv`, many hybrid methods of *GLM*, *kriging*, and *IDW* (*GLMkrigingIDW*) can be resulted (Table 10.2).

If `lambda` is < 1, more weight is placed on *kriging*; and if it is > 1, more weight is placed on *IDW*.

If `lambda` is 0 and the default `hybrid.parameter` is used, then *IDW* is not considered and the method resulted is *GLMkriging*. If `lambda` is 2 and the default `hybrid.parameter` is used, then the method resulted is *GLMIDW*.

When `hybrid.parameter` is 3, if `lambda` is 0, then the method resulted is equivalent to the average of *GLM*, *GLMkriging*, and *GLMkriging* (*GLMGLMkrigingGLMkriging*), and if `lambda` is 2, then the method resulted is equivalent to the average of *GLM*, *GLMIDW* and *GLMIDW* (*GLMGLMIDWGLMIDW*). According to Equations (10.8) and (10.9), *GLMGLMkriging-GLMkriging* can be expressed as *GLMkriging* with $\lambda_k = 2/3$, and *GLMGLMIDWGLMIDW* as *GLMIDW* with $\lambda_m = 2/3$.

With various specifications of `idp`, `nmaxidw`, `beta`, `block`, `lambda`, and `hybrid.parameter`, many hybrid methods of *GLM, kriging,* and *IDW* (*GLMkrigingIDW*) can be resulted (Table 10.2).

When `0 < lambda < 2`, then numerous hybrid methods of *GLM, kriging,* and *IDW* could be possibly derived according to Equation (10.5) and implemented via `glmkrigeidwcv`, such as *GLMOKIDW* with $\lambda_m = 0.88/3$ and $\lambda_k = 1.12/3$, and *GLMSKKNN* with $\lambda_m = 1.21/2$ and $\lambda_k = 0.79/2$.

TABLE 10.2: *GLMkrigingIDW* methods can be derived and implemented via `glmkrigeidwcv` in relation to `idp`, `nmaxidw`, `beta`, `block`, `lambda`, and `hybrid.parameter` (ϕ); and λ_m and λ_k are from Equations (10.5), (10.6) and (10.7).

No.	idp	nmaxidw	beta	block	lambda	ϕ	Hybrid methods
1	!= 0	> 0	NULL	0	1	2	*GLMOKGLMIDW*
2	0	> 1	NULL	0	1	2	*GLMOKGLMKNN*
3	0	1	NULL	0	1	2	*GLMOKGLMNN*
4	!= 0	> 0	not NULL	0	1	2	*GLMSKGLMIDW*
5	0	> 1	not NULL	0	1	2	*GLMSKGLMKNN*
6	0	1	not NULL	0	1	2	*GLMSKGLMNN*
7	!= 0	> 0	NULL	> 0	1	2	*GLMBOKGLMIDW*
8	0	> 1	NULL	> 0	1	2	*GLMBOKGLMKNN*
9	0	1	NULL	> 0	1	2	*GLMBOKGLMNN*
10	!= 0	> 0	not NULL	> 0	1	2	*GLMBSKGLMIDW*
11	0	> 1	not NULL	> 0	1	2	*GLMBSKGLMKNN*
12	0	1	not NULL	> 0	1	2	*GLMBSKGLMNN*
13	!= 0	> 0	NULL	0	0	2	*GLMOK*
14	!= 0	> 0	not NULL	0	0	2	*GLMSK*
15	!= 0	> 0	NULL	> 0	0	2	*GLMBOK*
16	!= 0	> 0	not NULL	> 0	0	2	*GLMBSK*
17	!= 0	> 0	NULL	0	2	2	*GLMIDW*
18	0	> 1	NULL	0	2	2	*GLMKNN*
19	0	1	NULL	0	2	2	*GLMNN*
20	!= 0	> 0	NULL	0	1	3	*GLMGLMOKGLMIDW*
21	0	> 1	NULL	0	1	3	*GLMGLMOKGLMKNN*
22	0	1	NULL	0	1	3	*GLMGLMOKGLMNN*
23	!= 0	> 0	not NULL	0	1	3	*GLMGLMSKGLMIDW*
24	0	> 1	not NULL	0	1	3	*GLMGLMSKGLMKNN*
25	0	1	not NULL	0	1	3	*GLMGLMSKGLMNN*
26	!= 0	> 0	NULL	> 0	1	3	*GLMGLMBOKGLMIDW*
27	0	> 1	NULL	> 0	1	3	*GLMGLMBOKGLMKNN*
28	0	1	NULL	> 0	1	3	*GLMGLMBOKGLMNN*
29	!= 0	> 0	not NULL	> 0	1	3	*GLMGLMBSKGLMIDW*
30	0	> 1	not NULL	> 0	1	3	*GLMGLMBSKGLMKNN*
31	0	1	not NULL	> 0	1	3	*GLMGLMBSKGLMNN*
32	!= 0	> 0	NULL	0	0	3	*GLMOK* with $\lambda_k = 2/3$
33	!= 0	> 0	not NULL	0	0	3	*GLMSK* with $\lambda_k = 2/3$
34	!= 0	> 0	NULL	> 0	0	3	*GLMBOK* with $\lambda_k = 2/3$
35	!= 0	> 0	not NULL	> 0	0	3	*GLMBSK* with $\lambda_k = 2/3$
36	!= 0	> 0	NULL	0	2	3	*GLMIDW* with $\lambda_m = 2/3$
37	0	> 1	NULL	0	2	3	*GLMKNN* with $\lambda_m = 2/3$
38	0	1	NULL	0	2	3	*GLMNN* with $\lambda_m = 2/3$

The hybrid methods from No. 1 to No. 31 in Table 10.2 are exact methods, and the rest are not exact methods.

All these methods can be implemented and assessed via `glmkrigeidwcv`. Furthermore, the most accurate method can be selected from all these hybrid methods based on predictive accuracy with `glmkrigeidwcv` as demonstrated below.

Variable selection for *GLMkrigingIDW* can be conducted using the same methods and procedures introduced in Chapter 7 for *GLM*. We will demonstrate how to estimate the optimal parameters for *IDW* and *kriging* components of the hybrid methods based on the residuals of the optimal *GLM* predictive model. This optimization can be done with `glmkrigeidwcv`, which has been introduced in Section 10.3.1 for *LMOKIDW*.

We can use the optimal `idp`, `nmaxidw`, `vgm.args`, and `nmaxkrige` in Sections 10.4 and 10.5 to conduct further *parameter optimization*. These optimal parameters were estimated or selected separately for *GLMIDW* and *GLMOK*. We can also estimate or select optimal parameters for these arguments using `glmkrigeidwcv` for *GLMOKIDW*, and the optimization can be easily done by following the examples provided for *parameter optimization* for *GLMIDW* and *GLMOK* in Sections 10.4 and 10.5, as well as the example for estimating the optimal `lambda` below.

Below is an example using `glmkrigeidwcv` with the optimal parameters to estimate the optimal value of `lambda` for *GLMOKIDW* with `hybrid.parameter = 2`.

```
lambda3 <- c(0:100)*0.02
lambda3opt <- NULL
for (i in 1:length(lambda3)) {
  set.seed(1234)
  glmkrigeidwcv1 <- glmkrigeidwcv(formula.glm = model, longlat = longlat,
      trainxy =  glm.opt2, y = y, family = "poisson", transformation = "none",
      formula.krige = res1 ~ 1, vgm.args = "Exc", nmaxkrige = 14, idp = 1, nmaxidw
      = 38, lambda = lambda3[i], hybrid.parameter = 2, validation = "CV", predacc
      = "VEcv")
  lambda3opt[i] <- glmkrigelmidwcv1
}
```

The optimal value of `lambda` that leads to the maximal predictive accuracy is

```
lambda3.opt <- which (lambda3opt == max(lambda3opt))
lambda3[lambda3.opt]
```

```
[1] 1.9
```

The maximal predictive accuracy is

```
max(lambda3opt)
```

```
[1] 35.62
```

The parameters estimated can also be used to optimize `lambda` for *GLMOKIDW* with `hybrid.parameter = 3` in the same way as *GLMOKIDW* with `hybrid.parameter = 2`. The results are in `lambda4opt`, and the optimal `lambda` and the maximal predictive accuracy are

```
lambda4.opt <- which (lambda4opt == max(lambda4opt))
lambda4[lambda4.opt]
```

```
[1] 2
```

```
max(lambda4opt)
```

```
[1] 34.37
```

The predictive accuracy of the hybrid model resulted is lower than that of *GLMOKIDW* with `hybrid.parameter = 2`. Hence, the *GLMOKIDW* with `hybrid.parameter = 2` and a `lambda = 1.9` will be used.

10.6.2 Predictive accuracy

We will use the `model` and `y` for `glm1` and the optimal `nmaxkrige`, `vgm.args`, `idp`, and `nmaxidw` to get the *stabilized predictive accuracy* of *GLMOKIDW* with $\lambda_m = 1.9/2$ and $\lambda_k = 0.1/2$ (see Equation (10.5)).

```
set.seed(1234)
n <- 100
glmkrigeglmidwvecv <- NULL
for (i in 1:n) {
  glmkrigeglmidwcv1 <- glmkrigeidwcv(formula.glm = model, longlat = longlat,
      trainxy = glm.opt2, y = y, family = "poisson", transformation = "none",
      formula.krige = res1 ~ 1, vgm.args = "Exc", nmaxkrige = 14, idp = 1, nmaxidw
      = 38, lambda = 1.9, hybrid.parameter = 2, validation = "CV", predacc = "
      VEcv")
  glmkrigeglmidwvecv[i] <- glmkrigeglmidwcv1
}
```

The median and range of predictive accuracy of the *GLMOKIDW* model based on 100 repetitions of 10-fold cross-validation using `glmkrigeidwcv` are

```
median(glmkrigeglmidwvecv)
```

```
[1] 32.21
```

```
range(glmkrigeglmidwvecv)
```

```
[1] 14.50 40.33
```

10.6.3 Predictions

The predictions of *GLMOKIDW* can be generated with `glmkrigeidwpred` for the hybrid methods, which has been introduced in Section 10.3.3.

We will use `glmkrigeidwpred` with $\lambda = 1.9$ and *hybrid.parameter* = 2 to show how to generate spatial predictions. The area based on the `sponge.grid` data set will be used as an example.

```
glmkrigeglmidwpred1 <- glmkrigeidwpred(formula.glm = model, longlat = glm.opt2[,
    1:2], trainxy = glm.opt2, y = glm.opt2$sponge, longlatpredx = glm.opt.grid[,
    1:2], predx = glm.opt.grid, family = "poisson", transformation = "none",
    formula.krige = res1 ~ 1, vgm.args = "Exc", nmaxkrige = 14, idp = 1, nmaxidw =
    38, lambda = 1.9, hybrid.parameter = 2)
```

```
names(glmkrigeglmidwpred1)
```

```
[1] "predictions"
```

```
range(glmkrigeglmidwpred1$predictions)
```

```
[1] 13.70 14.62
```

The spatial distribution of predictions by *GLMOKIDW* with $\lambda = 1.9$ is illustrated in Figure 10.7. Similar to *GLMIDW*, the abrupt changes in the predictions are resulted from the number used for `nmaxidw`.

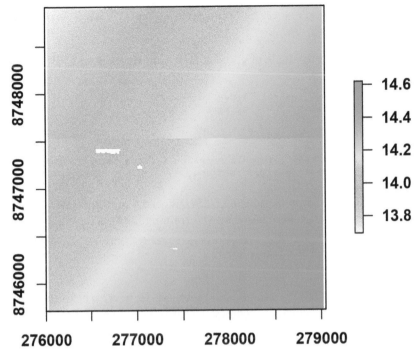

FIGURE 10.7: The spatial patterns of predictions using *GLMOKIDW* with `lambda = 1.9`.

10.7   Hybrid method of *GLS* and *IDW*

We will use the `petrel` data set to demonstrate the application of the hybrid method of *GLS* and *IDW* (*GLSIDW*). The predictive variables and parameters of `gls2` for the `petrel` data in Chapter 7 will be used to demonstrate the application of *GLSIDW*.

Firstly, we need to re-derive the relevant parameters that were used in `gls2` for `correlation` argument in the function `gls`.

```
library(spm)
data(petrel)

diag <- sqrt((diff(range(petrel$long)))^2+(diff(range(petrel$lat)))^2)
gravel <- petrel[, c(1, 2, 5)]
coordinates(gravel) = ~long+lat
gravel$loggravel <- log(gravel$gravel + 1)

vgm1 <- variogram(loggravel ~ 1, gravel)

psill1 = (max(vgm1$gamma) + median(vgm1$gamma)) / 2
range1 = 0.1 * diag
nugget1 = min(vgm1$gamma)
```

Secondly, the derived parameters and the predictive variables in `gls2` will be used for `gls` modeling.

```
library(nlme)
gls1 <- gls(log(gravel + 1) ~ long + lat + bathy + dist + I(long^2) + I(lat^2)
+ I(lat^3) + I(bathy^2) + I(bathy^3) + I(dist^2) + I(dist^3) + I(relief^2)
```

```
+ I(relief^3), petrel, corr = corExp(c(range1, nugget1), form = ~ lat + long,
    nugget = T))
```

Then we can produce the residuals of `gls1` for *IDW* and develop an *IDW* model based on the residuals.

```
gls1.pred <- predict(gls1, petrel, type="response")
res1 <- log(petrel$gravel + 1) - gls1.pred

gls.res <- data.frame(petrel, res1)

idw1.res <- gstat(id = "res1", formula = res1 ~ 1, locations = ~ long + lat, data
    = gls.res, set = list(idp = 2), nmax = 12)
```

Finally, we can produce the predictions of *GLSIDW* using `gls1` and `idw1`. We will still use the predictions `gls1.pred` that were based on the training data `petrel`.

```
idw.res.pred <- predict(idw1.res, gls.res)$res1.pred
glsidw.pred1 <- gls1.pred + idw.res.pred
glsidw.pred <- exp(glsidw.pred1) - 1
```

Similar to *LMIDW*, *GLSIDW* is also an *exact* method. However, the *goodness of fit* is not necessarily related to the predictive accuracy, which is also applicable to *GLSIDW*.

In addition, if `idp` is specified as 0 in model `idw1`, then *KNN* is used, and the hybrid method resulted would be *GLSKNN*. If `nmax` is specified as 1, then *NN* is used, and the hybrid method resulted would be *GLSNN*.

10.7.1   Variable selection and parameter optimization based on predictive accuracy

For *GLSIDW*, *variable selection* can be conducted using the procedures introduced in Section 7.5 for *GLS*. We will only demonstrate how to estimate the optimal parameters for the *GLSIDW* based on the residuals of the optimal *GLS* predictive model. This optimization can be done with the function `glsidwcv` in the `spm2` package. We will introduce `glsidwcv` first and then use it to conduct parameter optimization.

1. Function `glsidwcv`

For `glsidwcv`, the following arguments are required:
(1) `model`, a formula defining response variable and predictive variables;
(2) `longlat`, `trainxy`, `y`, `idp`, `nmaxidw`, `validation`, `cv.fold`, and `predacc`, are the same as those for `glmidwcv`; and
(3) `corr.args`, arguments for 'gls', which have been detailed in Section 7.5.1.
For details, see `?glsidwcv`.

2. Parameter optimization based on predictive accuracy

We can estimate the optimal `idp` and `nmax` for *GLSIDW* based on the residuals of the optimal *GLS* predictive model. We will use `glsidwcv` with the same formula as that for `gls1` model to demonstrate the *parameter optimization*.

```
gravel <- petrel[, c(1, 2, 6:9, 5)]
longlat <- petrel[, c(1, 2)]

model <- log(gravel + 1) ~ long + lat + bathy + dist + I(long^2) + I(lat^2) +
I(lat^3) + I(bathy^2) + I(bathy^3) + I(dist^2) + I(dist^3) + I(relief^2) + I(
    relief^3)
```

```
y <- log(gravel$gravel +1)
idp <- (1:20) * 0.2
nmax <- c(5:40)
glsidwopt <- array(0, dim = c(length(idp), length(nmax)))

for (i in 1:length(idp)) {
  for (j in 1:length(nmax)) {
    set.seed(1234)
    glsidwcv1 <- glsidwcv(model = model, longlat = longlat, trainxy = gravel, y =
        y, corr = corExp(c(range1, nugget1), form = ~ lat + long, nugget = T), idp
        = idp[i], nmaxidw = nmax[j], validation = "CV", predacc = "VEcv")
    glsidwopt[i, j] <- glsidwcv1
  }
}
```

The optimal `idp` and `nmax` that lead to the maximal predictive accuracy are

```
gls.para.opt <- which (glsidwopt == max(glsidwopt), arr.ind = T)
idp[gls.para.opt[, 1]]
```

`[1] 0.4`

```
nmax[gls.para.opt[, 2]]
```

`[1] 10`

These parameters should be used for the *GLSIDW* predictive model.

In addition, `glsidwcv` can also be used to assess the predictive accuracy of *GLSKNN* and *GLSNN* if `idp` is specified as 0 and `nmax` as 1 in `glsidwcv`, respectively.

10.7.2 Predictive accuracy

We can use `glsidwcv` to produce the *stabilized predictive accuracy* of the *GLSIDW* predictive model. We will use the `model`, `y`, and the optimal `idp` and `nmax` in the previous section to demonstrate the accuracy stabilization.

```
set.seed(1234)
n <- 100
glsidwvecv <- NULL
for (i in 1:n) {
  glsidwcv1 <- glsidwcv(model = model, longlat = longlat, trainxy = gravel, y = y,
      corr = corExp(c(range1, nugget1), form = ~ lat + long, nugget = T), idp =
      0.4, nmaxidw = 10, validation = "CV", predacc = "VEcv")

  glsidwvecv[i] <- glsidwcv1
}
```

The median and range of predictive accuracy of *GKSIDW* based on 100 repetitions of 10-fold cross-validation using `glsidwcv` are

```
median(glsidwvecv)
```

`[1] 35.47`

```
range(glsidwvecv)
```

`[1] 28.18 39.04`

10.7.3   Predictions

The predictions of *GLSIDW* can be generated with the function `glsidwpred` in the `spm2` package.

1. Function `glsidwpred`

For `glsidwpred`, the following arguments are required:
(1) `model`, `longlat`, `trainxy`, `y`, `corr.args`, `idp`, and `nmaxidw`, which are the same as those for `glmidwcv`, `gls`, and `glsidwcv`;
(2) `longlatpredx`, a dataframe contains the longitude and latitude of point locations (i.e., the centers of grids) to be predicted; and
(3) `predx`, a dataframe of predictive variables for the grids to be predicted.
For details, see `?glsidwpred`.

2. Generation of spatial predictions using `glsidwpred`

We will use the above `model`, `y`, and the optimal `idp` and `nmaxidw` to produce spatial predictions for the area based on the `petrel.grid` data set.

```
glsidwpred1 <- glsidwpred(model = model, longlat = longlat, trainxy = petrel,
y = y, longlatpredx = petrel.grid[, c(1:2)], predx = petrel.grid, corr = corExp(c(
    range1, nugget1), form = ~ lat + long, nugget = T), idp = 0.4, nmaxidw = 10)
```

The predictions need to be back-transformed to percentage data.

```
glsidwpred1$predictions <- exp(glsidwpred1$predictions) - 1
```

The range of the predictions are

```
range(glsidwpred1$predictions)
```

```
[1]   -0.8191 226.5991
```

A portion of the transformed predictions are beyond the range of percentage data and need to be corrected by resetting the faulty predictions to the nearest bound of the data range.

```
glsidwpred <- glsidwpred1
glsidwpred$predictions[glsidwpred$predictions <= 0] <- 0
glsidwpred$predictions[glsidwpred$predictions >= 100] <- 100
```

The spatial distribution of predictions by *GLSIDW* is illustrated in Figure 10.8.

10.8   Hybrid method of *GLS* and *OK*

For the hybrid method of *GLS* and *OK* (*GLSOK*), we will also use the `petrel` data set to demonstrate its application. To develop an optimal predictive model and generate spatial predictions, we need to apply *OK* to the residuals of `gls1` that are stored in `gls.res` in the previous section. We will conduct a variogram modeling for *GLSOK* first.

```
data.train <- gls.res
coordinates(data.train) = ~ long + lat
vgm1 <- variogram(object = res1 ~ 1, data.train)

model.1 <- fit.variogram(vgm1, vgm(mean(vgm1$gamma), "Sph", mean(vgm1$dist), min(
    vgm1$gamma) / 10))
```

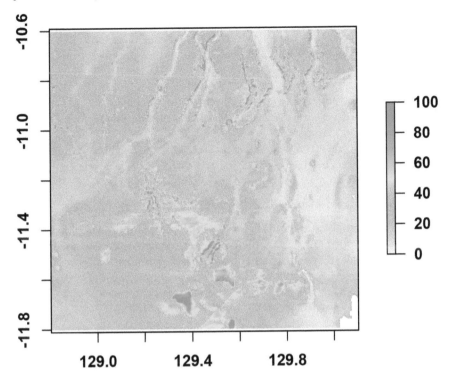

FIGURE 10.8: The spatial patterns of predictions using *GLSIDW*.

Then use `model.1` to make *OK* predictions based on `res1` in the training data set `data.train`, and finally make the predictions of *GLSOK*.

```
krige.res.pred <- krige(res1 ~ 1, data.train, data.train, model = model.1, nmax =
    12)
```

```
names(as.data.frame(krige.res.pred))
glsok.pred1 <- as.data.frame(krige.res.pred)[, 3] + gls1.pred
glsok.pred <- exp(glsok.pred1) - 1
```

Similar to *LMOK*, *GLSOK* is also an *exact* method. Again, the *goodness of fit* is not a measure of the predictive accuracy, which is also applicable to *GLSOK*.

In addition, if `beta` is specified for `krige.res.pred`, then *SK* is used, and the hybrid method resulted would be *GLSSK*. If `block` is specified, then *BK* is used, and the hybrid method resulted would be *GLSBOK* or *GLSBSK* depending on the specification of `beta`. For ways to specify these arguments, see relevant sections in Chapter 5.

If the `res1 ~ 1` in `vgm1` above is extended to include external variable(s) (e.g., `res1 ~ bathy`), then the hybrid method resulted would be *GLSKED*; but this extension should not be used because any possible external trends should have already been taken care of in *GLS* model, and as to such extension further information has been provided in Section 10.2.

10.8.1 Variable selection and parameter optimization based on predictive accuracy

With regard to *variable selection* for *GLSOK*, it can be conducted using the procedures introduced in Section 7.5 for *GLS*. Here we only need to demonstrate how to estimate the optimal parameters for the *GLSOK* based on the residuals of the optimal *GLS* predictive model. This optimization can be carried out with the function `glskrigecv` in the `spm2` package. We will introduce `glskrigecv` first and then use it to conduct parameter optimization.

1. Function `glskrigecv`

For `glskrigecv`, the following arguments may need to be specified:

(1) `model`, `longlat`, `trainxy`, `y`, `corr.args`, `validation`, `cv.fold`, and `predacc`, which are the same as those for `glmidwcv` and `gls`; and

(2) `transformation`, `delta`, `formula.krige`, `vgm.args`, `anis`, `alpha`, `block`, `beta`, and `nmaxkrige`, which are the same as those for `glmkrigecv`.

For details, see `?glskrigecv`.

2. Parameter optimization based on predictive accuracy

We can select the optimal `vgm.args` and estimate the optimal `nmaxkrige` for *GLSOK* based on the residuals of the optimal *GLS* predictive model. We will use `glskrigecv` with the same formula as in `gls1` model to demonstrate the *parameter optimization*.

```
nmax.ok <- c(5:40)
vgm.args <- c("Exp", "Gau", "Sph", "Exc", "Mat", "Ste", "Lin")
glsokopt <- matrix(0, length(nmax.ok), length(vgm.args))

for (i in 1:length(nmax.ok)) {
  for (j in 1:length(vgm.args)) {
    set.seed(1234)
    glskrigecv1 <- glskrigecv(model = model, longlat = longlat, trainxy = petrel,
        y = y, transformation = "none", formula.krige = res1 ~ 1, vgm.args = vgm.
        args[j], nmaxkrige = nmax.ok[i], corr.args = corSpher(c(range1, nugget1),
        form = ~ lat + long, nugget = T), validation = "CV", predacc = "VEcv")
    glsokopt[i, j] <- glskrigecv1
  }
}
```

The optimal `nmaxkrige` and `vgm.args` that lead to the maximal predictive accuracy are

```
glsok.opt <- which(glsokopt == max(glsokopt), arr.ind = T)
nmax.ok[glsok.opt[, 1]]
```

```
[1] 23
```

```
vgm.args[glsok.opt[, 2]]
```

```
[1] "Sph"
```

These parameters should be used for the *GLSOK* predictive model.

All other parameters (e.g., `transformation`, `alpha`) can also be optimized in the same way as shown in Chapter 5, so their optimization will not be presented here.

In addition, `glskrigewcv` can also be used to assess the predictive accuracy of other hybrid methods such as *GLSSK* and *GLSBSK* if `beta` or both `beta` and `block` are specified, respectively.

10.8.2 Predictive accuracy

We can use `glskrigecv` to produce the *stabilized predictive accuracy* of the *GLSOK* model. We will use the `model` and `y` in Section 10.7.1 and the optimal `nmaxkrige` and `vgm.args` to demonstrate the accuracy stabilization as follows:

```
set.seed(1234)
n <- 100
glsokvecv <- NULL
for (i in 1:n) {
  glsokcv1 <- glskrigecv(model = model, longlat = longlat, trainxy = petrel, y = y
      , transformation = "none", formula.krige = res1 ~ 1, vgm.args = "Sph",
      nmaxkrige = 23, corr.args = corSpher(c(range1, nugget1), form = ~ lat + long
      , nugget = T), validation = "CV", predacc = "VEcv")
  glsokvecv[i] <- glsokcv1
}
```

The median and range of predictive accuracy of *GLSOK* based on 100 repetitions of
glskrigecv, median(glsokvecv) are

```
median(glsokvecv)
```

```
[1] 35.14
```

```
range(glsokvecv)
```

```
[1] 29.75 40.50
```

10.8.3   Predictions

The predictions can be generated using the function glskrigepred in the spm2 package for
GLSOK.

1. Function glskrigepred

For glskrigepred, the following arguments may need to be specified:
(1) model, longlat, trainxy, predx, y, longlatpredx, and corr.args, which are the same as those
for gls, glsidwcv, and glsidwpred; and
(2) transformation, delta, formula.krige, vgm.args, anis, alpha, block, beta, and nmaxkrige, which
are the same as those for glmkrigecv.
For details, see ?glskrigepred.

2. Generation of spatial predictions using glskrigepred

We will use the model and y in Section 10.7.1 and the optimal vgm.args and nmaxkrige to
produce spatial predictions for the area based on the petrel.grid data set.

```
glskrigepred1 <- glskrigepred(model = model, longlat = longlat, trainxy = petrel,
    predx = petrel.grid, y = log(petrel$gravel +1), longlatpredx = petrel.grid[, c
    (1:2)], transformation = "none", formula.krige = res1 ~ 1, vgm.args = "Sph",
    nmaxkrige = 23, corr.args = corSpher(c(range1, nugget1), form = ~ lat + long,
    nugget = T))
```

The predictions need to be back-transformed to percentage data.

```
glskrigepred1$predictions <- exp(glskrigepred1$predictions) - 1
```

```
range(glskrigepred1$predictions)
```

```
[1]  -0.7752 189.1739
```

A portion of the transformed predictions are beyond the range of percentage data and need
to be corrected by resetting the faulty predictions to the nearest bound of the data range.
The spatial distribution of the corrected predictions of *GLSOK* is illustrated in Figure 10.9.

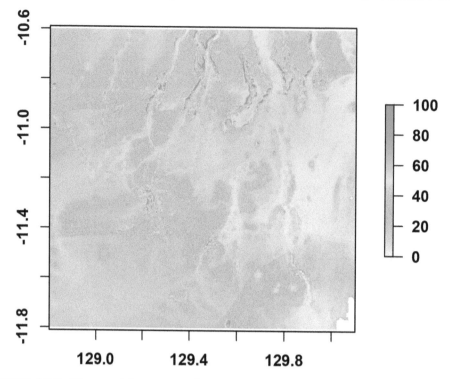

FIGURE 10.9: The spatial patterns of predictions using *GLSOK*.

10.9   Hybrid methods of *GLS*, *OK*, and *IDW*

For the hybrid methods of *GLS*, *OK*, and *IDW* (*GLSOKIDW*), we will also use the `petrel` data set to demonstrate their applications. To develop an optimal predictive model for and generate spatial predictions of *GLSOKIDW*, firstly, we need to develop an optimal *GLS* predictive model and then we apply *IDW* and *OK* to the residuals of the model, which have been done and stored in `gls1.pred`, `idw.res.pred`, and `krige.res.pred` in Sections 10.7 and 10.8 for *GLSIDW* and *GLSOK*. We will use them to demonstrate the generation of the predictions of *GLSOKIDW* methods.

1. Average of *GLSOK* and *GLSIDW*

The average of *GLSOK* and *GLSIDW* (*GLSOKGLSIDW*) is a *GLSOKIDW* method with $\lambda_m = \lambda_k = 1/2$ according to Equation (10.5). We will produce the predictions of *GLSOKGLSIDW* using `gls1.pred`, `idw.res.pred`, and `krige.res.pred` as below.

```
lambdai <- 1 / 2
lambdak <- 1 / 2
glsokglsidw.pred1 <- gls1.pred + lambdak * as.data.frame(krige.res.pred)[, 3] +
    lambdai * idw.res.pred

glsokglsidw.pred <- exp(glsokglsidw.pred1) - 1
```

Similar to *LMOKLMIDW* and *GLMOKGLMIDW*, *GLSOKGLSIDW* is also an *exact* method.

2. Average of *GLS*, *GLSOK*, and *GLSIDW*

The average of *GLS*, *GLSOK*, and *GLSIDW* (*GLSGLSOKGLSIDW*) is a *GLSOKIDW* method with $\lambda_m = \lambda_k = 1/3$ according to Equation (10.5). We will produce the predictions of *GLSGLSOKGLSIDW* using `gls1.pred`, `idw.res.pred`, and `krige.res.pred` as below.

```
lambdai <- 1 / 3
lambdak <- 1 / 3
glsglsokglsidw.pred1 <- gls1.pred + lambdak * as.data.frame(krige.res.pred)[, 3] +
    lambdai * idw.res.pred

glsglsokglsidw.pred <- exp(glsglsokglsidw.pred1) - 1
```

Due to $\lambda_m + \lambda_k = 2/3$, the fitted values are no longer the same as the observed values, thus *GLSGLSOKGLSIDW* is not an exact method.

In fact, λ_m and λ_k for *GLSOKIDW* can take any values within their range (i.e., 0 to 1) as long as their sum is 1 or 2/3. Their values can be determined according to the predictive accuracy of the hybrid method resulted as shown below.

10.9.1 Variable selection and parameter optimization based on predictive accuracy

For *GLSOKIDW*, *variable selection* can be conducted using the procedures introduced in Section 7.5 for *GLS*. Here we only need to demonstrate how to estimate or select the optimal parameters for *IDW* and *OK* components of the hybrid methods based on the residuals of the optimal *GLS* predictive model. This optimization can be done with the function `glskrigeidwcv` in the `spm2` package. We will introduce `glskrigeidwcv` first and then use it to conduct parameter optimization.

1. Function `glskrigeidwcv`

For `glskrigeidwcv`, the following arguments may need to be specified:
(1) `model`, `longlat`, `trainxy`, `y`, `corr.args`, `transformation`, `delta`, `formula.krige`, `vgm.args`, `anis`, `alpha`, `block`, `beta`, `nmaxkrige`, `idp`, `nmaxidw`, `validation`, `cv.fold`, and `predacc`, which are the same as those for `glmidwcv`, `glmkrigecv`, and `gls`;
(2) `hybrid.parameter`, either 2 or 3, with the default value of 2; and
(3) `lambda`, ranging from 0 to 2, with the default value of 1.
For details, see `?glskrigeidwcv`.

If `lambda` is < 1, more weight is placed on *kriging*; and if it is > 1, more weight is placed on *IDW*.

If `lambda` is 0 and the default `hybrid.parameter` is used, then *IDW* is not considered and the method resulted is *GLSkriging*. If `lambda` is 2 and the default `hybrid.parameter` is used, then the method resulted is *GLSIDW*.

When `hybrid.parameter` is 3, if `lambda` is 0, then the method resulted is equivalent to the average of *GLS*, *GLSkriging*, and *GLSkriging* (*GLSGLSkrigingGLSkriging*), and if `lambda` is 2, then the method resulted is equivalent to the average of *GLS*, *GLSIDW* and *GLSIDW* (*GLSGLSIDWGLSIDW*). According to Equations (10.8) and (10.9), *GLSGLSkrigingGLSkriging* can be expressed as *GLSkriging* with $\lambda_k = 2/3$, and *GLSGLSIDWGLSIDW* as *GLSIDW* with $\lambda_m = 2/3$.

With various specifications of `idp`, `nmaxidw`, `beta`, `block`, `lambda`, and `hybrid.parameter`, many hybrid methods of *GLS*, *kriging*, and *IDW* (*GLSkrigingIDW*) can be resulted (Table 10.3).

When $0 <$ `lambda` < 2, then numerous hybrid methods of *GLS*, *kriging*, and *IDW* could be possibly derived according to Equation (10.5) and implemented via `glskrigeidwcv`, such as *GLSOKIDW* with $\lambda_m = 0.88/3$ and $\lambda_k = 1.12/3$, and *GLSSKKNN* with $\lambda_m = 1.21/2$ and $\lambda_k = 0.79/2$.

The hybrid methods from No. 1 to No. 31 in Table 10.3 are exact methods, and the rest are not exact methods.

TABLE 10.3: *GLSkrigingIDW* methods can be derived and implemented via `glskrigeidwcv` in relation to `idp`, `nmaxidw`, `beta`, `block`, `lambda`, and `hybrid.parameter` (ϕ); and λ_m and λ_k are from Equations (10.5), (10.6) and (10.7).

No.	idp	nmaxidw	beta	block	lambda	ϕ	Hybrid methods
1	!= 0	> 0	NULL	0	1	2	*GLSOKGLSIDW*
2	0	> 1	NULL	0	1	2	*GLSOKGLSKNN*
3	0	1	NULL	0	1	2	*GLSOKGLSNN*
4	!= 0	> 0	not NULL	0	1	2	*GLSSKGLSIDW*
5	0	> 1	not NULL	0	1	2	*GLSSKGLSKNN*
6	0	1	not NULL	0	1	2	*GLSSKGLSNN*
7	!= 0	> 0	NULL	> 0	1	2	*GLSBOKGLSIDW*
8	0	> 1	NULL	> 0	1	2	*GLSBOKGLSKNN*
9	0	1	NULL	> 0	1	2	*GLSBOKGLSNN*
10	!= 0	> 0	not NULL	> 0	1	2	*GLSBSKGLSIDW*
11	0	> 1	not NULL	> 0	1	2	*GLSBSKGLSKNN*
12	0	1	not NULL	> 0	1	2	*GLSBSKGLSNN*
13	!= 0	> 0	NULL	0	0	2	*GLSOK*
14	!= 0	> 0	not NULL	0	0	2	*GLSSK*
15	!= 0	> 0	NULL	> 0	0	2	*GLSBOK*
16	!= 0	> 0	not NULL	> 0	0	2	*GLSBSK*
17	!= 0	> 0	NULL	0	2	2	*GLSIDW*
18	0	> 1	NULL	0	2	2	*GLSKNN*
19	0	1	NULL	0	2	2	*GLSNN*
20	!= 0	> 0	NULL	0	1	3	*GLSGLSOKGLSIDW*
21	0	> 1	NULL	0	1	3	*GLSGLSOKGLSKNN*
22	0	1	NULL	0	1	3	*GLSGLSOKGLSNN*
23	!= 0	> 0	not NULL	0	1	3	*GLSGLSSKGLSIDW*
24	0	> 1	not NULL	0	1	3	*GLSGLSSKGLSKNN*
25	0	1	not NULL	0	1	3	*GLSGLSSKGLSNN*
26	!= 0	> 0	NULL	> 0	1	3	*GLSGLSBOKGLSIDW*
27	0	> 1	NULL	> 0	1	3	*GLSGLSBOKGLSKNN*
28	0	1	NULL	> 0	1	3	*GLSGLSBOKGLSNN*
29	!= 0	> 0	not NULL	> 0	1	3	*GLSGLSBSKGLSIDW*
30	0	> 1	not NULL	> 0	1	3	*GLSGLSBSKGLSKNN*
31	0	1	not NULL	> 0	1	3	*GLSGLSBSKGLSNN*
32	!= 0	> 0	NULL	0	0	3	*GLSOK* with $\lambda_k = 2/3$
33	!= 0	> 0	not NULL	0	0	3	*GLSSK* with $\lambda_k = 2/3$
34	!= 0	> 0	NULL	> 0	0	3	*GLSBOK* with $\lambda_k = 2/3$
35	!= 0	> 0	not NULL	> 0	0	3	*GLSBSK* with $\lambda_k = 2/3$
36	!= 0	> 0	NULL	0	2	3	*GLSIDW* with $\lambda_m = 2/3$
37	0	> 1	NULL	0	2	3	*GLSKNN* with $\lambda_m = 2/3$
38	0	1	NULL	0	2	3	*GLSNN* with $\lambda_m = 2/3$

All these methods can be implemented and assessed via `glskrigeidwcv`. Furthermore, the most accurate method can be selected from all these hybrid methods based on predictive accuracy with `glskrigeidwcv` as demonstrated below.

2. Parameter optimization based on predictive accuracy

We can use the optimal `idp`, `nmaxidw`, `vgm.args`, and `nmaxkrige` in Sections 10.7 and 10.8 to do further parameter *optimization* for *GLSOKIDW*. These parameters were estimated or selected separately for *GLSIDW* and *GLSOK*. If it is believed that these parameters are not optimal for *GLSOKIDW*, then we need to re-do the optimization. This can be done using `glskrigeidwcv`, but will not be presented here, as it can be easily implemented by following the examples provided for parameter optimization for *GLSIDW* and *GLSOK* in Sections 10.7 and 10.8 as well as the example for the estimation of optimal `lambda` below.

We will provide an example to estimate the optimal value of `lambda` for *GLSOKIDW* with `hybrid.parameter = 2`.

```
lambda5 <- c(0:100)*0.02
lambda5opt <- NULL

for (i in 1:length(lambda5)) {
  set.seed(1234)
  glskrigeglsidwcv1 <- glskrigeidwcv(model = model, longlat = longlat, trainxy =
      petrel, y = log(petrel$gravel +1), corr = corExp(c(range1, nugget1), form =
      ~ lat + long, nugget = T), transformation = "none", formula.krige = res1 ~
      1, vgm.args = "Sph", nmaxkrige = 23, idp = 0.4, nmaxidw = 10, lambda =
      lambda[i], hybrid.parameter = 2, validation = "CV", predacc = "VEcv")
  lambda5opt[i] <- glskrigeglsidwcv1
}
```

The optimal `lambda` that leads to the maximal predictive accuracy is

```
lambda5.opt <- which (lambda5opt == max(lambda5opt))
lambda5[lambda5.opt]
```

```
[1] 1.06
```

The maximal predictive accuracy is

```
max(lambda5opt)
```

```
[1] 37.87
```

Similar to *GLSOKIDW* with `hybrid.parameter = 2`, the parameters estimated can also be used to optimize `lambda` for *GLSOKIDW* with `hybrid.parameter = 3`. The results are in `lambda6opt`, and the optimal `lambda` and maximal predictive accuracy are

```
lambda6.opt <- which (lambda6opt == max(lambda6opt))
lambda6[lambda6.opt]
```

```
[1] 1.74
```

```
max(lambda6opt)
```

```
[1] 37.94
```

The predictive accuracy of the hybrid model resulted is higher than that of *GLSOKIDW* with `hybrid.parameter = 2`. Hence, the *GLSOKIDW* with `hybrid.parameter = 3` and a `lambda = 1.74` will be used.

10.9.2   Predictive accuracy

We can use `glskrigeidwcv` to produce the *stabilized predictive accuracy* of the hybrid methods. We will use the above `model` and `y` in Section 10.7.1 and the optimal `nmaxkrige`, `vgm.args`, `idp`, and `nmaxidw` to stabilize the predictive accuracy of *GLSGLSOKGLSIDW* method with `lambda = 1.74` (i.e., *GLSOKIDW* with $\lambda_k = 0.26/3$ and $\lambda_m = 1.74/3$) as a demonstration.

```
set.seed(1234)
n <- 100
glsglskrigeglsidwvecv <- NULL
for (i in 1:n) {
  glskrigeidwcv1 <- glskrigeidwcv(model = model, longlat = longlat, trainxy =
      petrel, y = log(petrel$gravel +1), corr = corExp(c(range1, nugget1), form =
      ~ lat + long, nugget = T), transformation = "none", formula.krige = res1 ~
      1, vgm.args = "Sph", nmaxkrige = 23, idp = 0.4, nmaxidw = 10, lambda = 1.74,
      hybrid.parameter = 3, validation = "CV", predacc = "VEcv")
  glsglskrigeglsidwvecv[i] <- glskrigeidwcv1
}
```

The median and range of predictive accuracy of the *GLSOKIDW* method based on 100 repetitions of 10-fold cross-validation using `glmkrigeidwcv` are

```
median(glsglskrigeglsidwvecv)
```

```
[1] 36.01
```

```
range(glsglskrigeglsidwvecv)
```

```
[1] 29.62 39.66
```

10.9.3   Predictions

The predictions can be generated using the function `glskrigeidwpred` in the `spm2` package for the hybrid methods.

1. Function `glskrigeidwpred`

For `glskrigeidwpred`, the following arguments may need to be specified:
(1) `model`, `longlat`, `trainxy`, `predx`, `y`, `longlatpredx`, `corr.args`, `transformation`, `delta`, `formula.krige`, `vgm.args`, `anis`, `alpha`, `block`, `beta`, `nmaxkrige`, `idp`, and `nmaxidw`, which are the same as those for `glmkrigecv`, `glsidwcv`, `gls`, and `glsidwpred`; and
(2) `hybrid.parameter` and `lambda`, which are the same as those for `glskrigeidwcv`.
For details, see `?glskrigeidwpred`.

2. Generation of spatial predictions using `glskrigeidwpred`

We will use `model` and `y` in Section 10.7.1 and the optimal `vgm.args`, `nmaxkrige`, `idp`, and `nmaxidw` to produce spatial predictions for the area based on the `petrel.grid` data set. We will use a hybrid method, *GLSGLSOKGLSIDW* method with `lambda = 1.74`, to generate the predictions.

```
glskrigeidwpred1 <- glskrigeidwpred(model = model, longlat = longlat, trainxy =
    petrel, predx = petrel.grid, y = log(petrel$gravel +1), longlatpredx = petrel.
    grid[, c(1:2)], corr = corExp(c(range1, nugget1), form = ~ lat + long, nugget
    = T), transformation = "none", formula.krige = res1 ~ 1, vgm.args = "Sph",
    nmaxkrige = 23, idp = 0.4, nmaxidw = 10, lambda = 1.74, hybrid.parameter = 3)
```

The predictions need to be back-transformed to percentage data.

```
glskrigeidwpred1$predictions <- exp(glskrigeidwpred1$predictions) - 1
```

```
range(glskrigeidwpred1$predictions)
```

```
[1]   -0.8422 183.6652
```

A portion of the transformed predictions are beyond the range of percentage data and need to be corrected by resetting the faulty predictions to the nearest bound of the data range.

The spatial distribution of predictions by *GLSGLSOKGLSIDW* method with `lambda = 1.74` is illustrated in Figure 10.10.

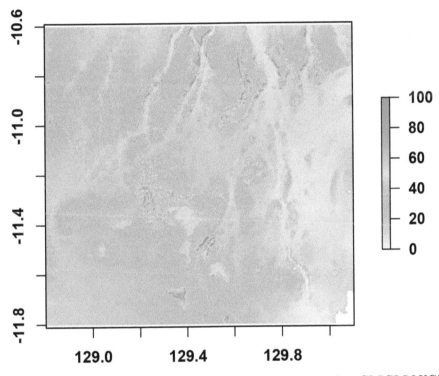

FIGURE 10.10: The spatial patterns of predictions using the *GLSGLSOKGLSIDW* method with `lambda = 1.74`.

11

Hybrids of machine learning methods with geostatistical methods

This chapter introduces the *hybrids* of *machine learning methods* with geostatistical methods, where the geostatistical methods refer to *mathematical methods, univariate geostatistical methods*, or both methods. The machine learning methods are the methods introduced in Chapters 8 and 9, while the mathematical and univariate geostatistical methods are the methods introduced in Chapters 4 and 5, respectively. In this chapter, we will concentrate on the hybrid methods of machine learning methods with *IDW* and/or *kriging*. *IDW* also includes *NN* and *KNN* as they can be regarded as special cases of *IDW*, while *kriging* methods include *OK, SK, BOK*, and *BSK*. They are sourced from previous studies (Li, Potter, Huang, Daniell, et al. 2010; Li, Heap, Potter, and Daniell 2011b; Li, Potter, Huang, and Heap 2012b). These hybrid methods are the

(1) hybrid method of *RF* and *IDW* (*RFIDW*),
(2) hybrid method of *RF* and *OK* (*RFOK*),
(3) hybrid methods of *RF, OK*, and *IDW* (*RFOKIDW*),
(4) hybrid method of *GBM* and *IDW* (*GBMIDW*),
(5) hybrid method of *GBM* and *OK* (*GBMOK*),
(6) hybrid methods of *GBM, OK*, and *IDW* (*GBMOKIDW*),
(7) hybrid method of *SVM* and *IDW* (*SVMIDW*),
(8) hybrid method of *SVM* and *OK* (*SVMOK*), and
(9) hybrid methods of *SVM, OK*, and *IDW* (*SVMOKIDW*).

We will also discuss various extensions of these hybrid methods (i.e., the hybrid methods of machine learning methods with *NN, KNN, SK, BOK*, and *BSK*). These methods are contained in:

(1) hybrid methods of *RF, kriging*, and *IDW* (*RFkrigingIDW*) (see Table 11.1),
(2) hybrid methods of *GBM, kriging*, and *IDW* (*GBMkrigingIDW*) (see Table 11.2), and
(3) hybrid methods of *SVM, kriging*, and *IDW* (*SVMkrigingIDW*) (see Table 11.3).

The predictions of these hybrid methods can also be generated according to Equations (10.1) to (10.7).

11.1 Hybrid method of *RF* and *IDW*

We will use the `sponge` data set in the `spm` package (Li 2019b) and `sponge2` data set in the `spm2` package (Li 2021a) to demonstrate the application of the hybrid method of *RF* and *IDW* (*RFIDW*). For the `sponge2` data, the predictive variables selected for *RF* are in `steprf`
`.3$steprfPredictorsAll` in Chapter 8. Firstly, we will use the predictive variables selected to construct the optimal *RF* model for *RFIDW*.

DOI: 10.1201/9781003091776-11

```
library(spm2)
data(sponge2)

variables.most.accurate <- c("easting", "northing", "bathy", "bs27", "bs36", "prof
    _cur_o", "tpi3") # from `steprf.3$steprfPredictorsAll[2][[1]]$variables.most.
    accurate`
species.richness <- sponge2[, 3]

rf.opt <- data.frame(sponge2[, variables.most.accurate], species.richness)

library(randomForest)
set.seed(1234)
rf1 <- randomForest(rf.opt[, 1:7], rf.opt[, 8], ntree = 500)

rf1.pred <- predict(rf1, rf.opt)
```

Then produce the residuals of `rf1` for *IDW* and develop an *IDW* model based on the residuals.

```
library(gstat)
rf.opt$res1 <- rf.opt[, 8] - rf1.pred

idw1 <- gstat(id = "res1", formula = res1 ~ 1, locations = ~ easting + northing,
    data = rf.opt, set = list(idp = 2), nmax = 12)
```

Finally, produce the predictions of *RFIDW* using `rf1` and `idw1`. We will still use the training data `rf.opt` to make the predictions.

```
idw.pred <- predict(idw1, rf.opt)$res1.pred
rfidw.pred <- rf1.pred + idw.pred
```

Similar to the hybrids of modern statistical methods with *IDW*, *RFIDW* is also an *exact* method. The *goodness of fit* is not necessarily related to the predictive accuracy (Li, Alvarez, et al. 2017), which is equally applicable to *RFIDW*.

In addition, if `idp` is specified as 0 in model `idw1`, then *KNN* is used, and the hybrid method resulted would be *RFKNN*. If `nmax` is specified as 1, then *NN* is used, and the hybrid method resulted would be *RFNN*.

11.1.1   Variable selection and parameter optimization based on predictive accuracy

For *RFIDW*, *variable selection* can be conducted using the methods and procedures introduced in Chapter 8 for *RF*. Here we will demonstrate how to estimate the optimal parameters for *RFIDW* based on the residuals of the optimal *RF* predictive model. This optimization can be done using the function `rfidwcv` in the `spm` package. We will briefly introduce `rfidwcv` first and then use it to estimate relevant parameters.

1. Function `rfidwcv`

For `rfidwcv`, the following arguments are essential:
(1) `longlat`, a dataframe contains the longitude and latitude of point samples;
(2) `trainx`, a dataframe or matrix contains the columns of predictive variables;
(3) `trainy`, a vector of response variable;
(4) `cv.fold`, integer; the number of folds for cross-validation;
(5) `idp`, a numeric number specifying the inverse distance weighting power;
(6) `nmax`, the number of nearest observations that should be used for a prediction; and

(7) `predacc`, can be either "VEcv" for `vecv` or "ALL" for all measures.
For details, see `?rfidwcv`.

2. Parameter optimization based on predictive accuracy

The optimal `idp` and `nmax` for *RFIDW* can be estimated based on the residuals of the optimal *RF* predictive model. This is going to be demonstrated based on the `sponge` data set.

We will use `rfidwcv` with all the predictive variables in the `sponge` data set to estimate optimal parameters.

```
idp <- (1:20) * 0.2
nmax <- c(5:40)
rfidwopt1 <- array(0, dim = c(length(idp), length(nmax)))

for (i in 1:length(idp)) {
  for (j in 1:length(nmax)) {
    set.seed(1234)
    rfidwcv1 <- rfidwcv(sponge[, 1:2], sponge[, c(1:2, 4:8)], sponge[, 3], idp =
        idp[i], nmax = nmax[j], predacc = "VEcv")
    rfidwopt1[i, j] <- rfidwcv1
  }
}
```

The optimal `idp` and `nmax` that lead to the maximal predictive accuracy are

```
para.rfidwopt1 <- which (rfidwopt1 == max(rfidwopt1), arr.ind = T)
idp[para.rfidwopt1[, 1]]
```

```
[1] 0.8
```

```
nmax[para.rfidwopt1[, 2]]
```

```
[1] 6
```

These parameters should be used to derive a stabilized predictive accuracy and generate spatial predictions as shown below.

In addition, `rfidwcv` can also be used to assess the predictive accuracy of *RFKNN* and *RFNN* if `idp` is specified as 0 and `nmax` as 1 in `rfidwcv`, respectively.

11.1.2   Predictive accuracy

We can use `rfidwcv` to produce a *stabilized predictive accuracy* of the *RFIDW* model. We will apply `rfidwcv` to `sponge` data with the optimal `idp` and `nmax` to demonstrate the *accuracy stabilization*.

```
set.seed(1234)
n <- 100
rfidwvecv1 <- NULL
for (i in 1:n) {
  rfidwcv1 <- rfidwcv(sponge[, 1:2], sponge[, c(1:2, 4:8)], sponge[, 3], idp =
      0.8, nmax = 6, predacc = "VEcv")
  rfidwvecv1[i] <- rfidwcv1
}
```

The median and range of predictive accuracy of *RFIDW* based on 100 repetitions of 10-fold cross-validation using `rfidwcv` are

```
median(rfidwvecv1)
```

```
[1] 45.84
```

```
range(rfidwvecv1)
```

```
[1] 32.49 50.57
```

11.1.3   Predictions

The predictions of *RFIDW* can be generated using the function `rfidwpred` in the `spm` package.

1. Function `rfidwpred`

For `rfidwpred`, the following arguments may need to be specified:
(1) `longlat`, `trainx`, `trainy`, `idp`, and `nmax`, the same as those for `rfidwcv`;
(2) `longlatpredx`, a dataframe contains the longitude and latitude of point locations (i.e., the centers of grids) to be predicted; and
(3) `predx`, a dataframe of predictive variables for the grids to be predicted.
For details, see `?rfidwpred`.

2. Generation of spatial predictions using `rfidwpred`

The predictions of *RFIDW* will be demonstrated by applying `rfidwpred` with the optimal `idp` and `nmax` to the area based on the `sponge.grid` data set.

```
data(sponge.grid)
```

```
set.seed(1234)
rfidwpred1 <- rfidwpred(sponge[, 1:2], sponge[, c(1:2, 4:8)], sponge[, 3], sponge.
    grid[, c(1,2)], sponge.grid, ntree = 500, idp = 0.8, nmax = 6)
```

The range of the predictions are

```
names(rfidwpred1)
```

```
[1] "LON"          "LAT"          "Predictions"
```

```
range(rfidwpred1$Predictions)
```

```
[1] 10.43 28.95
```

The spatial distribution of *RFIDW* predictions is illustrated in Figure 11.1.

11.1.4   A note on *RFIDW*

A function `rgidwcv` for cross-validation and a function `rgidwpred` for generating spatial predictions are available in the `spm` package for *RFIDW*. These functions are based on the function `ranger` (Wright, Wager, and Probst 2020) that is a fast implementation of *RF*, particularly for high-dimensional data. They can be easily applied according to the above examples for `rfidwcv` and `rfidwpred`, and their applications will not be presented here.

11.2   Hybrid method of *RF* and *OK*

For the hybrid method of *RF* and *OK* (*RFOK*), we will also use the `sponge` data set to demonstrate its application. To develop an optimal predictive model and generate spatial

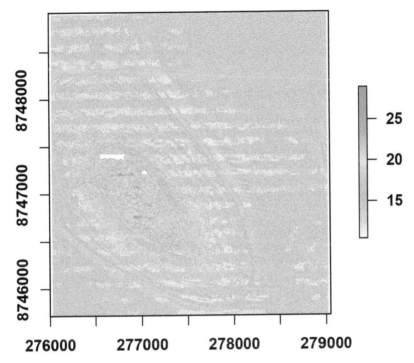

FIGURE 11.1: The spatial patterns of predictions using *RFIDW*.

predictions using *RFOK*, we need to apply *OK* to the residuals of a *RF* predictive model, such as `rf1` that is in `rf.opt` in Section 11.1. We will firstly conduct a variogram modeling for *RFOK* based on the residuals of `rf1`.

```
library(sp)
data.train <- rf.opt
coordinates(data.train) = ~ easting + northing
vgm1 <- variogram(object = res1 ~ 1, data.train)

model.1 <- fit.variogram(vgm1, vgm(mean(vgm1$gamma), "Sph", mean(vgm1$dist), min(
    vgm1$gamma) / 10))
```

Then use `model.1` and the training data `data.train` to make the predictions of *RFOK*.

```
pred.krige1 <- krige(res1 ~ 1, data.train, data.train, model = model.1, nmax = 12)

names(as.data.frame(pred.krige1))
rfok.pred  <- as.data.frame(pred.krige1)[, 3] + rf1.pred
```

Similar to the hybrids of modern statistical methods with *OK*, *RFOK* is also an *exact* method. Again, the *goodness of fit* is not necessarily a measure of the predictive accuracy of *RFOK* either.

In addition, if `beta` is specified for `pred.krige1`, then *SK* is used, and the hybrid method resulted would be *RFSK*. If `block` is specified, then *BK* is used, and the hybrid method resulted would be *RFBOK* or *RFBSK* depending on the specification of `beta`. For ways to specify these arguments, see relevant sections in Chapter 5.

If the `res1 ~ 1` in `vgm1` is extended to include external variable(s) (e.g., `res1 ~ tpi3`), then the hybrid method resulted would be *RFKED*; but this extension should not be used because any possible external trends should have already been considered in the *RF* model (e.g., rf1). Despite this note, the cross-validation function (i.e., rfkrigeidwcv) to be introduced in Section 11.3.1 could be used to assess the predictive accuracy of such hybrid methods.

11.2.1 Variable selection and parameter optimization based on predictive accuracy

Similar to *RFIDW*, *variable selection* for *RFOK* can be conducted using the methods and procedures introduced in Chapter 8 for *RF*. We will demonstrate how to optimize relevant parameters for the *RFOK* based on the residuals of the optimal *RF* predictive model. This optimization can be done with the function rfokcv in the spm package. We will introduce rfokcv first and then use it to conduct parameter optimization.

1. Function rfokcv

For rfokcv, the following arguments may need to be specified:
(1) longlat, trainx, trainy, cv.fold, nmax, and predacc, the same as those for rfidwcv;
(2) vgm.args, arguments for vgm, e.g., variogram model of response variable and anisotropy parameters; and by default, "Sph" is used; and
(3) block, block size.
For details, see ?rfokcv.

2. Parameter optimization based on predictive accuracy

We can optimize vgm.args and nmax for *RFOK* based on the residuals of the optimal *RF* predictive model. This is going to be demonstrated based on the sponge data set below.

```
nmax.ok <- c(5:40)
vgm.args <- c("Exp", "Gau", "Sph", "Exc", "Mat", "Ste", "Lin")
rfokopt <- matrix(0, length(nmax.ok), length(vgm.args))

for (i in 1:length(nmax.ok)) {
  for (j in 1:length(vgm.args)) {
    set.seed(1234)
    rfokcv1 <- rfokcv(sponge[, 1:2], sponge[, c(1:2, 4:8)], sponge[, 3], nmax =
        nmax.ok[i], vgm.args = vgm.args[j], predacc = "VEcv")
    rfokopt[i, j] <- rfokcv1
  }
}
```

The optimal nmax and vgm.args that lead to the maximal predictive accuracy are

```
rfok.opt <- which(rfokopt == max(rfokopt), arr.ind = T)
nmax.ok[rfok.opt[, 1]]
```

```
[1] 6
```

```
vgm.args[rfok.opt[, 2]]
```

```
[1] "Exc"
```

These parameters should be used for the *RFOK* predictive model.

As demonstrated in Chapter 5, all other parameters (e.g., transformation, alpha) for *RFOK* can also be optimized using rfokcv with lambda = 0. Their optimization will not be presented here.

11.2.2 Predictive accuracy

We will demonstrate how to use `rfokcv` to derive a stabilized predictive accuracy of the *RFOK* model based on the `sponge` data set with the optimal `nmax` and `vgm.args` below.

```
set.seed(1234)
n <- 100
rfokvecv <- NULL
for (i in 1:n) {
  rfokcv1 <- rfokcv(sponge[, 1:2], sponge[, c(1:2, 4:8)], sponge[, 3], nmax = 6,
     vgm.args = "Exc", predacc = "VEcv")
  rfokvecv[i] <- rfokcv1
}
```

The median and range of predictive accuracy of *RFOK* based on 100 repetitions of 10-fold cross-validation using `rfokcv` are

```
median(rfokvecv)
```

```
[1] 45.01
```

```
range(rfokvecv)
```

```
[1] 31.99 49.24
```

11.2.3 Predictions

The predictions can be generated using the function `rfokpred` in the `spm` package for *RFOK*.

1. Function `rfokpred`

For `rfokpred`, the following arguments may need to be specified:
(1) `longlat`, `trainx`, `trainy`, `vgm.args`, `block`, and `nmax`, the same as those for `rfidwcv` and `rfokcv`;
(2) `longlatpredx`, a dataframe contains the longitude and latitude of point locations (i.e., the centers of grids) to be predicted; and
(3) `predx`, a dataframe of predictive variables for the grids to be predicted.
For details, see `?rfokpred`.

2. Generation of spatial predictions using `rfokpred`

The spatial predictions of *RFOK* will be demonstrated by applying `rfokpred` with the optimal `vgm.args` and `nmax` to the area based on the `sponge.grid` data set.

```
set.seed(1234)
rfokpred1 <- rfokpred(sponge[, 1:2], sponge[, c(1:2, 4:8)], sponge[, 3], sponge.
   grid[, c(1,2)], sponge.grid, ntree = 500, nmax = 6, vgm.args = ("Exc"))
```

```
names(rfokpred1)
```

```
[1] "LON"         "LAT"         "Predictions" "Variances"
```

```
range(rfokpred1$Predictions)
```

```
[1]   9.916 28.460
```

The spatial distribution of *RFOK* predictions is illustrated in Figure 11.2.

Predictions

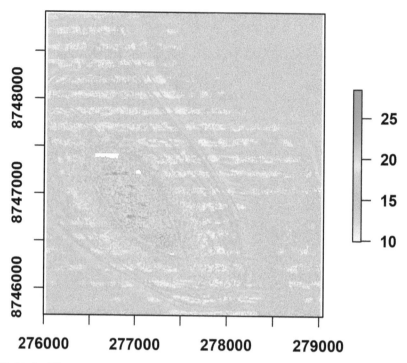

FIGURE 11.2: The spatial patterns of predictions using *RFOK*.

11.2.4   Notes on *RFOK*

A function `rgokcv` for cross-validation and a function `rgokpred` for making spatial predictions are available in the `spm` package for *RFOK*. They are based on the function `ranger` that is faster than the function `randomForest` (Breiman et al. 2018). They can be easily applied according to the above examples for `rfokcv` and `rfokpred`, and their applications will not be presented.

RFOK was initially published under the name of *RKrf* (Li, Potter, Huang, Daniell, et al. 2010; Li, Heap, Potter, Huang, et al. 2011). It is also called *Random forest RK* (Hengl et al. 2015) and *RFRK* (Ruiz-Álvarez, Alonso-Sarria, and Gomariz-Castillo 2019).

11.3   Hybrid methods of *RF*, *OK*, and *IDW*

To generate spatial predictions using the hybrid methods of *RF*, *OK*, and *IDW* (*RFOKIDW*), we need to develop an optimal *RFOKIDW* predictive model. First, we need to develop an optimal *RF* predictive model and then apply *IDW* and *OK* to the residuals of the model. Finally, the predictions of *RFOKIDW* can be generated based on the predictions of the *RF*, *IDW*, and *OK* models.

We will use the `sponge2` data set, `rf1.pred` and `idw.pred` produced for *RFIDW* in Section 11.1 and `pred.krige1` for *RFOK* in Section 11.2 to demonstrate the applications of *RFOKIDW*.

1. Average of *RFOK* and *RFIDW*

The average of *RFOK* and *RFIDW* (*RFOKRFIDW*) is a *RFOKIDW* method with $\lambda_m = \lambda_k = 1/2$ according to Equation (10.5). We will use `rf1.pred`, `idw.pred`, and `pred.krige1` to demonstrate how to produce the predictions of *RFOKRFIDW*.

```
lambdai <- 1 / 2
lambdak <- 1 / 2
rfokrfidw.pred <- rf1.pred + lambdak * as.data.frame(pred.krige1)[,3] + lambdai *
    idw.pred
```

RFOKRFIDW is an *exact* method.

2. Average of *RF*, *RFOK*, and *RFIDW*

The average of *RF*, *RFOK*, and *RFIDW* (*RFRFOKRFIDW*) is a *RFOKIDW* method with $\lambda_m = \lambda_k = 1/3$ according to Equation (10.5). We will produce the predictions of *RFR-FOKRFIDW* using `rf1.pred`, `idw.pred`, and `pred.krige1` as below.

```
lambdai <- 1 / 3
lambdak <- 1 / 3
rfrfokrfidw.pred  <- rf1.pred + lambdak * as.data.frame(pred.krige1)[,3] + lambdai
    * idw.pred
```

Due to $\lambda_m + \lambda_k = 2/3$ (see Equation (10.5)), the fitted values are no longer the same as the observed values, thus *RFRFOKRFIDW* is not an exact method.

In fact, λ_m and λ_k for *RFOKIDW* can take any values within their range (i.e., 0 to 1) as long as their sum is 1 or 2/3. Their values can be determined according to the predictive accuracy of the hybrid method resulted as shown below.

11.3.1 Variable selection and parameter optimization based on predictive accuracy

Regarding *variable selection* for *RFOKIDW*, it can be conducted using the methods and procedures introduced in Chapter 8 for *RF*. We will demonstrate how to optimize relevant parameters for *IDW* and *OK* based on the residuals of the optimal *RF* predictive model. This optimization can be carried out with the function `rfkrigeidwcv` in the spm2 package. We will briefly introduce `rfkrigeidwcv` first and then use it to conduct parameter optimization based on the `sponge` data set.

1. Function `rfkrigeidwcv`

For `rfkrigeidwcv`, the following arguments may need to be specified:
(1) `trainx`, `trainy`, `transformation`, `delta`, `formula`, `vgm.args`, `anis`, `alpha`, `block`, `beta`, `nmaxkrige`, `validation`, `cv.fold`, and `predacc`, the same as those for `rfokcv`, and `krigecv`;
(2) `longlat`, `mtry`, and `idp`, the same as those for `rfidwcv`;
(3) `nmaxidw`, the same as `nmax` for `rfidwcv`;
(4) `hybrid.parameter`, either 2 (default) or 3; and
(5) `lambda`, ranging from 0 to 2, with the default value of 1.
For details, see `?rfkrigeidwcv`.

If `lambda` is < 1, more weight is placed on *kriging*; and if it is > 1, more weight is placed on *IDW*.

If lambda is 0 and the default hybrid.parameter is used, *IDW* is not considered, and the method resulted is *RFkriging*. If lambda is 2 and the default hybrid.parameter is used, then the method resulted is *RFIDW*.

When hybrid.parameter is 3, if lambda is 0, then the method resulted is equivalent to the average of *RF*, *RFkriging*, and *RFkriging* (*RFRFkrigingRFkriging*), and if lambda is 2, then the method resulted is equivalent to the average of *RF*, *RFIDW*, and *RFIDW* (*RFRFIDWRFIDW*). According to Equations (10.8) and (10.9), *RFRFkrigingRFkriging* can be expressed as *RFkriging* with $\lambda_k = 2/3$, and *RFRFIDWRFIDW* as *RFIDW* with $\lambda_m = 2/3$.

With the changes in the specifications of relevant parameters (i.e., idp, nmaxidw, beta, block , lambda, and hybrid.parameter), many hybrid methods of *RF*, *kriging*, and *IDW* (*RFkrigingIDW*) can be resulted (Table 11.1).

When 0 < lambda < 2, then numerous hybrid methods of *RF*, *kriging*, and *IDW* could be possibly derived according to Equation (10.5) and implemented via rfkrigeidwcv, such as *RFOKIDW* with $\lambda_m = 0.22/3$ and $\lambda_k = 1.780/3$, and *RFSKKNN* with $\lambda_m = 1.11/2$ and $\lambda_k = 0.89/2$.

The hybrid methods from No. 1 to No. 31 in Table 11.1 are exact methods, and the rest are not exact methods.

In spite of the numerous hybrid methods, the most accurate method can be selected from all these hybrid methods based on predictive accuracy using rfkrigeidwcv as demonstrated below.

2. Parameter optimization based on predictive accuracy

We can use the optimal idp and nmax (for nmaxidw) in Section 11.1, and vgm.args and nmax (for nmaxkrige) in Section 11.2 based on the sponge data set for further *parameter optimization*.

These parameters were optimized separately for *RFIDW* and *RFOK*. If it is believed that these parameters are not optimal for *RFOKIDW* methods, then we need to re-do the optimization using rfkrigeidwcv, but we will not present it here, as it can be easily implemented by following the examples provided for parameter optimization for *RFIDW* in Section 11.1 and for *RFOK* in Section 11.2 as well as the example below for estimating the optimal lambda.

The optimal parameters of relevant arguments can be estimated or selected with rfkrigeidwcv. The examples provided for other functions can be adopted for such optimization. We will provide one example to estimate the optimal value of lambda for *RFOKIDW* with hybrid. parameter = 2 below.

```
data(sponge)
longlat <- sponge[, 1:2]
y = sponge[, 3]
trainx = sponge[, -3]

lambda <- c(0:100)*0.02
lambdaopt <- NULL

for (i in 1:length(lambda)) {
  set.seed(1234)
  rfkrigerfidwcv1 <- rfkrigeidwcv(longlat = longlat, trainx = trainx, trainy = y,
    transformation = "none", formula = res1 ~ 1, vgm.args = "Exc", nmaxkrige =
    6, idp = 0.8, nmaxidw = 6, lambda = lambda[i], hybrid.parameter = 2,
    validation = "CV",  predacc = "VEcv")
  lambdaopt[i] <- rfkrigerfidwcv1
}
```

TABLE 11.1: *RFkrigingIDW* methods can be derived and implemented via `rfkrigeidwcv` in relation to `idp`, `nmaxidw`, `beta`, `block`, `lambda`, and `hybrid.parameter` (ϕ); and λ_m and λ_k are from Equations (10.5), (10.6) and (10.7).

No.	idp	nmaxidw	beta	block	lambda	ϕ	Hybrid methods
1	!= 0	> 0	NULL	0	1	2	*RFOKRFIDW*
2	0	> 1	NULL	0	1	2	*RFOKRFKNN*
3	0	1	NULL	0	1	2	*RFOKRFNN*
4	!= 0	> 0	not NULL	0	1	2	*RFSKRFIDW*
5	0	> 1	not NULL	0	1	2	*RFSKRFKNN*
6	0	1	not NULL	0	1	2	*RFSKRFNN*
7	!= 0	> 0	NULL	> 0	1	2	*RFBOKRFIDW*
8	0	> 1	NULL	> 0	1	2	*RFBOKRFKNN*
9	0	1	NULL	> 0	1	2	*RFBOKRFNN*
10	!= 0	> 0	not NULL	> 0	1	2	*RFBSKRFIDW*
11	0	> 1	not NULL	> 0	1	2	*RFBSKRFKNN*
12	0	1	not NULL	> 0	1	2	*RFBSKRFNN*
13	!= 0	> 0	NULL	0	0	2	*RFOK*
14	!= 0	> 0	not NULL	0	0	2	*RFSK*
15	!= 0	> 0	NULL	> 0	0	2	*RFBOK*
16	!= 0	> 0	not NULL	> 0	0	2	*RFBSK*
17	!= 0	> 0	NULL	0	2	2	*RFIDW*
18	0	> 1	NULL	0	2	2	*RFKNN*
19	0	1	NULL	0	2	2	*RFNN*
20	!= 0	> 0	NULL	0	1	3	*RFRFOKRFIDW*
21	0	> 1	NULL	0	1	3	*RFRFOKRFKNN*
22	0	1	NULL	0	1	3	*RFRFOKRFNN*
23	!= 0	> 0	not NULL	0	1	3	*RFRFSKRFIDW*
24	0	> 1	not NULL	0	1	3	*RFRFSKRFKNN*
25	0	1	not NULL	0	1	3	*RFRFSKRFNN*
26	!= 0	> 0	NULL	> 0	1	3	*RFRFBOKRFIDW*
27	0	> 1	NULL	> 0	1	3	*RFRFBOKRFKNN*
28	0	1	NULL	> 0	1	3	*RFRFBOKRFNN*
29	!= 0	> 0	not NULL	> 0	1	3	*RFRFBSKRFIDW*
30	0	> 1	not NULL	> 0	1	3	*RFRFBSKRFKNN*
31	0	1	not NULL	> 0	1	3	*RFRFBSKRFNN*
32	!= 0	> 0	NULL	0	0	3	*RFOK* with $\lambda_k = 2/3$
33	!= 0	> 0	not NULL	0	0	3	*RFSK* with $\lambda_k = 2/3$
34	!= 0	> 0	NULL	> 0	0	3	*RFBOK* with $\lambda_k = 2/3$
35	!= 0	> 0	not NULL	> 0	0	3	*RFBSK* with $\lambda_k = 2/3$
36	!= 0	> 0	NULL	0	2	3	*RFIDW* with $\lambda_m = 2/3$
37	0	> 1	NULL	0	2	3	*RFKNN* with $\lambda_m = 2/3$
38	0	1	NULL	0	2	3	*RFNN* with $\lambda_m = 2/3$

The optimal value that leads to the maximal predictive accuracy is

```
lambda.opt <- which (lambdaopt == max(lambdaopt))
lambda[lambda.opt]
```

[1] 2

The maximal predictive accuracy is

```
max(lambdaopt)
```

```
[1] 48.46
```

Since the value estimated for `lambda` is 2, the hybrid method resulted is, in fact, *RFIDW* according to Equation (10.5).

Similar to *RFOKIDW* with `hybrid.parameter = 2`, the parameters estimated can also be used to optimize `lambda` for *RFOKIDW* with `hybrid.parameter = 3`. The results are in `lambda2opt`, and the optimal value and the maximal predictive accuracy are

```
lambda2.opt <- which (lambda2opt == max(lambda2opt))
lambda2[lambda2.opt]
```

```
[1] 2
```

```
max(lambda2opt)
```

```
[1] 47.56
```

Since the value estimated for `lambda` is also 2, the hybrid method resulted is actually a *RFIDW* but with $\lambda_m = 2/3$ according to Equation (10.6).

As demonstrated in Chapter 5, all other parameters (e.g., `transformation`, `alpha`) for *RFOK* can also be optimized using `rfkrigeidwcv` with `lambda = 0`. Their optimization will not be presented here.

The predictive accuracy of the hybrid model resulted is lower than that of *RFOKIDW* with `hybrid.parameter = 2`. Hence, the *RFOKIDW* with `hybrid.parameter = 2` and a `lambda = 2`, that is, *RFIDW* will be used. Since an example has been provided for the application of *RFIDW*, we will use *RFIDW* with $\lambda_m = 2/3$ as example to demonstrate the application of *RFOKIDW* below.

11.3.2   Predictive accuracy

The function `rfkrigeidwcv` can be used to produce a *stabilized predictive accuracy* of the *RFkrigingIDW* model. We will use `rfkrigeidwcv` with the optimal `nmaxkrige` (i.e., `nmax` for *RFOK*), `vgm.args`, `idp`, and `nmaxidw` (i.e., `nmax` for *RFIDW*) to demonstrate the *accuracy stabilization* for *RFIDW* with $\lambda_m = 2/3$ below.

```
set.seed(1234)
n <- 100
rfkrigeidwvecv <- NULL
for (i in 1:n) {
  rfkrigeidwcv1 <- rfkrigeidwcv(longlat = longlat, trainx = trainx, trainy = y,
      transformation = "none", formula = res1 ~ 1, idp = 0.8, nmaxidw = 6, lambda
      = 2, hybrid.parameter = 3, validation = "CV",  predacc = "VEcv")
  rfkrigeidwvecv[i] <- rfkrigeidwcv1
}
```

Given that this is about *RFIDW* with $\lambda_m = 2/3$, the specifications of `vgm.args` and `nmaxkrige` have no effects on the results, so the defaults have been used.

The median and range of predictive accuracy of *RFIDW* with $\lambda_m = 2/3$ based on 100 repetitions of 10-fold cross-validation using `rfkrigeidwcv` are

```
median(rfkrigeidwvecv)
```

```
[1] 45.12
```

```
range(rfkrigeidwvecv)
```

```
[1] 32.91 49.34
```

In addition, the function `rfokrfidwcv` in the `spm` package can also be used to assess the predictive accuracy of *RFOKRFIDW*.

11.3.3 Predictions

The predictions can be generated using the function `rfkrigeidwpred` in the `spm2` package for the hybrid methods.

1. Function `rfkrigeidwpred`

For `rfkrigeidwpred`, the following arguments may need to be specified:
(1) `trainx`, `trainy`, `transformation`, `delta`, `formula`, `vgm.args`, `anis`, `alpha`, `block`, and `beta`, the same as those for `rfokcv` and `krigecv`;
(2) `nmaxidw`, the same as `nmax` for `rfidwcv`;
(3) `nmaxkrige`, the same as `nmax` for `krigecv`;
(4) `longlat`, `longlatpredx`, `predx`, and `idp`, the same as those for `rfidwpred`; and
(5) `hybrid.parameter` and `lambda`, the same as those for `rfkrigeidwcv`.
For details, see `?rfkrigeidwpred`.

2. Generation of spatial predictions using `rfkrigeidwpred`

We can use `rfkrigeidwpred` to generate the spatial predictions of *RFkrigingIDW*. The hybrid method, *RFIDW* with $\lambda_m = 2/3$, will be used to demonstrate the generation of the predictions. The function `rfkrigeidwpred` with the optimal `vgm.args`, `nmaxkrige` (i.e., `nmax` for *RFOK*), `idp` and `nmaxidw` (i.e., `nmax` for *RFIDW*) will be applied to the `sponge` data set and the `petrel.grid` data set as below.

```
data(sponge)
data(sponge.grid)
```

```
set.seed(1234)
rfkrigeidwpred1 <- rfkrigeidwpred(longlat = sponge[, 1:2], trainx = sponge[, -3],
    predx = sponge.grid, trainy = sponge[, 3], longlatpredx = sponge.grid[, c
    (1:2)], formula.krige = res1 ~ 1, idp = 0.8, nmaxidw = 6, lambda = 2, hybrid.
    parameter = 3)
```

Since `lambda = 2` is specified, the specifications of `vgm.args` and `nmaxkrige` have no effects on the results of the hybrid method that is *RFIDW* with $\lambda_m = 2/3$, so the defaulted `vgm.args` and `nmaxkrige` have been used.

```
names(rfkrigeidwpred1)
```

```
[1] "long"        "lat"         "predictions"
```

```
range(rfkrigeidwpred1$predictions)
```

```
[1]  9.736 28.243
```

The spatial distribution of predictions by *RFIDW* with $\lambda_m = 2/3$ (see Equation (10.6)) is illustrated in Figure 11.3.

In addition, the function `rfokrfidwpred` in the `spm` package can also be used to generate spatial predictions for *RFOKRFIDW*.

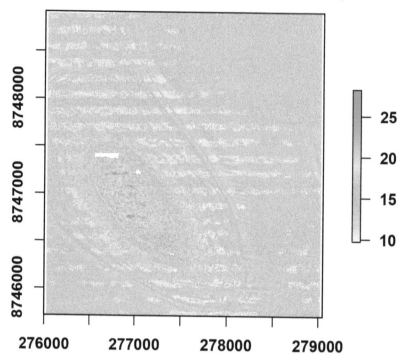

FIGURE 11.3: The spatial patterns of predictions using *RFIDW* with `lambda_m = 2/3`.

11.3.4 A note on *RFOKRFIDW*

A function `rgokrgidwcv` for cross-validation and a function `rgokrgidwpred` for generating spatial predictions are available in the `spm` package for *RFOKRFIDW*. These functions are based on the function `ranger` that is faster than the function `randomForest`. These functions can be easily applied according to the above examples for `rfkrigeidwcv` and `rfkrigeidwpred`, and their applications will not be presented here.

11.4 Hybrid method of *GBM* and *IDW*

We will use the `sponge2` data set to demonstrate the application of the hybrid method of *GBM* and *IDW* (*GBMIDW*) for developing an optimal predictive model and generating spatial predictions. For the `sponge2` data set, the predictive variables selected for *GBM* are in `stepgbm2$stepgbmPredictorsAll[[2]]` in Chapter 8. We will use these variables to construct the optimal *GBM* model for the hybrid method, *GBMIDW*, first.

```
library(spm2)
data(sponge2)

variables.most.accurate <- c("easting", "sand", "prof_cur_o", "tpi3", "relief_5",
    "homo3", "bathy_lmi7", "relief_3", "entro7") # from `stepgbm2$
    stepgbmPredictorsAll[[2]]$variables.most.accurate`
species.richness <- sponge2[, 3]

gbm.opt <- data.frame(sponge2[, variables.most.accurate], species.richness)
nc <- ncol(gbm.opt)-1; nr <- nrow(gbm.opt)
```

The gbm.opt can also be generated using gbm.opt <- data.frame(sponge2[, stepgbm2$stepgbmPredictorsAll[[2]]$variables.most.accurate], species.richness), which can be done only if stepgbm2$stepgbmPredictorsAll[[2]]$variables.most.accurate is made available.

```
library(gbm)
set.seed(1234)

gbm1 <- gbm(species.richness ~ ., distribution = "poisson", gbm.opt, weights = rep
    (1, nr), var.monotone = rep(0, nc), n.trees = 6000, interaction.depth = 5, n.
    minobsinnode = 10, shrinkage = 0.001, bag.fraction = 0.5, cv.fold = 10, keep.
    data = FALSE, verbose = TRUE, n.cores = 8)

best.iter <- gbm.perf(gbm1, method = "cv")
print(best.iter)
gbm1.pred <- predict.gbm(gbm1, gbm.opt, n.trees = best.iter, type = "response")
```

Then, derive the residuals of gbm1 for *IDW* and develop an *IDW* model based on the residuals.

```
library(gstat)
gbm.opt$res1 <- gbm.opt$species.richness - gbm1.pred
gbm.opt$northing <- sponge2$northing

idw1 <- gstat(id = "res1", formula = res1 ~ 1, locations = ~ easting + northing,
    data = gbm.opt, set = list(idp = 2), nmax = 12)
```

Finally, produce the predictions of *GBMIDW* using gbm1 and idw1. We will still use the training data gbm.opt to make the predictions.

```
idw.pred <- predict(idw1, gbm.opt)$res1.pred
gbmidw.pred <- gbm1.pred + idw.pred
```

Similar to *RFIDW*, *GBMIDW* is also an *exact* method. However, the *goodness of fit* is not necessarily related to the predictive accuracy of *GBMIDW*.

In addition, if idp is specified as 0 in model idw1, then *KNN* is used, and the hybrid method resulted would be *GBMKNN*. If nmax is specified as 1, then *NN* is used, and the hybrid method resulted would be *GBMNN*.

11.4.1 Variable selection and parameter optimization based on predictive accuracy

For *GBMIDW*, *variable selection* can be undertaken using the methods and procedures introduced in Chapter 8 for *GBM*. We will demonstrate how to optimize relevant parameters for *GBMIDW* based on the residuals of the optimal *GBM* predictive model. This optimization can be carried out using the function gbmidwcv in the spm package. We will introduce gbmidwcv first and then use it to conduct parameter optimization.

1. Function gbmidwcv

For gbmidwcv, the following arguments may need to be specified:
(1) longlat, a dataframe contains the longitude and latitude of point samples;
(2) trainx, trainy, var.monotone, family, n.trees, learning.rate, interaction.depth, bag.fraction, train.fraction, n.minobsinnode, cv.fold, predacc, and n.cores, the same as those for gbm and gbmcv;
(3) idp, a numeric number specifying the inverse distance weighting power; and
(4) nmax, the number of nearest observations that should be used for a prediction.
For details, see ?gbmidwcv.

2. Parameter optimization based on predictive accuracy

The optimal `idp` and `nmax` for *GBMIDW* can be estimated based on the residuals of the optimal *GBM* predictive model.

We will use `gbmidwcv` with all the predictive variables in the `sponge` data set to estimate optimal parameters.

```
idp <- (1:20) * 0.2
nmax <- c(5:40)
gbmidwopt1 <- array(0, dim = c(length(idp), length(nmax)))

for (i in 1:length(idp)) {
  for (j in 1:length(nmax)) {
    set.seed(1234)
    gbmidwcv1 <- gbmidwcv(sponge[, c(1,2)], sponge[, -3], sponge[, 3], idp = idp[i
        ], nmax = nmax[j], cv.fold = 10, interaction.depth = 3, family = "poisson"
        , n.cores = 8, predacc = "VEcv")
    gbmidwopt1[i, j] <- gbmidwcv1
  }
}
```

The optimal estimations of `idp` and `nmax` that lead to the maximal predictive accuracy are

```
para.gbmidwopt1 <- which (gbmidwopt1 == max(gbmidwopt1), arr.ind = T)
idp[para.gbmidwopt1[, 1]]
```

```
[1] 0.8
```

```
nmax[para.gbmidwopt1[, 2]]
```

```
[1] 11
```

These parameters will be used for the *GBMIDW* predictive model.

In addition, `gbmidwcv` can also be used to assess the predictive accuracy of *GBMKNN* and *GBMNN* if `idp` is specified as 0 and `nmax` as 1 in `gbmidwcv`, respectively.

11.4.2   Predictive accuracy

The function `gbmidwcv` can be used to produce a *stabilized predictive accuracy* of the *GB-MIDW* model. We will apply *GBMIDW* with the optimal `idp` and `nmax` to `sponge` data to demonstrate the *accuracy stabilization* below.

```
set.seed(1234)
n <- 100
gbmidwvecv1 <- NULL
for (i in 1:n) {
  gbmidwcv1 <- gbmidwcv(sponge[, c(1,2)], sponge[, -3], sponge[, 3], idp = 0.8,
      nmax = 11, cv.fold = 10, interaction.depth = 3, family = "poisson", n.cores
      = 8, predacc = "VEcv")
  gbmidwvecv1[i] <- gbmidwcv1
}
```

The median and range of predictive accuracy of *GBMIDW* based on 100 repetitions of 10-fold cross-validation using `gbmidwcv` are

```
median(gbmidwvecv1)
```

```
[1] 41.92
```

```
range(gbmidwvecv1)
```

```
[1] 30.55 49.42
```

11.4.3 Predictions

The predictions of *GBMIDW* can be generated with the function `gbmidwpred` in the `spm` package (Li 2019b).

1. Function `gbmidwpred`

For `gbmidwpred`, the following arguments may need to be specified:

(1) `longlat`, `trainx`, `trainy`, `family`, `n.trees`, `learning.rate`, `interaction.depth`, `cv.fold`, `idp`, and `nmax`, the same as those for `gbm`, `gbmcv`, and `gbmidwcv`;

(2) `longlatpredx`, a dataframe contains the longitude and latitude of point locations (i.e., the centers of grids) to be predicted; and

(3) `predx`, a dataframe of predictive variables for the grids to be predicted.

For details, see `?gbmidwpred`.

2. Generation of spatial predictions using `gbmidwpred`

The generation of the predictions of *GBMIDW* will be demonstrated by applying `gbmidwpred` with the optimal `idp` and `nmax` for the area based on the `sponge.grid` data set as shown below.

```
data(sponge.grid)
```

```
set.seed(1234)
gbmidwpred1 <- gbmidwpred(sponge[, 1:2], sponge[, -3], sponge[, 3], sponge.grid[,
    c(1,2)], sponge.grid, n.trees = 5000, interaction.depth = 3, cv.fold = 10, idp
    = 0.8, nmax = 11, family = "poisson")
```

The argument `cv.fold` needs to be specified for the function `gbm.perf` that is used within `gbmidwpred` to find the best iterations for generating spatial predictions. The range of the predictions are

```
names(gbmidwpred1)
```

```
[1] "LON"          "LAT"          "Predictions"
```

```
range(gbmidwpred1$Predictions)
```

```
[1] 10.64 47.02
```

The spatial distribution of *GBMIDW* predictions is illustrated in Figure 11.4.

```
gbmidwpred <- gbmidwpred1
gridded(gbmidwpred) = ~ LON + LAT
```

11.5 Hybrid method of *GBM* and *OK*

For the hybrid method of *GBM* and *OK* (*GBMOK*), we will also use the `sponge2` data set to demonstrate its application. To develop an optimal predictive model and generate spatial predictions using *GBMOK*, we need to apply *OK* to the residuals of a *GBM* predictive model, such as `gbm1` in `gbm.opt` in Section 11.4. We will firstly conduct a variogram modeling based on the residuals of `gbm1` for *GBMOK*.

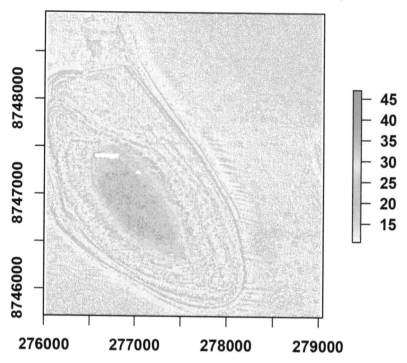

FIGURE 11.4: The spatial patterns of predictions using *GBMIDW*.

```
data.train <- gbm.opt
coordinates(data.train) = ~ easting + northing
vgm1 <- variogram(object = res1 ~ 1, data.train)

model.1 <- fit.variogram(vgm1, vgm(mean(vgm1$gamma), "Sph", mean(vgm1$dist), min(
    vgm1$gamma) / 10))
```

Then use `model.1` and the training data `data.train` to make the predictions of *GBMOK*.

```
pred.krige1 <- krige(res1 ~ 1, data.train, data.train, model = model.1, nmax = 12)

names(as.data.frame(pred.krige1))
gbmok.pred <- as.data.frame(pred.krige1)[, 3] + gbm1.pred
```

Similar to *RFOK*, *GBMOK* is also an *exact* method. Again, the *goodness of fit* is not necessarily a measure of the predictive accuracy of *GBMOK* either.

In addition, if `beta` is specified for `pred.krige1`, then *SK* is used, and the hybrid method resulted would be *GBMSK*. If `block` is specified, then *BK* is used, and the hybrid method resulted would be *GBMBOK* or *GBMBSK* depending on the specification of `beta`. For ways to specify these arguments, see relevant sections in Chapter 5.

If the `res1 ~ 1` in `vgm1` is extended to include external variable(s) (e.g., `res1 ~ tpi3`), then the hybrid method resulted would be *GBMKED*; but this extension should not be used because any possible external trends should have already been considered in the *GBM* model (e.g., `gbm1`). Despite this note, the cross-validation function (i.e., `gbmkrigeidwcv`) to be introduced in Section 11.6.1 could be used to assess the predictive accuracy of such hybrid methods.

11.5.1 Variable selection and parameter optimization based on predictive accuracy

Similar to *GBMIDW*, *variable selection* for *GBMOK* can be conducted using the methods and procedures introduced in Chapter 8 for *GBM*. We will demonstrate how to optimize relevant parameters for the *GBMOK* based on the residuals of the optimal *GBM* predictive model. This optimization can be done with the function `gbmokcv` in the `spm` package. We will introduce `gbmokcv` first and then use it to optimize relevant parameters.

1. Function `gbmokcv`

For `gbmokcv`, the following arguments may need to be specified:

(1) `longlat`, `trainx`, `trainy`, `var.monotone`, `family`, `n.trees`, `learning.rate`, `interaction.depth`, `bag.fraction`, `train.fraction`, `n.minobsinnode`, `cv.fold`, `nmax`, `predacc`, and `n.cores`, the same as those for `gbm`, `gbmcv`, and `gbmidwcv`;

(2) `vgm.args`, arguments for `vgm`, e.g., variogram model of response variable and anisotropy parameters; and by default, "Sph" is used; and

(3) `block`, block size.

For details, see `?gbmokcv`.

2. Parameter optimization based on predictive accuracy

The `vgm.args` and `nmax` for *GBMOK* can be optimized based on the residuals of the optimal *GBM* predictive model. This is going to be demonstrated based on the `sponge` data set.

```
nmax.ok <- c(5:40)
vgm.args <- c("Exp", "Gau", "Sph", "Exc", "Mat", "Ste", "Lin")
gbmokopt <- matrix(0, length(nmax.ok), length(vgm.args))

for (i in 1:length(nmax.ok)) {
  for (j in 1:length(vgm.args)) {
    set.seed(1234)
    gbmokcv1 <- gbmokcv(sponge[, c(1,2)], sponge[, -c(3)], sponge[, 3], cv.fold =
        10, interaction.depth = 3, family = "poisson", n.cores = 8, nmax = nmax.ok
        [i], vgm.args = vgm.args[j],  predacc = "VEcv")
    gbmokopt[i, j] <- gbmokcv1
  }
}
```

The optimal `nmax` and `vgm.args` that lead to the maximal predictive accuracy are

```
gbmok.opt <- which(gbmokopt == max(gbmokopt), arr.ind = T)
nmax.ok[gbmok.opt[, 1]]
```

```
[1] 17
```

```
vgm.args[gbmok.opt[, 2]]
```

```
[1] "Exc"
```

These parameters should be used for the *GBMOK* predictive model.

As shown in Chapter 5, all other parameters (e.g., `transformation`, `alpha`) for *GBMOK* can also be optimized using `gbmokcv` with `lambda = 0`. Their optimization will not be repeated here.

11.5.2 Predictive accuracy

The function `gbmokcv` can be used to produce a *stabilized predictive accuracy* of the *GBMOK* model. We will apply `gbmokcv` with the optimal `nmax` and `vgm.args` to the `sponge` data set to demonstrate the *accuracy stabilization* below.

```
set.seed(1234)
n <- 100
gbmokvecv <- NULL
for (i in 1:n) {
  gbmokcv1 <- gbmokcv(sponge[, c(1,2)], sponge[, -c(3)], sponge[, 3], cv.fold =
      10, interaction.depth = 3, family = "poisson", n.cores = 8, nmax = 17, vgm.
      args = "Exc",  predacc = "VEcv")
  gbmokvecv[i] <- gbmokcv1
}
```

The median and range of predictive accuracy of *GBMOK* based on 100 repetitions of 10-fold cross-validation using `gbmokcv` are

```
median(gbmokvecv)
```

```
[1] 40.89
```

```
range(gbmokvecv)
```

```
[1] 28.50 49.07
```

11.5.3 Predictions

The predictions can be generated using the function `gbmokpred` in the `spm` package for *GBMOK*.

1. Function `gbmokpred`

For `gbmokpred`, the following arguments may need to be specified:

(1) `longlat`, `trainx`, `trainy`, `var.monotone`, `family`, `n.trees`, `learning.rate`, `interaction.depth`, `bag.fraction`, `train.fraction`, `n.minobsinnode`, `cv.fold`, `vgm.args`, `block`, and `nmax`, the same as those for `gbm`, `gbmcv`, `gbmidwcv`, and `gbmokcv`; and
(2) `longlatpredx` and `predx`, the same as those for `gbmidwpred`.
For details, see `?gbmokpred`.

2. Generation of spatial predictions using `gbmokpred`

The predictions of *GBMOK* will be demonstrated with the optimal `vgm.args` and `nmax` to produce spatial predictions for the area based on the `sponge.grid` data set as below.

```
set.seed(1234)
gbmokpred1 <- gbmokpred(sponge[, 1:2], sponge[, -3], sponge[, 3], sponge.grid[, c
    (1,2)], sponge.grid, cv.fold = 10, interaction.depth = 3, family = "poisson",
    n.cores = 8, ntree = 5000, nmax = 17, vgm.args = ("Exc"))
```

```
names(gbmokpred1)
```

```
[1] "LON"          "LAT"          "Predictions" "Variances"
```

```
range(gbmokpred1$Predictions)
```

```
[1] 12.32 33.90
```

The spatial distribution of *GBMOK* predictions is illustrated in Figure 11.5.

Predictions

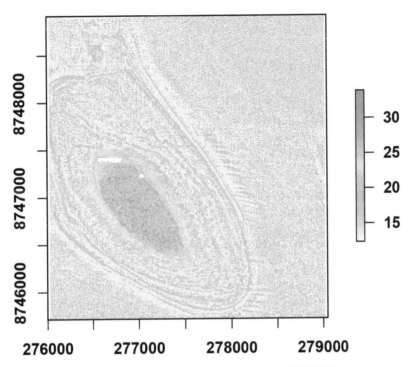

FIGURE 11.5: The spatial patterns of predictions using *GBMOK*.

11.6 Hybrid methods of *GBM*, *OK*, and *IDW*

To generate spatial predictions using the hybrid methods of *GBM*, *OK*, and *IDW* (*GBMOKIDW*), we need to develop an optimal *GBMOKIDW* predictive model. Firstly, we need to develop an optimal *GBM* predictive model, and then apply *IDW* and *OK* to the residuals of the model. Finally, the predictions of *GBMOKIDW* can be generated based on the predictions of the *GBM*, *IDW*, and *OK* models.

We will use the sponge and sponge2 data sets, and gbm1.pred and idw.pred in Section 11.4 and pred.krige1 in Section 11.5 to demonstrate the applications of *GBMOKIDW* methods.

1. Average of *GBMOK* and *GBMIDW*

The average of *GBMOK* and *GBMIDW* (*GBMOKGBMIDW*) is a *GBMOKIDW* method with $\lambda_m = \lambda_k = 1/2$ according to Equation (10.5). We will produce the predictions of *GBMOKGBMIDW* using gbm1.pred, idw.pred, and pred.krige1 as an example below.

```
lambdai <- 1 / 2
lambdak <- 1 / 2
gbmokgbmidw.pred <- gbm1.pred + lambdak * as.data.frame(pred.krige1)[,3] + lambdai
     * idw.pred
```

Similar to *RFOKRFIDW*, *GBMOKGBMIDW* is also an *exact* method.

2. Average of *GBM*, *GBMOK*, and *GBMIDW*

The average of *GBM*, *GBMOK*, and *GBMIDW* (*GBMGBMOKGBMIDW*) is a *GBMOKIDW* method with $\lambda_m = \lambda_k = 1/3$ according to Equation (10.5). We will produce the predictions of *GBMGBMOKGBMIDW* using gbm1.pred, idw.pred, and pred.krige1 as an example below.

```
lambdai <- 1 / 3
lambdak <- 1 / 3
gbmgbmokgbmidw.pred <- gbm1.pred + lambdak * as.data.frame(pred.krige1)[,3] +
    lambdai * idw.pred
```

Due to $\lambda_m + \lambda_k = 2/3$ (see Equation (10.5)), the fitted values are no longer the same as the observed values, thus *GBMGBMOKGBMIDW* is not an exact method.

In fact, λ_m and λ_k for *GBMOKIDW* can take any values within their range (i.e., 0 to 1) as long as their sum is 1 or 2/3. Their values can be determined according to the predictive accuracy of the hybrid method resulted as shown below.

11.6.1 Variable selection and parameter optimization based on predictive accuracy

With regard to *variable selection* for *GBMOKIDW*, it can be carried out using the methods and procedures introduced in Chapter 8 for *GBM*. We will demonstrate how to optimize relevant parameters for *IDW* and *OK* based on the residuals of the optimal *GBM* predictive model. This optimization can be done with the function gbmkrigeidwcv in the spm2 package. We will introduce gbmkrigeidwcv first and then use it to conduct parameter optimization based on the sponge data set.

1. Function gbmkrigeidwcv

For gbmkrigeidwcv, the following arguments may need to be specified:
(1) trainx, trainy, transformation, delta, formula, vgm.args, anis, alpha, block, beta, nmaxkrige, validation, cv.fold, and predacc, the same as those for krigecv;
(2) nmaxkrige, the same as nmax for krigecv;
(3) longlat, var.monotone, family, n.trees, learning.rate, interaction.depth, bag.fraction, train.fraction, n.minobsinnode, idp, and n.cores, the same as those for gbm, gbmcv, and gbmidwcv;
(4) nmaxidw, the same as nmax for gbmidwcv;
(5) hybrid.parameter, either 2 (default) or 3; and
(6) lambda, ranging from 0 to 2, with the default value of 1.
For details, see ?gbmkrigeidwcv

If lambda is < 1, more weight is placed on *kriging*; and if it is > 1, more weight is placed on *IDW*.

When the default hybrid.parameter is used and lambda is 0, *IDW* is not considered and the method resulted is *GBMkriging*. When the default hybrid.parameter is used and lambda is 2, then the method resulted is *GBMIDW*.

When hybrid.parameter is 3, if lambda is 0, then the method resulted is equivalent to the average of *GBM*, *GBMkriging*, and *GBMkriging* (*GBMGBMkrigingGBMkriging*), and if lambda is 2, then the method resulted is equivalent to the average of *GBM*, *GBMIDW*, and *GBMIDW* (*GBMGBMIDWGBMIDW*). According to Equations (10.8) and (10.9), *GBMGBMkriging-GBMkriging* can be expressed as *GBMkriging* with $\lambda_k = 2/3$, and *GBMGBMIDWGBMIDW* as *GBMIDW* with $\lambda_m = 2/3$.

With the changes in the specifications of `idp`, `nmaxidw`, `beta`, `block`, `lambda`, and `hybrid.parameter`, many hybrid methods of *GBM*, *kriging*, and *IDW* (*GBMkrigingIDW*) can be resulted (Table 11.2).

TABLE 11.2: *GBMkrigingIDW* methods can be derived and implemented via `gbmkrigeidwcv` in relation to `idp`, `nmaxidw`, `beta`, `block`, `lambda`, and `hybrid.parameter` (ϕ); and λ_m and λ_k are from Equations (10.5), (10.6) and (10.7).

No.	idp	nmaxidw	beta	block	lambda	ϕ	Hybrid methods
1	!= 0	> 0	NULL	0	1	2	*GBMOKGBMIDW*
2	0	> 1	NULL	0	1	2	*GBMOKGBMKNN*
3	0	1	NULL	0	1	2	*GBMOKGBMNN*
4	!= 0	> 0	not NULL	0	1	2	*GBMSKGBMIDW*
5	0	> 1	not NULL	0	1	2	*GBMSKGBMKNN*
6	0	1	not NULL	0	1	2	*GBMSKGBMNN*
7	!= 0	> 0	NULL	> 0	1	2	*GBMBOKGBMIDW*
8	0	> 1	NULL	> 0	1	2	*GBMBOKGBMKNN*
9	0	1	NULL	> 0	1	2	*GBMBOKGBMNN*
10	!= 0	> 0	not NULL	> 0	1	2	*GBMBSKGBMIDW*
11	0	> 1	not NULL	> 0	1	2	*GBMBSKGBMKNN*
12	0	1	not NULL	> 0	1	2	*GBMBSKGBMNN*
13	!= 0	> 0	NULL	0	0	2	*GBMOK*
14	!= 0	> 0	not NULL	0	0	2	*GBMSK*
15	!= 0	> 0	NULL	> 0	0	2	*GBMBOK*
16	!= 0	> 0	not NULL	> 0	0	2	*GBMBSK*
17	!= 0	> 0	NULL	0	2	2	*GBMIDW*
18	0	> 1	NULL	0	2	2	*GBMKNN*
19	0	1	NULL	0	2	2	*GBMNN*
20	!= 0	> 0	NULL	0	1	3	*GBMGBMOKGBMIDW*
21	0	> 1	NULL	0	1	3	*GBMGBMOKGBMKNN*
22	0	1	NULL	0	1	3	*GBMGBMOKGBMNN*
23	!= 0	> 0	not NULL	0	1	3	*GBMGBMSKGBMIDW*
24	0	> 1	not NULL	0	1	3	*GBMGBMSKGBMKNN*
25	0	1	not NULL	0	1	3	*GBMGBMSKGBMNN*
26	!= 0	> 0	NULL	> 0	1	3	*GBMGBMBOKGBMIDW*
27	0	> 1	NULL	> 0	1	3	*GBMGBMBOKGBMKNN*
28	0	1	NULL	> 0	1	3	*GBMGBMBOKGBMNN*
29	!= 0	> 0	not NULL	> 0	1	3	*GBMGBMBSKGBMIDW*
30	0	> 1	not NULL	> 0	1	3	*GBMGBMBSKGBMKNN*
31	0	1	not NULL	> 0	1	3	*GBMGBMBSKGBMNN*
32	!= 0	> 0	NULL	0	0	3	*GBMOK* with $\lambda_k = 2/3$
33	!= 0	> 0	not NULL	0	0	3	*GBMSK* with $\lambda_k = 2/3$
34	!= 0	> 0	NULL	> 0	0	3	*GBMBOK* with $\lambda_k = 2/3$
35	!= 0	> 0	not NULL	> 0	0	3	*GBMBSK* with $\lambda_k = 2/3$
36	!= 0	> 0	NULL	0	2	3	*GBMIDW* with $\lambda_m = 2/3$
37	0	> 1	NULL	0	2	3	*GBMKNN* with $\lambda_m = 2/3$
38	0	1	NULL	0	2	3	*GBMNN* with $\lambda_m = 2/3$

When `0 < lambda < 2`, then numerous hybrid methods of *GBM*, *OK*, and *IDW* could be possibly derived according to Equation (10.5) and implemented via `gbmkrigeidwcv`, such as

GBMOKIDW with $\lambda_m = 0.6/3$ and $\lambda_k = 1.4/3$, and *GBMSKKNN* with $\lambda_m = 1.55/2$ and $\lambda_k = 0.45/2$.

The hybrid methods from No. 1 to No. 31 in Table 11.2 are exact methods and the remaining methods are not exact methods.

Despite the numerous hybrid methods, the most accurate method can be selected from all these hybrid methods based on predictive accuracy using gbmkrigeidwcv as demonstrated below.

2. Parameter optimization based on predictive accuracy

We can use the optimal idp and nmax (for nmaxidw) in Section 11.4, and vgm.args and nmax (for nmaxkrige) in Section 11.5 based on the sponge data set to conduct further parameter optimization. These parameters were estimated or selected separately for *GBMIDW* and *GBMOK*. If it is believed that these parameters are not optimal for *GBMOKIDW* methods, then we need to re-do the optimization using gbmkrigeidwcv, but we will not repeat it here, as it can be easily implemented by following the examples provided for parameter optimization for *GBMIDW* in Section 11.4 and *GBMOK* in Section 11.5 as well as the example below for estimating the optimal lambda.

The optimal parameters of relevant arguments can be determined with gbmkrigeidwcv. The examples provided for other functions can be adopted for such optimization. We will provide an example to estimate the optimal value of lambda for *GBMOKIDW* with hybrid.parameter = 2 as follows.

```
data(sponge)
longlat <- sponge[, 1:2]
y = sponge[, 3]
trainx = sponge[, -3]

lambda3 <- c(0:100)*0.02
lambda3opt <- NULL

for (i in 1:length(lambda3)) {
  set.seed(1234)
  gbmkrigegbmidwcv1 <- gbmkrigeidwcv(longlat = longlat, trainx = trainx, trainy =
    y, family = "poisson", interaction.depth = 3, transformation = "none",
    formula = res1 ~ 1, vgm.args = "Exc", nmaxkrige = 17, idp = 0.8, nmaxidw =
    11, hybrid.parameter = 2, lambda = lambda3[i], validation = "CV", predacc =
    "VEcv", , n.cores = 8)
  lambda3opt[i] <- gbmkrigegbmidwcv1
}
```

The optimal lambda that leads to the maximal predictive accuracy is

```
lambda3.opt <- which (lambda3opt == max(lambda3opt))
lambda3[lambda3.opt]
```

```
[1] 0
```

The maximal predictive accuracy is

```
max(lambda3opt)
```

```
[1] 45
```

Since value estimated for lambda is 0, the hybrid method resulted is, in fact, *GBMOK* according to Equation (10.5).

Similar to *GBMOKIDW* with `hybrid.parameter = 2`, the parameters estimated can also be used to optimize `lambda` for *GBMOKIDW* with `hybrid.parameter = 3`. The results are in `lambda4opt`, and the optimal value and the maximal predictive accuracy are

```
lambda4.opt <- which (lambda4opt == max(lambda4opt))
lambda4[lambda4.opt]
```

```
[1] 0
```

```
max(lambda4opt)
```

```
[1] 43.21
```

This value estimated for `lambda` is again 0, which suggests that the hybrid method resulted is actually *GBMOK* with $\lambda_k = 2/3$ according to Equation (10.7).

As shown in Chapter 5, all other parameters (e.g., `transformation`, `alpha`) for *GBMOK* can also be optimized using `gbmkrigeidwcv` with `lambda = 0`. Their optimization will not be repeated here.

The predictive accuracy of the hybrid model resulted is lower than that of *GBMOKIDW* with `hybrid.parameter = 2`. Hence, the *GBMOKIDW* with `hybrid.parameter = 2` and `lambda = 0`, that is, *GBMOK* will be used. Since an example has already been provided for the application of *GBMOK*, we will use *GBMOK* with $\lambda_k = 2/3$ as an example to demonstrate the application of *GBMOKIDW* below.

11.6.2 Predictive accuracy

We can use `gbmkrigeidwcv` to produce a stabilized predictive accuracy of the *GBMkrigingIDW* model. We will apply `gbmkrigeidwcv` with the optimal `nmaxkrige` (i.e., `nmax` for *GBMOK*), `vgm.args`, `idp`, and `nmaxidw` (i.e., `nmax` for *GBMIDW*) to `sponge` data to demonstrate the accuracy stabilization for *GBMOK* with $\lambda_k = 2/3$ (see Equation (10.7)).

```
set.seed(1234)
n <- 100
gbmkrigeidwvecv <- NULL
for (i in 1:n) {
  gbmkrigegbmidwcv1 <- gbmkrigeidwcv(longlat = longlat, trainx = trainx, trainy =
      y, family = "poisson", interaction.depth = 3, transformation = "none",
      formula = res1 ~ 1, vgm.args = "Exc", nmaxkrige = 17, hybrid.parameter = 3,
      lambda = 0, validation = "CV", predacc = "VEcv", , n.cores = 8)
  gbmkrigeidwvecv[i] <- gbmkrigegbmidwcv1
}
```

Given that this is about *GBMOK* with $\lambda_k = 2/3$, the specifications of `idp` and `nmaxidw` have no effects on the results, so the defaults have been used.

The median and range of predictive accuracy of *GBMOK* with $\lambda_k = 2/3$ based on 100 repetitions of 10-fold cross-validation using `gbmkrigeidwcv` are

```
median(gbmkrigeidwvecv)
```

```
[1] 41.16
```

```
range(gbmkrigeidwvecv)
```

```
[1] 30.54 48.18
```

In addition, the function `gbmokgbmidwcv` in the `spm` package can also be used to assess the predictive accuracy of *GBMOKGBMIDW*.

11.6.3   Predictions

The predictions can be generated using the function `gbmkrigeidwpred` in the `spm2` package for *GBMkrigingIDW*.

1. Function `gbmkrigeidwpred`

For `gbmkrigeidwpred`, the following arguments may need to be specified:

(1) `trainx`, `trainy`, `transformation`, `delta`, `formula`, `vgm.args`, `anis`, `alpha`, `block`, and `beta`, the same as those for `krigepred`;

(2) `nmaxidw`, the same as `nmax` for `gbmidwcv`;

(3) `nmaxkrige`, the same as `nmax` for `krigecv`;

(4) `longlat`, `predx`, `longlatpredx`, `family`, `n.trees`, `learning.rate`, `interaction.depth`, and `idp`, the same as those for `gbmidwpred`; and

(5) `hybrid.parameter` and `lambda`, the same as those for `gbmkrigeidwcv`.

For details, see `?gbmkrigeidwpred`.

2. Generation of spatial predictions using `gbmkrigeidwpred`

To demonstrate the generation of the predictions, the hybrid method, *GBMOK* with $\lambda_k = 2/3$, will be used. The function `gbmkrigeidwpred` with the optimal `vgm.args`, `nmaxkrige` (i.e., `nmax` for *GBMOK*), `idp`, and `nmaxidw` (i.e., `nmax` for *GBMIDW*) will be applied to the `sponge` data set to produce spatial predictions for the area based on the `petrel.grid` data set as follows.

```
data(sponge)
data(sponge.grid)

set.seed(1234)
gbmkrigeidwpred1 <- gbmkrigeidwpred(longlat = sponge[, 1:2], trainx = sponge[,
    -3], predx = sponge.grid, trainy = sponge[, 3], longlatpredx = sponge.grid[, c
    (1:2)], family = "poisson", interaction.depth = 3, transformation = "none",
    formula.krige = res1 ~ 1, vgm.args = "Exc", nmaxkrige = 17, hybrid.parameter =
    3, lambda = 0)
```

Since `lambda = 0` is specified, the specifications of `idp` and `nmaxidw` have no effects on the results of the hybrid method that is *GBMOK* with $\lambda_k = 2/3$, so the defaulted `idp` and `nmaxidw` have been used.

```
names(gbmkrigeidwpred1)

[1] "predictions"

range(gbmkrigeidwpred1$predictions)

[1] 11.75 33.34
```

The spatial distribution of predictions by *GBMOK* with $\lambda_k = 2/3$ is illustrated in Figure 11.6.

In addition, the function `gbmokgbmidwpred` in the `spm` package can also be used to generate spatial predictions for *GBMOKGBMIDW*.

11.7   Hybrid method of *SVM* and *IDW*

We will use the `sponge` and `sponge2` data sets to demonstrate the application of the hybrid method of *SVM* and *IDW* (*SVMIDW*) for developing an optimal predictive model and

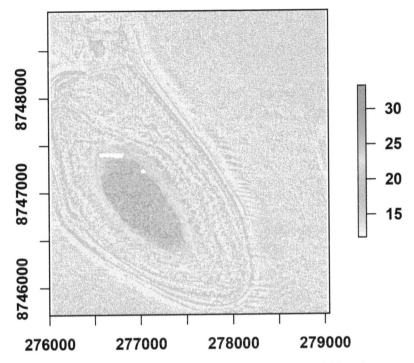

FIGURE 11.6: The spatial patterns of predictions using *GBMOK* with `lambda_k = 2/3`.

generating spatial predictions. For `sponge2`, the most accurate model is `svmvecv1` as shown in Chapter 9, where all predictive variables were used. We will use these predictive variables and the optimal parameters to construct the optimal *SVM* predictive model for *SVMIDW* first.

```
library(spm2)
data(sponge2)
svm.opt <- sponge2[, -4]

library(e1071)
svm1 <- svm(species.richness ~ ., data = svm.opt, cost = 4, gamma = 0.009)

svm1.pred <- predict(svm1, svm.opt)
```

Then produce the residuals of `svm1` for *IDW* and develop an *IDW* model based on the residuals.

```
library(gstat)
svm.opt$res1 <- sponge2$species.richness - svm1.pred

idw1 <- gstat(id = "res1", formula = res1 ~ 1, locations = ~ easting + northing,
    data = svm.opt, set = list(idp = 2), nmax = 12)
```

Finally, produce the predictions of *SVMIDW* using `svm1` and `idw1`. We will still use the training data set `svm.opt` to make the predictions.

```
idw.pred <- predict(idw1, svm.opt)$res1.pred
svmidw.pred <- svm1.pred + idw.pred
```

Similar to *RFIDW* and *GBMIDW*, *SVMIDW* is also an *exact* method; however, the *goodness of fit* is not necessarily related to the predictive accuracy of *SVMIDW* either.

In addition, if `idp` is specified as 0 in model `idw1`, then *KNN* is used, and the hybrid method resulted would be *SVMKNN*. If `nmax` is specified as 1, then *NN* is used, and the hybrid method resulted would be *SVMNN*.

11.7.1   Variable selection and parameter optimization based on predictive accuracy

For *SVMIDW*, *variable selection* can be conducted using the methods and procedures introduced in Chapter 9 for *SVM*. We will demonstrate how to optimize relevant parameters for the *SVMIDW* based on the residuals of the optimal *SVM* predictive model. This optimization can be done with the function `svmidwcv` in the `spm2` package. We will introduce `svmidwcv` first and then use it to conduct parameter optimization.

1. Function `svmidwcv`

For `svmidwcv`, the following arguments may need to be specified:

(1) `formula`, `trainxy`, `y`, `scale`, `type`, `kernel`, `degree`, `gamma`, `cost`, `nu`, `tolerance`, `epsilon`, `validation`, `cv.fold`, and `predacc`, the same as those for `svm` and `svmcv`;
(2) `longlat`, a dataframe contains the longitude and latitude of point samples;
(3) `idp`, a numeric number specifying the inverse distance weighting power; and
(4) `nmaxidw`, the number of nearest observations that should be used for a prediction.
For details, see `?svmidwcv`.

2. Parameter optimization based on predictive accuracy

We can estimate the optimal `idp` and `nmaxidw` for *SVMIDW* using `svmidwcv` based on the residuals of the optimal *SVM* predictive. The parameters estimated for `cost` and `gamma` based on `sponge` data in Chapter 9 will be used for `svmidwcv` in the demonstration below.

```
data(sponge)
model <- sponge ~ .
longlat <- sponge[, 1:2]

idp <- (1:20) * 0.2
nmax <- c(5:40)
svmidwopt <- array(0, dim = c(length(idp), length(nmax)))

for (i in 1:length(idp)) {
  for (j in 1:length(nmax)) {
    set.seed(1234)
    svmidwcv1 <- svmidwcv(formula = model, longlat = longlat, trainxy = sponge,
 y = sponge[, 3], gamma = 0.039, cost = 6.5, scale = TRUE, idp = idp[i], nmaxidw =
      nmax[j],
 validation = "CV", predacc = "VEcv")
    svmidwopt[i, j] <- svmidwcv1
  }
}
```

The optimal values of `idp` and `nmaxidw` that lead to the maximal predictive accuracy are

```
para.svmidwopt <- which (svmidwopt == max(svmidwopt), arr.ind = T)
idp[para.svmidwopt[, 1]]
```

```
[1] 0.6
```

```
nmax[para.svmidwopt[, 2]]
```

```
[1] 10
```

These parameters should be used for the *SVMIDW* predictive model.

In addition, `svmidwcv` can also be used to assess the predictive accuracy of *SVMKNN* and *SVMNN* if `idp = 0` is specified and `nmaxidw = 1` is used, respectively.

11.7.2   Predictive accuracy

We can use `svmidwcv` to produce a stabilized predictive accuracy of the *SVMIDW* model. We will apply `svmidwcv` with the above `model`, `y`, and the optimal `idp` and `nmaxidw` to `sponge` data set to demonstrate the accuracy stabilization.

```
set.seed(1234)
n <- 100
svmidwvecv <- NULL

for (i in 1:n) {
  svmidwcv1 <- svmidwcv(formula = model, longlat = longlat, trainxy = sponge,
  y = sponge[, 3], gamma = 0.039, cost = 6.5, scale = TRUE, idp = 0.6, nmaxidw =
    10,
  validation = "CV", predacc = "VEcv")
  svmidwvecv[i] <- svmidwcv1
}
```

The median and range of predictive accuracy of *SVMIDW* based on 100 repetitions of 10-fold cross-validation using `svmidwcv` are

```
median(svmidwvecv)
```

```
[1] 39.85
```

```
range(svmidwvecv)
```

```
[1] 27.57 47.43
```

11.7.3   Predictions

The predictions of *SVMIDW* can be generated using the function `svmidwpred` in the `spm2` package.

1. Function `svmidwpred`

For `svmidwpred`, the following arguments may need to be specified:
(1) `formula`, `longlat`, `trainxy`, `y`, `scale`, `type`, `kernel`, `degree`, `gamma`, `cost`, `nu`, `tolerance`, `epsilon`, `idp`, and `nmaxidw`, the same as those for `svm`. `svmcv`, and `svmidwcv`;
(2) `longlatpredx`, a dataframe contains the longitude and latitude of point locations (i.e., the centers of grids) to be predicted; and
(3) `predx`, a dataframe of predictive variables for the grids to be predicted.
For details, see `?svmidwpred`.

2. Generation of spatial predictions using `svmidwpred`

The generation of predictions of *SVMIDW* will be demonstrated by applying `svmidwpred` with the above `model`, `longlat`, and the optimal `gamma`, `cost`, `idp`, and `nmaxidw` for the area based on the `sponge.grid` data set as shown below.

```
svmidwpred1 <- svmidwpred(formula = model, longlat = longlat, trainxy = sponge,
y = sponge[, 3], longlatpredx = sponge.grid[, c(1:2)], predx = sponge.grid, gamma
    = 0.039, cost = 6.5, idp = 0.6, nmaxidw = 10)
```

The range of the predictions are

```
range(svmidwpred1$predictions)
```

```
[1] -12.83  25.71
```

A portion of the predictions are negative and beyond the range of count data and need to be corrected by resetting the faulty predictions to 0.

The spatial distribution of the corrected *SVMIDW* predictions is illustrated in Figure 11.7.

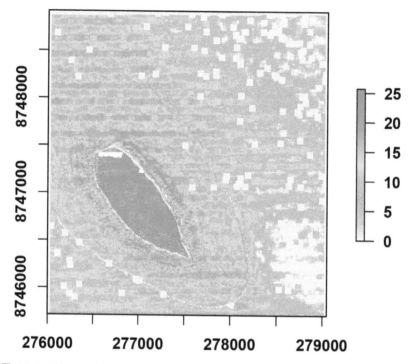

FIGURE 11.7: The spatial patterns of predictions using *SVMIDW*.

11.8   Hybrid method of *SVM* and *OK*

For the hybrid method of *SVM* and *OK* (*SVMOK*), we will continue using the sponge2 data set to demonstrate its application. To develop an optimal predictive model and generate spatial predictions using *SVMOK*, we need to apply *OK* to the residuals of a *SVM* predictive model like svm1 in svm.opt. We will firstly conduct a variogram modeling for *SVMOK* based on the residuals of svm1.

```
data.train <- svm.opt
coordinates(data.train) = ~ easting + northing
vgm1 <- variogram(object = res1 ~ 1, data.train)

model.1 <- fit.variogram(vgm1, vgm(mean(vgm1$gamma), "Sph", mean(vgm1$dist), min(
    vgm1$gamma) / 10))
```

Then use model.1 and the training data data.train to make predictions.

```
pred.krige1 <- krige(res1 ~ 1, data.train, data.train, model = model.1, nmax = 12)

names(as.data.frame(pred.krige1))
svmok.pred <- as.data.frame(pred.krige1)[, 3] + svm1.pred
```

Similar to *RFOK* and *GBMOK*, *SVMOK* is also an *exact* method; and the *goodness of fit* is not a measure of the predictive accuracy of *SVMOK* either.

In addition, if `beta` is specified for `pred.krige1`, then *SK* is used, and the hybrid method resulted would be *SVMSK*. If `block` is specified, then *BK* is used, and the hybrid method resulted would be *SVMBOK* or *SVMBSK* depending on the specification of `beta`. For ways to specify these arguments, see relevant sections in Chapter 5.

If the `res1 ~ 1` in `vgm1` is extended to include external variable(s) (e.g., `res1 ~ bathy + tpi3`), then the hybrid method resulted would be *SVMKED*; but this extension should not be used because any possible external trends should have already been dealt with in the *SVM* model. In spite of this note, the cross-validation function (i.e., `svmkrigecv`) to be introduced below can be used to assess the predictive accuracy of such hybrid methods.

11.8.1   Variable selection and parameter optimization based on predictive accuracy

With regard to *variable selection* for *SVMOK*, it can be conducted using the methods and procedures introduced in Chapter 9 for *SVM*. We will demonstrate how to optimize relevant parameters for the *SVMOK* based on the residuals of the optimal *SVM* predictive model. This optimization can be done with the function `svmkrigecv` in the `spm2` package. We will introduce `svmkrigecv` first and then use it to conduct parameter optimization.

1. Function `svmkrigecv`

For `svmkrigecv`, the following arguments are essential:
(1) `formula.svm`, the same as `formula` for `svmidwcv`,
(2) `longlat`, `trainxy`, `y`, `scale`, `type`, `kernel`, `degree`, `gamma`, `cost`, `nu`, `tolerance`, `epsilon`, `validation`, `cv.fold`, and `predacc`, the same as those for `svm` and `svmidwcv`;
(3) `transformation`, `delta`, `vgm.args`, `anis`, `alpha`, `block`, and `beta`, the same as those for `krigecv`; and
(4) `formula.krige` and `nmaxkrige`, the same as `formula` and `nmax` for `krigecv`, respectively.
For details, see `?svmkrigecv`.

2. Parameter optimization based on predictive accuracy

We can seek the optimal `vgm.args` and `nmaxkrige` for *SVMOK* using `svmkrigecv`, which will be based on the residuals of the optimal *SVM* predictive model. We will use `svmkrigecv` and the optimal `gamma` and `cost` for `sponge` data in Section 9 to demonstrate the parameter optimization as follows.

```
data(sponge)
model <- sponge ~ .
longlat <- sponge[, 1:2]

nmax.ok <- c(5:40)
vgm.args <- c("Exp", "Gau", "Sph", "Exc", "Mat", "Ste", "Lin")
svmokopt <- matrix(0, length(nmax.ok), length(vgm.args))

for (i in 1:length(nmax.ok)) {
  for (j in 1:length(vgm.args)) {
    set.seed(1234)
```

```
   svmkrigecv1 <- svmkrigecv(formula.svm = model, longlat = longlat, trainxy =
       sponge, y = sponge[, 3], gamma = 0.039, cost = 6.5, formula.krige = res1 ~
       1, nmaxkrige = nmax.ok[i], vgm.args = vgm.args[j], validation = "CV",
       predacc = "VEcv")
   svmokopt[i, j] <- svmkrigecv1
 }
}
```

The optimal `nmaxkrige` and `vgm.args` that lead to the maximal predictive accuracy are

```
svmok.opt <- which(svmokopt == max(svmokopt), arr.ind = T)
nmax.ok[svmok.opt[, 1]]
```

```
[1] 10
```

```
vgm.args[svmok.opt[, 2]]
```

```
[1] "Exc"
```

These parameters should be used for the *SVMOK* predictive model.

As shown in Chapter 5, all other parameters (e.g., `transformation`, `alpha`) can also be optimized and their optimization will not be performed here.

Furthermore, `svmkrigewcv` can also be used to assess the predictive accuracy of other hybrid methods, such as *SVMSK* and *SVMBSK* if `beta` or both `beta` and `block` are specified, respectively.

11.8.2　Predictive accuracy

The function `svmkrigecv` can be used to produce a stabilized predictive accuracy of the *SVMOK* model. We will apply `svmkrigecv` with the `model` and `longlat` in Section 11.8.1 and the optimal `cost`, `gamma`, `nmaxkrige`, and `vgm.args` to `sponge` data set to demonstrate the accuracy stabilization.

```
set.seed(1234)
n <- 100
svmokvecv <- NULL

for (i in 1:n) {
   svmkrigecv1 <- svmkrigecv(formula.svm = model, longlat = longlat, trainxy =
       sponge, y = sponge[, 3], gamma = 0.039, cost = 6.5, formula.krige = res1 ~
       1, nmaxkrige = 10, vgm.args = "Exc", validation = "CV", predacc = "VEcv")
   svmokvecv[i] <- svmkrigecv1
}
```

The median and range of predictive accuracy of *SVMOK* based on 100 repetitions of 10-fold cross-validation using `svmkrigecv` are

```
median(svmokvecv)
```

```
[1] 41.02
```

```
range(svmokvecv)
```

```
[1] 26.07 49.89
```

11.8.3 Predictions

The predictions can be generated with the function svmkrigepred in the spm2 package for *SVMOK*.

1. Function svmkrigepred

For svmkrigepred, the following arguments are required:
(1) formula.svm, longlat, trainxy, y, scale, type, kernel, degree, gamma, cost, nu, tolerance, epsilon, transformation, delta, formula.krige, vgm.args, anis, alpha, block, beta, and nmaxkrige, the same as those for svmkrigecv;
(2) predx, a dataframe of predictive variables for the grids to be predicted; and
(3) longlatpredx, a dataframe contains longitude and latitude of point locations (i.e., the centers of grids) to be predicted.
For details, see ?svmkrigepred.

2. Generation of spatial predictions using svmkrigepred

The generation of predictions of *SVMOK* will be demonstrated by applying svmkrigepred with the model and longlat in Section 11.8.1 and with the optimal gamma, cost, vgm.args, and nmaxkrige to sponge data set for the area based on the sponge.grid data set.

```
svmkrigepred1 <- svmkrigepred(formula.svm = model, longlat = longlat, trainxy =
    sponge,
y = sponge[, 3], predx = sponge.grid, longlatpredx = sponge.grid[, c(1:2)], gamma
    = 0.039, cost = 6.5, transformation = "none", formula.krige = res1 ~ 1, vgm.
    args = "Exc", nmaxkrige = 10)
```

```
range(svmkrigepred1$predictions)
```

```
[1]  -12.48   26.08
```

A portion of the predictions are negative and beyond the range of count data and need to be corrected by resetting the faulty predictions to 0.

The spatial distribution of the corrected *SVMOK* predictions is illustrated in Figure 11.8.

11.9 Hybrid methods of *SVM*, *OK*, and *IDW*

To generate spatial predictions using the hybrid methods of *SVM*, *OK*, and *IDW* (*SVMOKIDW*), we need to develop an optimal *SVMOKIDW* predictive model. First, we need to develop an optimal *SVM* predictive model and then apply *IDW* and *OK* to the residuals of the model. Finally, the predictions of *SVMOKIDW* can be generated based on the predictions of the *SVM*, *IDW*, and *OK* models.

We will use svm1.pred and idw.pred from Section 11.7, pred.krige1 from Section 11.8 and the sponge and sponge2 data sets to demonstrate the applications of *SVMOKIDW*.

1. Average of *SVMOK* and *SVMIDW*

The average of *SVMOK* and *SVMIDW* (*SVMOKSVMIDW*) is a *SVMOKIDW* method with $\lambda_m = \lambda_k = 1/2$ according to Equation (10.5). We will produce the predictions of *SVMOKSVMIDW* using svm1.pred, idw.pred, and pred.krige1.

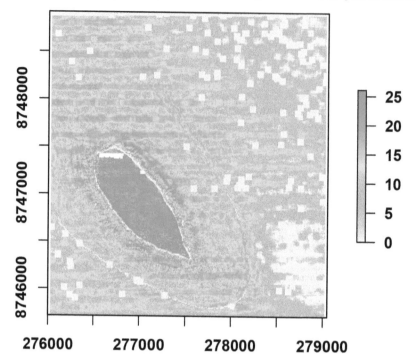

FIGURE 11.8: The spatial patterns of predictions using *SVMOK*.

```
lambdai <- 1 / 2
lambdak <- 1 / 2
svmoksvmidw.pred <- svm1.pred + lambdak * as.data.frame(pred.krige1)[, 3] +
    lambdai * idw.pred
```

Similar to *RFOKRFIDW* and *GBMOKGBMIDW*, *SVMOKSVMIDW* is also an *exact* method.

2. Average of *SVM*, *SVMOK*, and *SVMIDW*

The average of *SVM*, *SVMOK*, and *SVMIDW* (*SVMSVMOKSVMIDW*) is a *SVMOKIDW* method with $\lambda_m = \lambda_k = 1/3$ according to Equation (10.5). We will produce the predictions of *SVMSVMOKSVMIDW* using svm1.pred, idw.pred, and pred.krige1.

```
lambdai <- 1 / 3
lambdak <- 1 / 3
svmsvmoksvmidw.pred <- svm1.pred + lambdak * as.data.frame(pred.krige1)[, 3] +
    lambdai * idw.pred
```

Due to $\lambda_m + \lambda_k = 2/3$ (see Equation (10.5)), the fitted values are the same as the observed values, thus *SVMSVMOKSVMIDW* is not an exact method.

In fact, λ_m and λ_k for *SVMOKIDW* can take any values within their range (i.e., 0 to 1) as long as their sum is 1 or 2/3. Their values can be determined according to the predictive accuracy of the hybrid method resulted as shown below.

11.9.1 Variable selection and parameter optimization based on predictive accuracy

For *SVMOKIDW*, *variable selection* can be conducted using the methods and procedures introduced in Chapter 9 for *SVM*. We will demonstrate how to optimize relevant parameters for *IDW* and *OK* components of the hybrid methods based on the residuals of the optimal *SVM* predictive model here. This optimization can be done with the function svmkrigeidwcv in the spm2 package. We will introduce svmkrigeidwcv first and then use it to conduct parameter optimization.

1. Function svmkrigeidwcv

For svmkrigeidwcv, the following arguments may need to be specified:
(1) formula.svm, longlat, trainxy, y, scale, type, kernel, degree, gamma, cost, nu, tolerance, epsilon, transformation, delta, formula.krige, vgm.args, anis, alpha, block, beta, nmaxkrige, validation, cv.fold, and predacc, the same as those for svmidwcv and svmkrigecv;
(2) hybrid.parameter, either 2 (default) or 3; and
(3) lambda, ranging from 0 to 2, with the default value of 1.
For details, see ?svmkrigeidwcv.

If lambda is < 1, more weight is placed on *kriging*; and if it is > 1, more weight is placed on *IDW*.

When the default hybrid.parameter is used and lambda is 0, *IDW* is not considered, and the methods resulted is *SVMkriging*. When the default hybrid.parameter is used and lambda is 2, then the method resulted is *SVMIDW*.

When hybrid.parameter is 3, if lambda is 0, then the method resulted is equivalent to the average of *SVM*, *SVMkriging*, and *SVMkriging* (*SVMSVMkrigingSVMkriging*), and if lambda is 2, then the method resulted is equivalent to the average of *SVM*, *SVMIDW* and *SVMIDW* (*SVMSVMIDWSVMIDW*). According to Equations (10.8) and (10.9), *SVMSVMkrigingSVMkriging* can be expressed as *SVMkriging* with $\lambda_k = 2/3$, and *SVMSVMIDWSVMIDW* as *SVMIDW* with $\lambda_m = 2/3$.

With the changes in the specifications of idp, nmaxidw, beta, block, lambda, and hybrid.parameter, many hybrid methods of *SVM*, *kriging*, and *IDW* (*SVMkrigingIDW*) can be resulted (Table 11.3).

When $0 <$ lambda < 2, then numerous hybrid methods of *SVM*, *kring*, and *IDW* could be possibly derived according to Equation (10.5) and implemented via svmkrigeidwcv, such as *SVMOKIDW* with $\lambda_m = 0.85/3$ and $\lambda_k = 1.15/3$, and *SVMSKKNN* with $\lambda_m = 1.66/2$ and $\lambda_k = 0.34/2$.

The hybrid methods from No. 1 to No. 31 in Table 11.3 are exact methods, and the rest are not exact methods.

In spite of the numerous hybrid methods, the most accurate method can be selected from all these hybrid methods based on predictive accuracy using svmkrigeidwcv as demonstrated below.

2. Parameter optimization based on predictive accuracy

We can use the optimal idp, nmaxidw, vgm.args, and nmaxkrige in Sections 11.7 and 11.8 for further *parameter optimization*. These parameters were optimized separately for *SVMIDW* and *SVMOK*. If it is believed that these parameters are not optimal for *SVMOKIDW* methods, then we need to re-do the optimization. This can be done with svmkrigeidwcv, but we will not present it here, as it can be easily implemented by following the examples provided for parameter optimization for *SVMIDW* in Section 11.7 and *SVMOK* in Section 11.8 as well as the example below for estimating the optimal `lambda`.

TABLE 11.3: *SVMKrigingIDW* methods can be derived and implemented via `svmkrigeidwcv` in relation to `idp`, `nmaxidw`, `beta`, `block`, `lambda`, and `hybrid.parameter` (ϕ); and λ_m and λ_k are from Equations (10.5), (10.6) and (10.7).

No.	idp	nmaxidw	beta	block	lambda	ϕ	Hybrid methods
1	!= 0	> 0	NULL	0	1	2	*SVMOKSVMIDW*
2	0	> 1	NULL	0	1	2	*SVMOKSVMKNN*
3	0	1	NULL	0	1	2	*SVMOKSVMNN*
4	!= 0	> 0	not NULL	0	1	2	*SVMSKSVMIDW*
5	0	> 1	not NULL	0	1	2	*SVMSKSVMKNN*
6	0	1	not NULL	0	1	2	*SVMSKSVMNN*
7	!= 0	> 0	NULL	> 0	1	2	*SVMBOKSVMIDW*
8	0	> 1	NULL	> 0	1	2	*SVMBOKSVMKNN*
9	0	1	NULL	> 0	1	2	*SVMBOKSVMNN*
10	!= 0	> 0	not NULL	> 0	1	2	*SVMBSKSVMIDW*
11	0	> 1	not NULL	> 0	1	2	*SVMBSKSVMKNN*
12	0	1	not NULL	> 0	1	2	*SVMBSKSVMNN*
13	!= 0	> 0	NULL	0	0	2	*SVMOK*
14	!= 0	> 0	not NULL	0	0	2	*SVMSK*
15	!= 0	> 0	NULL	> 0	0	2	*SVMBOK*
16	!= 0	> 0	not NULL	> 0	0	2	*SVMBSK*
17	!= 0	> 0	NULL	0	2	2	*SVMIDW*
18	0	> 1	NULL	0	2	2	*SVMKNN*
19	0	1	NULL	0	2	2	*SVMNN*
20	!= 0	> 0	NULL	0	1	3	*SVMSVMOKSVMIDW*
21	0	> 1	NULL	0	1	3	*SVMSVMOKSVMKNN*
22	0	1	NULL	0	1	3	*SVMSVMOKSVMNN*
23	!= 0	> 0	not NULL	0	1	3	*SVMSVMSKSVMIDW*
24	0	> 1	not NULL	0	1	3	*SVMSVMSKSVMKNN*
25	0	1	not NULL	0	1	3	*SVMSVMSKSVMNN*
26	!= 0	> 0	NULL	> 0	1	3	*SVMSVMBOKSVMIDW*
27	0	> 1	NULL	> 0	1	3	*SVMSVMBOKSVMKNN*
28	0	1	NULL	> 0	1	3	*SVMSVMBOKSVMNN*
29	!= 0	> 0	not NULL	> 0	1	3	*SVMSVMBSKSVMIDW*
30	0	> 1	not NULL	> 0	1	3	*SVMSVMBSKSVMKNN*
31	0	1	not NULL	> 0	1	3	*SVMSVMBSKSVMNN*
32	!= 0	> 0	NULL	0	0	3	*SVMOK* with $\lambda_k = 2/3$
33	!= 0	> 0	not NULL	0	0	3	*SVMSK* with $\lambda_k = 2/3$
34	!= 0	> 0	NULL	> 0	0	3	*SVMBOK* with $\lambda_k = 2/3$
35	!= 0	> 0	not NULL	> 0	0	3	*SVMBSK* with $\lambda_k = 2/3$
36	!= 0	> 0	NULL	0	2	3	*SVMIDW* with $\lambda_m = 2/3$
37	0	> 1	NULL	0	2	3	*SVMKNN* with $\lambda_m = 2/3$
38	0	1	NULL	0	2	3	*SVMNN* with $\lambda_m = 2/3$

The optimal parameters of relevant arguments can be estimated or selected with `svmkrigeidwcv`. The examples provided for other functions can be adopted for such optimization. We will demonstrate how to estimate optimal value of `lambda` for *SVMOKIDW* with `hybrid.parameter = 2` below.

```
data(sponge)
model <- sponge ~ .
longlat <- sponge[, 1:2]
y = sponge[, 3]

lambda5 <- c(0:100)*0.02
lambda5opt <- NULL

for (i in 1:length(lambda5)) {
  set.seed(1234)
  svmkrigesvmidwcv1 <- svmkrigeidwcv(formula.svm = model, longlat = longlat,
      trainxy = sponge, y = y, gamma = 0.039, cost = 6.5, transformation = "none",
      formula.krige = res1 ~ 1, vgm.args = "Exc", nmaxkrige = 10, idp = 0.6,
      nmaxidw = 10, lambda = lambda5[i], hybrid.parameter = 2, validation = "CV",
      predacc = "VEcv")
  lambda5opt[i] <- svmkrigesvmidwcv1
}
```

This value estimated for `lambda` that leads to the maximal predictive accuracy is

```
lambda5.opt <- which (lambda5opt == max(lambda5opt))
lambda5[lambda5.opt]
```

```
[1] 0.3
```

The maximal predictive accuracy is

```
max(lambda5opt)
```

```
[1] 43.76
```

Similar to *SVMOKIDW* with `hybrid.parameter = 2`, the parameters estimated can also be used to optimize `lambda` for *SVMOKIDW* with `hybrid.parameter = 3`. The results are in `lambda6opt`, and the optimal value and the maximal predictive accuracy are

```
lambda6.opt <- which (lambda6opt == max(lambda6opt))
lambda6[lambda6.opt]
```

```
[1] 0
```

```
max(lambda6opt)
```

```
[1] 44.22
```

The value estimated for `lambda` is 0, suggesting that the hybrid method resulted is actually *SVMOK* with $\lambda_k = 2/3$ according to Equation (10.7).

As shown in Chapter 5, all other parameters (e.g., `transformation`, `alpha`) for *SVMOK* can also be optimized using `svmkrigeidwcv` with `lambda = 0`. Their optimization will not be repeated here.

The predictive accuracy of the hybrid model resulted is higher than that of *SVMOKIDW* with `hybrid.parameter = 2`. Hence, the *SVMOKIDW* with `hybrid.parameter = 3` and `lambda = 0`, that is, *SVMOK* with $\lambda_k = 2/3$ according to Equation (10.7) will be used.

11.9.2   Predictive accuracy

The function svmkrigeidwcv can be used to produce a stabilized predictive accuracy of the *SVMkrigingIDW* model. We will use svmkrigeidwcv with the above model, y, longlat, gamma, cost and the optimal nmaxkrige, vgm.args, idp, and nmaxidw to stabilize the predictive accuracy of *SVMOK* with $\lambda_k = 2/3$.

```
set.seed(1234)
n <- 100
svmkrigesvmidwvecv <- NULL
for (i in 1:n) {
  svmkrigesvmidwcv1 <- svmkrigeidwcv(formula.svm = model, longlat = longlat,
      trainxy = sponge, y = y, gamma = 0.039, cost = 6.5, transformation = "none",
      formula.krige = res1 ~ 1, vgm.args = "Exc", nmaxkrige = 10, idp = 0.6,
      nmaxidw = 10, lambda = 0, hybrid.parameter = 3, validation = "CV",  predacc
      = "VEcv")
  svmkrigesvmidwvecv[i] <- svmkrigesvmidwcv1
}
```

The median and range of predictive accuracy of *SVMOK* with $\lambda_k = 2/3$ based on 100 repetitions of 10-fold cross-validation using svmkrigeidwcv are

```
median(svmkrigesvmidwvecv)
```

```
[1] 40.81
```

```
range(svmkrigesvmidwvecv)
```

```
[1] 29.16 48.04
```

11.9.3   Predictions

The predictions can be generated with the function svmkrigeidwpred in the spm2 package for the hybrid methods.

1. Function svmkrigeidwpred

For svmkrigeidwpred, the following arguments may need to be specified:

(1) formula.svm, longlat, trainxy, predx, y, longlatpredx, scale, type, kernel, degree, gamma, cost, nu, tolerance, epsilon, transformation, delta, formula.krige, vgm.args, anis, alpha, block, beta, and nmaxkrige, the same as the arguments for svmkrigepred; and

(2) idp, nmaxidw, hybrid.parameter, and lambda, the same as the arguments for svmkrigeidwcv. For details, see ?svmkrigeidwpred.

2. Generation of spatial predictions using svmkrigeidwpred

A special hybrid method, *SVMOK* with $\lambda_k = 2/3$, will be used to demonstrate the generation of spatial predictions for *SVMOKIDW*. We will apply svmkrigeidwpred with the above model, y, longlat, gamma, cost, and with the optimal vgm.args, nmaxkrige, idp, and nmaxidw to sponge data set to produce spatial predictions for the area based on the sponge.grid data set.

```
svmkrigesvmidwpred1 <- svmkrigeidwpred(formula.svm = model, longlat = longlat,
    trainxy = sponge, predx = sponge.grid, y = y, longlatpredx = sponge.grid[, c
    (1:2)], gamma = 0.039, cost = 6.5, transformation = "none", formula.krige =
    res1 ~ 1, vgm.args = "Exc", nmaxkrige = 10, idp = 0.6, nmaxidw = 10, lambda =
    0, hybrid.parameter = 3)
```

```
range(svmkrigesvmidwpred1$predictions)
```

`[1]  -12.41   26.14`

A portion of the predictions are negative and need to be corrected by resetting the faulty predictions to 0.

The spatial distribution of the corrected predictions by $SVMIDW$ with $\lambda_k = 2/3$ is illustrated in Figure 11.9.

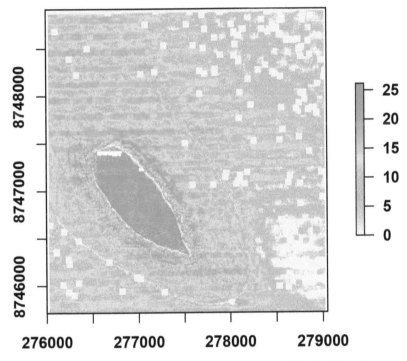

FIGURE 11.9: The spatial patterns of predictions using $SVMIDW$ with `lambda_k = 2/3`.

12

Applications and comparisons of spatial predictive methods

The spatial predictive methods that have been introduced in Chapters from 4 to 11 are ranging from *mathematical methods, geostatistical methods, modern statistical methods, machine learning methods*, the *hybrids* of modern statistical with mathematical and/or univariate geostatistical methods, and the hybrids of machine learning methods with mathematical and/or univariate geostatistical methods. Some of these methods have been applied to various data types and compared in terms of various error and/or accuracy measures (e.g., Li, Potter, Huang, Daniell, et al. 2010; Li and Heap 2011; Li, Heap, Potter, and Daniell 2011b; Li, Heap, Potter, Huang, et al. 2011; Li, Potter, Huang, and Heap 2012b; Sanabria et al. 2013; Stephens and Diesing 2014; Hengl et al. 2015; Appelhans et al. 2015; Ließ, Schmidt, and Glaser 2016; Zhang et al. 2017; Hinz, Grigoryev, and Novikov 2020; Xia et al. 2020; Fox, Ver Hoef, and Olsen 2020). In this chapter, we provide further examples on the applications of the following spatial predictive methods and also provide an example on systematically comparing the performance of these methods.

(1) *IDW*,
(2) *OK*,
(3) *KED*,
(4) *GLM*,
(5) *GLMkrigingIDW*,
(6) *RF*,
(7) *RFkrigingIDW*,
(8) *GBM*,
(9) *GBMkrigingIDW*,
(10) *SVM*, and
(11) *SVMkrigingIDW*.

The following three response variables (i.e., `zinc`, `species.richness`, and `gravel`) are used to demonstrate the application and comparison of the performance of these methods:
(1) `zinc` in the `meuse` data set in the `sp` package (Pebesma and Bivand 2020), which is continuous data;
(2) `species.richness` in the `sponge2` data set in the `spm2` package (Li 2021a), which is count data; and
(3) `gravel` in the `petrel` data set in the `spm` package (Li 2019b), which is percentage data.

We optimize relevant parameters and select predictive variables for each method with each data set based on predictive accuracy and then stabilize the predictive accuracy using the parameters optimized and predictive variables selected. Finally, we compare the performance of these methods.

The R code for parameter optimization, variable selection, and assessment of predictive accuracy are provided for `zinc`. For `species.richness` and `gravel`, the code is only presented if

DOI: 10.1201/9781003091776-12

the code for `zinc` cannot be easily adopted due to the amount of modification required, but the results are displayed for all three variables.

12.1 Parameter optimization and/or variable selection

12.1.1 *IDW*

We will estimate `idp` from 0 to 4 and `nmax` from 1 to 40 using `idwcv`.

1. `zinc` in `meuse`

```
library(sp)
data(meuse)
zn <- meuse[, c(1,2, 7:8, 14, 6)]

library(spm)

idp <- (0:20) * 0.2
nmax <- c(1:40)
idwopt1 <- array(0, dim = c(length(idp), length(nmax)))

for (i in 1:length(idp)) {
  for (j in 1:length(nmax)) {
      set.seed(1234)
      idwcv1 <- idwcv(zn[, c(1, 2)], zn[, 7], nmax = nmax[j], idp = idp[i],
          predacc = "VEcv" )
      idwopt1[i, j] <- idwcv1
  }
}
```

The optimal estimations of `idp` and `namx` are

```
para.opt1 <- which (idwopt1 == max(idwopt1), arr.ind = T)
idp[para.opt1[, 1]]
```

```
[1] 1
```

```
nmax[para.opt1[, 2]]
```

```
[1] 4
```

2. `species.richness` in `sponge2`

```
library(spm2)
data(sponge2)
```

The optimal `idp` and `namx` are

```
[1] 1
```

```
[1] 38
```

3. `gravel` in `petrel`

```
library(spm)
data(petrel)
```

The optimal `idp` and `namx` are

```
[1] 0.6
```

```
[1] 9
```

12.1.2   *OK*

1. zinc in meuse

```
zn$logzn <- log(zn$zinc)
```

```
library(sp)
```

```
coordinates(zn)=~x+y
```

```
library(intamap)
```

```
variogram1 <- estimateAnisotropy(zn, "logzn", logzn ~ 1)
```

```
variogram1$doRotation
```

```
[1] TRUE
```

```
dir1 <- 90 - variogram1$direction
ratio1 <- 1 / variogram1$ratio
```

We assume that the data transformation used is optimal and will use okcv to seek optimal model and nmax to maximize the predictive accuracy of *OK*.

```
zn <- meuse[, c(1,2, 7:8, 14, 6)]
zn$logzn <- log(zn$zinc)
```

```
nmax.ok <- c(5:40)
vgm.args.ok <- c("Exp", "Gau", "Sph", "Exc", "Mat", "Ste", "Lin")
okopt1 <- matrix(0, length(nmax.ok), length(vgm.args.ok))
```

```
for (i in 1:length(nmax.ok)) {
  for (j in 1:length(vgm.args.ok)) {
    set.seed(1234)
    okcv1 <- okcv(zn[, c(1, 2)], zn[, 7], nmax = nmax.ok[i], vgm.args = vgm.args.
        ok[j], anis = c(dir1, ratio1), alpha = c((0:8)*20), predacc = "VEcv")
    okopt1[i, j] <- okcv1
  }
}
```

The optimal nmax and vgm are

```
ok.opt1 <- which (okopt1 == max(okopt1), arr.ind = T)
nmax.ok[ok.opt1[, 1]]
```

```
[1] 9
```

```
vgm.args.ok[ok.opt1[, 2]]
```

```
[1] "Gau"
```

2. species.richness in sponge2

For species.richness, first anis and vgm need to be optimized for the original and transformed data, and then transformation and nmax can be optimized. Since the initial selection of vgm by estimateAnisotropy may not be the optimal one according to Section 5.2, vgm will be further selected.

Estimation of anis *and selection of* vgm

Non-transformed data

```
species <- sponge2

coordinates(species) = ~ easting + northing

variogram2.nonesp <- estimateAnisotropy(species, "species.richness", species.
    richness ~ 1)

variogram2.nonesp$doRotation
```

```
[1] TRUE
```

```
dir2.none <- 90 - variogram2.nonesp$direction
ratio2.none <- 1 / variogram2.nonesp$ratio

library(automap)
model2.nonesp <- autofitVariogram(species.richness ~ 1, species)$var_model

model2.nonesp
```

```
  model psill range kappa
1   Nug 63.85     0     0
2   Ste 51.55 36956    10
```

Sqrt-transformed data

```
species$sqrtsp <- sqrt(species$species.richness)

variogram2.sqrtsp <- estimateAnisotropy(species, "sqrtsp", sqrtsp ~ 1)

variogram2.sqrtsp$doRotation
```

```
[1] TRUE
```

```
dir2.sqrtsp <- 90 - variogram2.sqrtsp$direction
ratio2.sqrtsp <- 1 / variogram2.sqrtsp$ratio

model2.sqrtsp <- autofitVariogram(sqrtsp ~ 1, species)$var_model

model2.sqrtsp
```

```
  model psill range kappa
1   Nug 1.212     0     0
2   Ste 1.362 30967     2
```

Log-transformed data

```
species$logsp <- log(species$species.richness)

variogram2.logsp <- estimateAnisotropy(species, "logsp", logsp ~ 1)

variogram2.logsp$doRotation
```

```
[1] TRUE
```

```
dir2.logsp <- 90 - variogram2.logsp$direction
ratio2.logsp <- 1 / variogram2.logsp$ratio

model2.logsp <- autofitVariogram(logsp ~ 1, species)$var_model

model2.logsp
```

```
  model  psill range kappa
1   Nug 0.4654     0   0.0
2   Ste 1.2152 69965   0.3
```

Selection of `transformation`

We can select `transformation` using `krigecv` based on the predictive accuracy of *OK*. We will use the optimal parameters above for `anis` (i.e., `dir`, `ratio`) and `vgm`, as well as the `alpha` values as suggested in Section 5.2.2.

```
library(spm2)

transf1 <- c("sqrt", "log", "none")
anis.dir1 <- c(dir2.sqrtsp, dir2.logsp, dir2.none)
anis.ratio1 <- c(ratio2.sqrtsp, ratio2.logsp, ratio2.none)
alpha.tr1 <- list (c((0:8)*20), c((0:8)*20), 0)
okopt.tr1 <- NULL

for (i in 1:length(transf1)) {
  set.seed(1234)
  okcv1 <- krigecv(longlat = sponge2[, c(1, 2)], trainy = sponge2[, 3], nmax = 12,
      transformation = transf1[i], vgm.args = "Ste", anis = c(anis.dir1[i], anis.
      ratio1[i]), alpha = alpha.tr1[[i]], predacc = "VEcv")
  okopt.tr1[i] <- okcv1
}
```

The optimal `transformation` type selected is

```
okopt.transf1 <- which (okopt.tr1 == max(okopt.tr1))
transf1[okopt.transf1] # optimal `transformation` type
```

```
[1] "sqrt"
```

Estimation of `nmax` and further selection of `vgm`

We assume that the data transformation selected above is optimal for `nmax` and `vgm`, and will use `okcv` to optimize `nmax` to maximize the predictive accuracy of *OK*. The `vgm` (i.e., `vgm.args`) will be further optimized in case it changes with `nmax` as observed in Chapter 5.

```
nmax.ok <- c(5:40)
vgm.args.ok <- c("Exp", "Gau", "Sph", "Exc", "Mat", "Ste", "Lin")
okopt2 <- matrix(0, length(nmax.ok), length(vgm.args.ok))

for (i in 1:length(nmax.ok)) {
  for (j in 1:length(vgm.args.ok)) {
    set.seed(1234)
    okcv1 <- okcv(sponge2[, c(1, 2)], sponge2[, 3], nmax = nmax.ok[i],
        transformation = "sqrt", vgm.args = vgm.args.ok[j], anis = c(dir2.sqrtsp,
        ratio2.sqrtsp), alpha = c((0:8)*20), predacc = "VEcv")
    okopt2[i, j] <- okcv1
  }
}
```

The optimal `nmax` and `vgm` are

```
ok.opt2 <- which (okopt2 == max(okopt2, na.rm=T), arr.ind = T)
nmax.ok[ok.opt2[, 1]]
```

```
[1] 17
```

```
vgm.args.ok[ok.opt2[, 2]]
```

```
[1] "Exc"
```

Since NAs are in okopt2 for "EXP", na.rm=T is used. The results suggest that *Exc* model should be used for *OK* instead of *Ste* when nmax = 17.

3. gravel in petrel

The parameter optimization for gravel in petrel has been done in Section 5.2.2. We will use those parameters for the application of *OK* to gravel.

12.1.3 KED

For each of three response variables, we need to estimate anis first for each detrend model (e.g., logzn ~ dist.m + dist), then select predictive variables, and finally seek optimal vgm and nmax.

1. zinc in meuse

Estimation of anis

```
zn <- meuse[, c(1,2, 7:8, 14, 6)]
zn$logzn <- log(zn$zinc)
```

```
library(sp)
```

```
coordinates(zn) = ~ x + y
```

```
library(intamap)
```

```
variogram3.1 <- estimateAnisotropy(zn, "logzn", logzn ~ dist.m)
variogram3.2 <- estimateAnisotropy(zn, "logzn", logzn ~ dist.m + dist)
variogram3.3 <- estimateAnisotropy(zn, "logzn", logzn ~ dist.m + dist + elev)
variogram3.4 <- estimateAnisotropy(zn, "logzn", logzn ~ dist.m + dist + elev + x)
variogram3.5 <- estimateAnisotropy(zn, "logzn", logzn ~ dist.m + dist + elev + x +
    y)
```

```
variogram3.1$doRotation
```

```
[1] TRUE
```

```
variogram3.2$doRotation
```

```
[1] TRUE
```

```
variogram3.3$doRotation
```

```
[1] TRUE
```

```
variogram3.4$doRotation
```

```
[1] TRUE
```

```
variogram3.5$doRotation
```

```
[1] TRUE

dir3.1 <- 90 - variogram3.1$direction; ratio3.1 <- 1 / variogram3.1$ratio
dir3.2 <- 90 - variogram3.2$direction; ratio3.2 <- 1 / variogram3.2$ratio
dir3.3 <- 90 - variogram3.3$direction; ratio3.3 <- 1 / variogram3.3$ratio
dir3.4 <- 90 - variogram3.4$direction; ratio3.4 <- 1 / variogram3.4$ratio
dir3.5 <- 90 - variogram3.5$direction; ratio3.5 <- 1 / variogram3.5$ratio
```

Variable selection

We will perform variable selection for *KED* by assuming `vgm.args = "Sph"` and `nmax = 12`, which will be optimized later.

```
zn <- meuse[, c(1,2, 7:8, 14, 6)]
zn$logzn <- log(zn$zinc)

formula.ked <- c(var1 ~ dist.m,
                 var1 ~ dist.m + dist,
                 var1 ~ dist.m + dist + elev,
                 var1 ~ dist.m + dist + elev + x,
                 var1 ~ dist.m + dist + elev + x + y)

ked.dir <- c(dir3.1, dir3.2, dir3.3, dir3.4, dir3.5)
ked.ratio <- c(ratio3.1, ratio3.2, ratio3.3, ratio3.4, ratio3.5)

kedopt1 <- NULL

for (i in 1:length(formula.ked)) {
  set.seed(1234)
  kedcv1 <- krigecv(zn[, 1:2], zn[, 7], zn[, 1:5], nmax = 12, transformation = "
      none", formula = formula.ked[[i]], vgm.args = "Sph", anis = c(ked.dir[i],
      ked.ratio[i]), alpha = c(0:8)*20, predacc = "VEcv")
  kedopt1[i] <- kedcv1
}

ked.opt1 <- which (kedopt1 == max(kedopt1))
formula.ked[ked.opt1]

[[1]]
var1 ~ dist.m + dist + elev
```

The results indicate that `var1 ~ dist.m + dist + elev` should be used for *KED*.

Selection of `vgm` and estimation of `nmax`

We assume that the `log` transformation used is optimal and will use `krigecv` to find optimal `vgm` (i.e., `vgm.args`) and `nmax` to maximize the predictive accuracy of *KED*.

```
nmax.ked <- c(5:50)
vgm.args.ked <- c("Exp", "Gau", "Sph", "Exc", "Mat", "Ste", "Lin")
kedopt1.2 <- matrix(0, length(nmax.ked), length(vgm.args.ked))

for (i in 1:length(nmax.ked)) {
  for (j in 1:length(vgm.args.ked)) {
    set.seed(1234)
    kedcv1 <- krigecv(zn[, 1:2], zn[, 7], zn[, 1:5], nmax = nmax.ked[i],
        transformation = "none", formula = formula.ked[[3]], vgm.args = vgm.args.
        ked[j], anis = c(ked.dir[3], ked.ratio[3]), alpha = c(0:8)*20, predacc = "
        VEcv")
    kedopt1.2[i, j] <- kedcv1
  }
}
```

The transformation for zinc has already been done in zn data set, so in krigecv the default transformation was used.

The optimal nmax and vgm are

```
ked.opt1.2 <- which (kedopt1.2 == max(kedopt1.2), arr.ind = T)
nmax.ked[ked.opt1.2[, 1]]
```

```
[1] 40
```

```
vgm.args.ked[ked.opt1.2[, 2]]
```

```
[1] "Exp"
```

The variables selected are based on vgm.args = "Sph" and nmax = 12 and assumed to be optimal for vgm.args = "Exp" and nmax = 40.

If needed, variable selection can be repeated based on the optimal parameters such as vgm .args = "Exp" and nmax = 40 until the variables selected and the parameters optimized are stable. This rule applies to all other response variables. We will further select the predictive variables for vgm.args = "Exp" and nmax = 40 below.

```
kedopt1.3 <- NULL
```

```
for (i in 1:length(formula.ked)) {
  set.seed(1234)
  kedcv1 <- krigecv(zn[, 1:2], zn[, 7], zn[, 1:5], nmax = 40, transformation = "
      none", formula = formula.ked[[i]], vgm.args = "Exp", anis = c(ked.dir[i],
      ked.ratio[i]), alpha = c(0:8)*20, predacc = "VEcv")
  kedopt1.3[i] <- kedcv1
}
```

```
ked.opt1.3 <- which (kedopt1.3 == max(kedopt1.3))
formula.ked[ked.opt1.3]
```

```
[[1]]
var1 ~ dist.m + dist + elev
```

It apparent that for zinc, the variables selected based on vgm.args = "Sph" and nmax = 12 are optimal for vgm.args = "Exp" and nmax = 40, so both the variables selected and the parameters estimated are optimal.

2. species.richness in sponge2

We will use the variables selected in Section 7.4.3 based on bestglm as initial predictive variables for *KED*.

Estimation of anis

```
spng <- sponge2
spng$sqrtsp <- sqrt(sponge2$species.richness)
spng$easting2 <- sponge2$easting^2
library(sp)
```

```
coordinates(spng) = ~ easting + northing
```

```
library(intamap)
```

```
variogram4.1 <- estimateAnisotropy(spng, "sqrtsp", sqrtsp ~ dist.coast)
variogram4.2 <- estimateAnisotropy(spng, "sqrtsp", sqrtsp ~ dist.coast + easting)
variogram4.3 <- estimateAnisotropy(spng, "sqrtsp", sqrtsp ~ dist.coast + easting +
      easting2)
variogram4.4 <- estimateAnisotropy(spng, "sqrtsp", sqrtsp ~ easting + easting2)
```

```
variogram4.1$doRotation

[1] TRUE

variogram4.2$doRotation

[1] TRUE

variogram4.3$doRotation

[1] TRUE

variogram4.4$doRotation

[1] TRUE

dir4.1 <- 90 - variogram4.1$direction; ratio4.1 <- 1 / variogram4.1$ratio
dir4.2 <- 90 - variogram4.2$direction; ratio4.2 <- 1 / variogram4.2$ratio
dir4.3 <- 90 - variogram4.3$direction; ratio4.3 <- 1 / variogram4.3$ratio
dir4.4 <- 90 - variogram4.4$direction; ratio4.4 <- 1 / variogram4.4$ratio
```

Variable selection

We will perform variable selection for *KED* by assuming vgm.args = "Sph" and nmax = 12, which will be optimized afterwards.

```
spng <- sponge2
spng$sqrtsp <- sqrt(sponge2$species.richness)
spng$easting2 <- sponge2$easting^2

formula.ked2 <- c(var1 ~ dist.coast,
                  var1 ~ dist.coast + easting,
                  var1 ~ dist.coast + easting + easting2,
                  var1 ~ easting + easting2)

ked.dir2 <- c(dir4.1, dir4.2, dir4.3, dir4.4)
ked.ratio2 <- c(ratio4.1, ratio4.2, ratio4.3, ratio4.4)

kedopt2 <- NULL

for (i in 1:length(formula.ked2)) {
  set.seed(1234)
  kedcv1 <- krigecv(spng[, 1:2], spng[, 82], spng[, c(1:5, 48, 83)], nmax = 12,
      transformation = "none", formula = formula.ked2[[i]], vgm.args = "Sph", anis
      = c(ked.dir2[i], ked.ratio2[i]), alpha = c(0:8)*20, predacc = "VEcv")
  kedopt2[i] <- kedcv1
}

ked.opt2 <- which (kedopt2 == max(kedopt2))
formula.ked2[ked.opt2]


[[1]]
var1 ~ dist.coast
```

The results show that var1 ~ dist.coast should be used for *KED*.

Selection of vgm and estimation of nmax

We assume that the data transformation used is optimal and will use krigecv to find optimal vgm (i.e., vgm.args) and nmax to maximize the predictive accuracy of *KED*.

```
nmax.ked <- c(5:50)
vgm.args.ked <- c("Exp", "Gau", "Sph", "Exc", "Mat", "Ste", "Lin")
kedopt2.2 <- matrix(0, length(nmax.ked), length(vgm.args.ked))

for (i in 1:length(nmax.ked)) {
  for (j in 1:length(vgm.args.ked)) {
    set.seed(1234)
    kedcv1 <- krigecv(spng[, 1:2], spng[, 82], spng[, c(1:5, 48, 83)], nmax = nmax
        .ked[i], transformation = "none", formula = formula.ked2[[1]], vgm.args =
        vgm.args.ked[j], anis = c(ked.dir2[1], ked.ratio2[1]), alpha = c(0:8)*20,
        predacc = "VEcv")
    kedopt2.2[i, j] <- kedcv1
  }
}
```

The optimal `nmax` and `vgm` are

```
ked.opt2.2 <- which (kedopt2.2 == max(kedopt2.2), arr.ind = T)
nmax.ked[ked.opt2.2[, 1]];
```

```
[1] 20
```

```
vgm.args.ked[ked.opt2.2[, 2]]
```

```
[1] "Exc"
```

3. `gravel` in `petrel`

We will use the parameters optimized and variables selected in Section 6.3.2 for `gravel` in `petrel`.

12.1.4   *GLM*

For *GLM*, *variable selection* needs to be performed, and `bestglm` will be used to select the predictive variables.

1. `zinc` in `meuse`

```
X1 <- data.frame(subset(meuse, select = c(elev, dist, dist.m)),
elev2 = meuse$elev^2,
dist2 = meuse$dist^2,
dist.m2 = meuse$dist.m^2)

logzn <- log(meuse$zinc)

Xy1 <- as.data.frame(cbind(X1, logzn))

library(bestglm)

set.seed(1234)

glm.cv1 <- bestglm(Xy1, IC = "CV", family = gaussian)

glm.cv1$BestModel$coefficients
```

```
(Intercept)         elev       dist.m      dist.m2
 8.65465467  -0.24743033  -0.00401001   0.00000307
```

The results show that the optimal model is with three variables: `elev`, `dist.m`, and `dist.m2`.

2. species.richness in sponge2

We will use the variables selected in Section 7.4.5 for species.richness.

3. gravel in petrel

```
gravelpredx <- petrel[, c(1, 2, 6:9)]

gravel <- petrel$gravel / 100

X3 <- data.frame(gravelpredx,
long2 = petrel$long^2,
lat2 = petrel$lat^2,
bathy2 = petrel$bathy^2,
dist2 = petrel$dist^2,
relief2 = petrel$relief^2,
slope2 = petrel$slope^2)

Xy3 <- as.data.frame(cbind(X3, gravel))

set.seed(1234)

glm.cv3 <- bestglm(Xy3, IC = "CV", family = binomial(link=logit)) # this not
    working.

glm.cv3$BestModel$coefficients
```

```
(Intercept)        bathy       relief
  -1.18445      0.01504      0.02724
```

The results show that the optimal model is with two variables: bathy and relief.

12.1.5 *GLMkrigingIDW*

We will use glmidwcv, glmkrigecv, and glmkrigeidwcv to select the most accurate *GLMkrigingIDW* method (that can be any method in Table 10.2) for zinc, species.richness, and gravel. The predictive variables selected for *GLM* will be used for *GLMkrigingIDW*.

1. zinc in meuse

We will use the predictive variables selected for *GLM* in glm.cv1$BestModel in Section 12.1.4 to seek optimal parameters.

Estimation of idp and nmaxidw

```
modelglm <- log(zinc) ~ elev + dist.m + I(dist.m^2)

longlat <- meuse[, c(1, 2)]
y <- log(meuse$zinc)

idp.glm1 <- (1:20) * 0.2
nmax.glm1 <- c(5:40)
glmidwopt1 <- array(0, dim = c(length(idp.glm1), length(nmax.glm1)))

for (i in 1:length(idp.glm1)) {
  for (j in 1:length(nmax.glm1)) {
    set.seed(1234)
    glmidwopt1[i, j] <- glmidwcv(formula = modelglm, longlat = longlat, trainxy =
        meuse, y = y, family = "gaussian", idp = idp.glm1[i], nmaxidw = nmax.glm1[
        j], validation = "CV", predacc = "VEcv")
  }
}
```

The optimal `idp` and `nmaxidw` are

```
para.glmidwopt1 <- which (glmidwopt1 == max(glmidwopt1), arr.ind = T)
idp.glm1[para.glmidwopt1[, 1]]
```

```
[1] 1.2
```

```
nmax.glm1[para.glmidwopt1[, 2]]
```

```
[1] 8
```

Estimation of `namxkrige` and selection of `vgm.args`

```
modelglm <- log(zinc) ~ elev + dist.m + I(dist.m^2)
```

```
longlat <- meuse[, c(1, 2)]
y <- log(meuse$zinc)
```

```
nmax.glmok1 <- c(5:40)
vgm.args <- c("Exp", "Gau", "Sph", "Exc", "Mat", "Ste", "Lin")
glmokopt1 <- matrix(0, length(nmax.glmok1), length(vgm.args))
```

```
for (i in 1:length(nmax.glmok1)) {
  for (j in 1:length(vgm.args)) {
    set.seed(1234)
    glmokopt1[i, j] <- glmkrigecv(formula.glm = modelglm, longlat = longlat,
        trainxy = meuse, y = y, family = "gaussian", nmaxkrige = nmax.glmok1[i],
        formula.krige = res1 ~ 1, vgm.args = vgm.args[j], validation = "CV",
        predacc = "VEcv")
  }
}
```

The optimal `nmaxkrige` and `vgm.args` are

```
glmok.opt1 <- which(glmokopt1 == max(glmokopt1), arr.ind = T)
nmax.glmok1[glmok.opt1[, 1]]
```

```
[1] 31
```

```
vgm.args[glmok.opt1[, 2]]
```

```
[1] "Lin"
```

Estimation of `lambda`

We will use the optimal `idp`, `nmaxidw`, `namxkrige`, and `vgm.args` to estimate `lambda` for *GLMk-rigingIDW* with `hybrid.parameter = 2`.

```
modelglm <- log(zinc) ~ elev + dist.m + I(dist.m^2)
longlat <- meuse[, c(1, 2)]
y <- log(meuse$zinc)
```

```
lambda.glmokidw1 <- c(0:100)*0.02
lambda.glmokidw1opt <- NULL
```

```
for (i in 1:length(lambda.glmokidw1)) {
  set.seed(1234)

  lambda.glmokidw1opt[i] <- glmkrigeidwcv(formula.glm = modelglm, longlat =
      longlat, trainxy = meuse, y = y, family = "gaussian", formula.krige = res1
      ~ 1, vgm.args = "Lin", nmaxkrige = 31, idp = 1.2, nmaxidw = 8, lambda =
      lambda.glmokidw1[i], hybrid.parameter = 2, validation = "CV", predacc = "
      VEcv")
}
```

The optimal `lambda` and the maximal predictive accuracy are

```
lambda.glmokidw1.opt <- which (lambda.glmokidw1opt == max(lambda.glmokidw1opt))
lambda.glmokidw1[lambda.glmokidw1.opt]
```

```
[1] 0
```

```
round(max(lambda.glmokidw1opt), 2)
```

```
[1] 80.06
```

The value estimated for `lambda` suggests that the hybrid method resulted from `glmkrigeidwcv` is, in fact, *GLMOK* according to Equation (10.7).

Similar to *GLMkrigingIDW* with `hybrid.parameter = 2`, the optimal `idp`, `nmaxidw`, `var.args`, and `namxkrige` can also be used to estimate `lambda` for *GLMkrigingIDW* with `hybrid.parameter = 3`. The results are in `lambda.glmokidw1.2opt`, and the optimal value and the maximal predictive accuracy are

```
lambda.glmokidw1.2.opt <- which (lambda.glmokidw1.2opt == max(lambda.glmokidw1.2
    opt))
lambda.glmokidw1.2[lambda.glmokidw1.2.opt]
```

```
[1] 0
```

```
round(max(lambda.glmokidw1.2opt), 2)
```

```
[1] 79.02
```

The value estimated for `lambda` is 0, suggesting that the hybrid method resulted from `glmkrigeidwcv` is, in fact, *GLMOK* but with $\lambda_k = 2/3$ according to Equation (10.7).

The predictive accuracy of the hybrid model resulted is lower than that of *GLMkrigingIDW* with `hybrid.parameter = 2`. Hence, the *GLMkrigingIDW* with `hybrid.parameter = 2` and `lambda = 0`, that is, *GLMOK*, will be used as the hybrid method for `zinc`.

2. `species.richness` in `sponge2`

We will use the predictive variables selected for *GLM* in `formula.glm2[glm.opt2]` in Section 7.4.3 to optimize relevant parameters for `species.richness`.

Estimation of `idp` and `nmaxidw`

```
X2 <- sponge2[, c(1:3)]
X2$easting2 = sponge2$easting^2

modelglm2 <- species.richness ~ easting + easting2

longlat <- X2[, c(1, 2)]
y <- X2[, 3]

idp.glm2 <- (1:20) * 0.2
nmax.glm2 <- c(5:40)
glmidwopt2 <- array(0, dim = c(length(idp.glm2), length(nmax.glm2)))

for (i in 1:length(idp.glm2)) {
  for (j in 1:length(nmax.glm2)) {
    set.seed(1234)
    glmidwopt2[i, j] <- glmidwcv(formula = modelglm2, longlat = longlat, trainxy =
        X2, y = y, family = "poisson", idp = idp.glm2[i], nmaxidw = nmax.glm2[j],
        validation = "CV", predacc = "VEcv")
  }
}
```

The optimal `idp` and `nmaxidw` are

```
para.glmidwopt2 <- which (glmidwopt2 == max(glmidwopt2), arr.ind = T)
idp.glm2[para.glmidwopt2[, 1]]
```

```
[1] 1
```

```
nmax.glm2[para.glmidwopt2[, 2]]
```

```
[1] 38
```

These parameters will be used for *GLMkrigingIDW*.

Estimation of `nmaxkrige` *and selection of* `vgm.args`

```
X2 <- sponge2[, c(1:3)]
X2$easting2 = sponge2$easting^2
```

```
modelglm2 <- species.richness ~ easting + easting2
```

```
longlat <- X2[, c(1, 2)]
y <- X2[, 3]
```

```
nmax.glmok2 <- c(5:40)
vgm.args <- c("Exp", "Gau", "Sph", "Exc", "Mat", "Ste", "Lin")
glmokopt2 <- matrix(0, length(nmax.glmok2), length(vgm.args))
```

```
for (i in 1:length(nmax.glmok2)) {
  for (j in 1:length(vgm.args)) {
    set.seed(1234)
    glmokopt2[i, j] <- glmkrigecv(formula.glm = modelglm2, longlat = longlat,
        trainxy = X2, y = y, family = "poisson", nmaxkrige = nmax.glmok2[i],
        formula.krige = res1 ~ 1, vgm.args = vgm.args[j], validation = "CV",
        predacc = "VEcv")
  }
}
```

The optimal `nmaxkrige` and `vgm.args` are

```
glmok.opt2 <- which(glmokopt2 == max(glmokopt2), arr.ind = T)
nmax.glmok2[glmok.opt2[, 1]]
```

```
[1] 14
```

```
vgm.args[glmok.opt2[, 2]]
```

```
[1] "Exc"
```

Estimation of `lambda`

We will use the optimal `idp`, `nmaxidw`, `namxkrige`, and `vgm.args` to estimate `lambda` for *GLMkrigingIDW* with `hybrid.parameter = 2`.

```
X2 <- sponge2[, c(1:3)]
X2$easting2 = sponge2$easting^2
```

```
modelglm2 <- species.richness ~ easting + easting2
```

```
longlat <- X2[, c(1, 2)]
y <- X2[, 3]
```

```
lambda.glmokidw2 <- c(0:100)*0.02
lambda.glmokidw2opt <- NULL
```

```
for (i in 1:length(lambda.glmokidw2)) {
  set.seed(1234)

  lambda.glmokidw2opt[i] <- glmkrigeidwcv(formula.glm = modelglm2, longlat =
      longlat, trainxy =  X2, y = y, family = "poisson", formula.krige = res1 ~ 1,
      vgm.args = "Exc", nmaxkrige = 14, idp = 1, nmaxidw = 38, lambda = lambda.
      glmokidw2[i], hybrid.parameter = 2, validation = "CV", predacc = "VEcv")
}
```

The optimal `lambda` and the maximal predictive accuracy are

```
lambda.glmokidw2.opt <- which (lambda.glmokidw2opt == max(lambda.glmokidw2opt))
lambda.glmokidw2[lambda.glmokidw2.opt]
```

```
[1] 1.88
```

```
round(max(lambda.glmokidw2opt), 2)
```

```
[1] 35.58
```

Similar to *GLMkrigingIDW* with `hybrid.parameter = 2`, the optimal `idp`, `nmaxidw`, `var.args`, and `namxkrige` can also be used to estimate `lambda` for *GLMkrigingIDW* with `hybrid.parameter = 3`. The results are in `lambda.glmokidw2.2opt`, and the optimal value of `lambda` and the maximal predictive accuracy are

```
lambda.glmokidw2.2.opt <- which (lambda.glmokidw2.2opt == max(lambda.glmokidw2.2
    opt))
lambda.glmokidw2.2[lambda.glmokidw2.2.opt]
```

```
[1] 2
```

```
round(max(lambda.glmokidw2.2opt), 2)
```

```
[1] 34.36
```

The predictive accuracy of the hybrid model resulted is lower than that of *GLMkrigingIDW* with `hybrid.parameter = 2`. Hence, the *GLMkrigingIDW* (i.e., *GLMOKIDW*) with `hybrid.parameter = 2` and `lambda = 1.88` will be used as the hybrid method for `species.richness`.

3. `gravel` **in** `petrel`

We will use the predictive variables selected for *GLM* in `glm.cv3$BestModel` in Section 12.1.4 to optimize relevant parameters for `gravel`.

Estimation of `idp` *and* `nmaxidw`

```
modelglm3 <- gravel / 100 ~ bathy + relief
```

```
longlat <- petrel[, c(1, 2)]
y <- petrel$gravel / 100
```

```
idp.glm3 <- (1:20) * 0.2
nmax.glm3 <- c(5:40)
glmidwopt3 <- array(0, dim = c(length(idp.glm3), length(nmax.glm3)))
```

```
for (i in 1:length(idp.glm3)) {
  for (j in 1:length(nmax.glm3)) {
    set.seed(1234)
    glmidwopt3[i, j] <- glmidwcv(formula = modelglm3, longlat = longlat, trainxy =
        petrel, y = y, family = binomial(link=logit), idp = idp.glm3[i], nmaxidw
        = nmax.glm3[j], validation = "CV", predacc = "VEcv")
  }
}
```

The optimal `idp` and `nmaxidw` are

```
para.glmidwopt3 <- which (glmidwopt3 == max(glmidwopt3), arr.ind = T)
idp.glm3[para.glmidwopt3[, 1]]
```

```
[1] 0.4
```

```
nmax.glm3[para.glmidwopt3[, 2]]
```

```
[1] 9
```

Estimation of `nmaxkrige` and selection of `vgm.args`

```
modelglm3 <- gravel / 100 ~ bathy + relief
```

```
longlat <- petrel[, c(1, 2)]
y <- petrel$gravel / 100
```

```
nmax.glmok3 <- c(5:40)
vgm.args <- c("Exp", "Gau", "Sph", "Exc", "Mat", "Ste", "Lin")
glmokopt3 <- matrix(0, length(nmax.glmok3), length(vgm.args))

for (i in 1:length(nmax.glmok3)) {
  for (j in 1:length(vgm.args)) {
    set.seed(1234)
    glmokopt3[i, j] <- glmkrigecv(formula.glm = modelglm3, longlat = longlat,
        trainxy = petrel, y = y, family = binomial(link=logit), nmaxkrige = nmax.
        glmok3[i], formula.krige = res1 ~ 1, vgm.args = vgm.args[j], validation =
        "CV", predacc = "VEcv")
  }
}
```

The optimal `nmaxkrige` and `vgm.args` are

```
glmok.opt3 <- which(glmokopt3 == max(glmokopt3), arr.ind = T)
nmax.glmok3[glmok.opt3[, 1]]
```

```
[1] 39
```

```
vgm.args[glmok.opt3[, 2]]
```

```
[1] "Lin"
```

Estimation of `lambda`

We will use the optimal `idp`, `nmaxidw`, `namxkrige`, and `vgm.args` to estimate `lambda` for *GLMk-rigingIDW* with `hybrid.parameter = 2`.

```
modelglm3 <- gravel / 100 ~ bathy + relief
```

```
longlat <- petrel[, c(1, 2)]
y <- petrel$gravel / 100
```

```
lambda.glmokidw3 <- c(0:100)*0.02
lambda.glmokidw3opt <- NULL

for (i in 1:length(lambda.glmokidw3)) {
  set.seed(1234)

  lambda.glmokidw3opt[i] <- glmkrigeidwcv(formula.glm = modelglm3, longlat =
      longlat, trainxy = petrel, y = y, family = binomial(link=logit), formula.
      krige = res1 ~ 1, vgm.args = "Lin", nmaxkrige = 39, idp = 0.4, nmaxidw = 9,
      lambda = lambda.glmokidw3[i], hybrid.parameter = 2, validation = "CV",
      predacc = "VEcv")
}
```

The value estimated for `lambda` and the maximal predictive accuracy are

```
lambda.glmokidw3.opt <- which (lambda.glmokidw3opt == max(lambda.glmokidw3opt))
lambda.glmokidw3[lambda.glmokidw3.opt]
```

```
[1] 0.32
```

```
round(max(lambda.glmokidw3opt), 2)
```

```
[1] 37.44
```

Similar to *GLMkrigingIDW* with `hybrid.parameter = 2`, the parameters estimated or selected can also be used to optimize `lambda` for *GLMkrigingIDW* with `hybrid.parameter = 3`. The results are in `lambda.glmokidw3.2opt`, and the optimal `lambda` and the maximal predictive accuracy are

```
lambda.glmokidw3.2.opt <- which (lambda.glmokidw3.2opt == max(lambda.glmokidw3.2
    opt))
lambda.glmokidw3.2[lambda.glmokidw3.2.opt]
```

```
[1] 0.2
```

```
round(max(lambda.glmokidw3.2opt))
```

```
[1] 36
```

The predictive accuracy of the hybrid model resulted is lower than that of *GLMkrigingIDW* with `hybrid.parameter = 2`. Hence, the *GLMkrigingIDW* (i.e., *GLMOKIDW*) with `hybrid.parameter = 2` and `lambda = 0.32` should be used as a hybrid method for `gravel`.

12.1.6   *RF*

For *RF*, variable selection needs to be performed, and `steprf` with `method = "KIAVI2"` will be used to select the predictive variables.

1. `zinc` in `meuse`

```
library(steprf)
```

```
set.seed(1234)
```

```
zn2 <- meuse[, -c(3:5, 9)]
zn2 <- na.omit(zn2)
```

```
steprf1 <- steprf(trainx = zn2[, -3], trainy = zn2[, 3], method = "KIAVI2", rpt =
    20, predacc = "VEcv", importance = TRUE, nsim = 20, delta.predacc = 0.005)
```

The predictive variables selected are

```
steprf1$steprfPredictorsAll[1][[1]]$variables.most.accurate
```

```
[1] "x"      "y"      "elev"   "dist"   "ffreq"  "dist.m"
```

The results show that `ffreq`, a categorical variable, is also selected as a predictor. The inclusion of categorical predictor was observed to prevent the selection of other important numerical predictive variables and then results in a less accurate predictive model (Li, Alvarez, et al. 2017), but it is not the case for `zinc` data because all numerical variables are selected for `zinc`.

2. `species.richness` in `sponge2`

```
set.seed(1234)

steprf2 <- steprf(trainx = sponge2[, -c(3:4)], trainy = sponge2[, 3], method = "
    KIAVI2", rpt = 20, predacc = "VEcv", importance = TRUE, nsim = 20, delta.
    predacc = 0.005)
```

The predictive variables selected are

```
steprf2$steprfPredictorsAll[2][[1]]$variables.most.accurate
```

```
[1] "easting"   "northing"   "bathy"      "bs27"       "bs36"
[6] "prof_cur_o" "tpi3"
```

3. `gravel` in `petrel`

```
set.seed(1234)

steprf3 <- steprf(trainx = petrel[, -c(3:5)], trainy = petrel[, 5], method = "
    KIAVI2", rpt = 20, predacc = "VEcv", importance = TRUE, nsim = 20, delta.
    predacc = 0.005)
```

The variables selected are

```
steprf3$steprfPredictorsAll[1][[1]]$variables.most.accurate
```

```
[1] "long"   "lat"   "bathy"  "dist"   "relief"
```

12.1.7 *RFkrigingIDW*

We will use `rfidwcv`, `rfokcv`, and `rfkrigeidwcv` to select the most accurate *RFkrigingIDW* method (that can be any method in Table 11.1) for zinc, `species.richness`, and `gravel`. The predictive variables selected for *RF* will be used for *RFkrigingIDW*.

1. `zinc` in `meuse`

We will use the predictive variables selected for *RF* in `steprf1$steprfPredictorsAll[1][[1]]` `$variables.most.accurate` in Section 12.1.6 to optimize relevant parameters, which will be carried out with `rfidwcv` and `rfokcv`.

Estimation of `idp` and `nmaxidw`

```
idp1 <- (1:20) * 0.2
nmax1 <- c(80:100)
rfidwopt1 <- array(0, dim = c(length(idp1), length(nmax1)))

for (i in 1:length(idp1)) {
  for (j in 1:length(nmax1)) {
    set.seed(1234)
    rfidwcv1 <- rfidwcv(meuse[, 1:2], meuse[, c(1:2, 7:8, 10, 14)], meuse[, 6],
        idp = idp1[i], nmax = nmax1[j], predacc = "VEcv")
    rfidwopt1[i, j] <- rfidwcv1
  }
}
```

The optimal `idp` and `nmaxidw` are

```
para.rfidwopt1 <- which (rfidwopt1 == max(rfidwopt1), arr.ind = T)
idp1[para.rfidwopt1[, 1]]
```

```
[1] 3.2
```

```
nmax1[para.rfidwopt1[, 2]]
```

```
[1] 97
```

Estimation of `nmaxkrige` and selection of `vgm.args`

```
nmax.ok <- c(5:40)
vgm.args <- c("Exp", "Gau", "Sph", "Exc", "Mat", "Ste", "Lin")
rfokopt1 <- matrix(0, length(nmax.ok), length(vgm.args))

for (i in 1:length(nmax.ok)) {
  for (j in 1:length(vgm.args)) {
    set.seed(1234)
    rfokcv1 <- rfokcv(meuse[, 1:2], meuse[, c(1:2, 7:8, 10, 14)], meuse[, 6], nmax
        = nmax.ok[i], vgm.args = vgm.args[j], predacc = "VEcv")
    rfokopt1[i, j] <- rfokcv1
  }
}
```

The optimal `nmaxkrige` and `vgm.args` are

```
para.rfokopt1 <- which(rfokopt1 == max(rfokopt1), arr.ind = T)
nmax.ok[para.rfokopt1[, 1]]
```

```
[1] 8
```

```
vgm.args[para.rfokopt1[, 2]]
```

```
[1] "Gau"
```

Estimation of `lambda`

We will use the optimal `idp`, `nmaxidw`, `namxkrige`, and `vgm.args` to estimate `lambda` for *RFkrigingIDW* with `hybrid.parameter = 2`.

```
longlat <- meuse[, 1:2]
y = meuse[, 6]
trainx = meuse[, c(1:2, 7:8, 10, 14)]
```

```
lambda1 <- c(0:100)*0.02
lambda1opt <- NULL

for (i in 1:length(lambda1)) {
  set.seed(1234)
  rfkrigerfidwcv1 <- rfkrigeidwcv(longlat = longlat, trainx = trainx, trainy = y,
      transformation = "none", formula = res1 ~ 1, vgm.args = "Gau", nmaxkrige =
      8, idp = 3.2, nmaxidw = 97, lambda = lambda1[i], hybrid.parameter = 2,
      validation = "CV", predacc = "VEcv")
  lambda1opt[i] <- rfkrigerfidwcv1
}
```

The optimal `lambda` and the maximal predictive accuracy are

```
lambda1.opt <- which (lambda1opt == max(lambda1opt))
lambda1[lambda1.opt]
```

```
[1] 2
```

```
round(max(lambda1opt), 2)
```

```
[1] 77.92
```

The value estimated for `lambda` suggests that the hybrid method resulted from `rfkrigeidwcv` is, in fact, *RFIDW* according to Equation (10.6).

Similar to *RFkrigingIDW* with `hybrid.parameter = 2`, the parameters estimated or selected can also be used to optimize `lambda` for *RFkrigingIDW* with `hybrid.parameter = 3`. The results are in `lambda2opt`, and the optimal `lambda` and the maximal predictive accuracy are

```
lambda2.opt <- which (lambda2opt == max(lambda2opt))
lambda2[lambda2.opt]
```

```
[1] 2
```

```
round(max(lambda2opt), 2)
```

```
[1] 77.62
```

The value estimated for `lambda` suggests that the hybrid method resulted from `rfkrigeidwcv` is also *RFIDW* but with $\lambda_m = 2/3$ according to Equation (10.6).

The predictive accuracy of the hybrid model resulted is lower than that of *RFkrigingIDW* with `hybrid.parameter = 2`. Hence, *RFkrigingIDW* with `hybrid.parameter = 2` and `lambda = 2`, that is, *RFIDW*, should be used as a hybrid method for `zinc`.

2. `species.richness` in `sponge2`

We will use the predictive variables selected for *RF* in `steprf2$steprfPredictorsAll[2][[1]]` `$variables.most.accurate` in Section 12.1.6 to optimize relevant parameters for `species.richness`.

Estimation of `idp` *and* `nmaxidw`

```
idp2 <- (1:20) * 0.2
nmax2 <- c(5:40)
rfidwopt2 <- array(0, dim = c(length(idp2), length(nmax2)))

for (i in 1:length(idp2)) {
  for (j in 1:length(nmax2)) {
    set.seed(1234)
    rfidwcv1 <- rfidwcv(sponge2[, 1:2], sponge2[, c(1:2, 7, 25, 34, 45, 52)],
        sponge2[, 3], idp = idp2[i], nmax = nmax2[j], predacc = "VEcv")
    rfidwopt2[i, j] <- rfidwcv1
  }
}
```

The optimal `idp` and `nmaxidw` are

```
para.rfidwopt2 <- which (rfidwopt2 == max(rfidwopt2), arr.ind = T)
idp2[para.rfidwopt2[, 1]]
```

```
[1] 0.6
```

```
nmax2[para.rfidwopt2[, 2]]
```

```
[1] 14
```

Estimation of `nmaxkrige` and selection of `vgm.args`

```
nmax.ok <- c(5:40)
vgm.args <- c("Exp", "Gau", "Sph", "Exc", "Mat", "Ste", "Lin")
rfokopt2 <- matrix(0, length(nmax.ok), length(vgm.args))

for (i in 1:length(nmax.ok)) {
  for (j in 1:length(vgm.args)) {
    set.seed(1234)
    rfokcv1 <- rfokcv(sponge2[, 1:2], sponge2[, c(1:2, 7, 25, 34, 45, 52)],
        sponge2[, 3], nmax = nmax.ok[i], vgm.args = vgm.args[j], predacc = "VEcv")
    rfokopt2[i, j] <- rfokcv1
  }
}
```

The optimal `nmaxkrige` and `vgm.args` are

```
para.rfokopt2 <- which(rfokopt2 == max(rfokopt2), arr.ind = T)
nmax.ok[para.rfokopt2[, 1]]
```

```
[1] 14
```

```
vgm.args[para.rfokopt2[, 2]]
```

```
[1] "Exc"
```

Estimation of `lambda`

We will use the optimal `idp`, `nmaxidw`, `namxkrige`, and `vgm.args` to estimate `lambda` for *RFk-rigingIDW* with `hybrid.parameter = 2`.

```
longlat <- sponge2[, 1:2]
y = sponge2[, 3]
trainx = sponge2[, c(1:2, 7, 25, 34, 45, 52)]

lambda3 <- c(0:100)*0.02
lambda3opt <- NULL

for (i in 1:length(lambda3)) {
  set.seed(1234)
  rfkrigerfidwcv1 <- rfkrigeidwcv(longlat = longlat, trainx = trainx, trainy = y,
      transformation = "none", formula = res1 ~ 1, vgm.args = "Exc", nmaxkrige =
      14, idp = 0.6, nmaxidw = 14, lambda = lambda3[i], hybrid.parameter = 2,
      validation = "CV", predacc = "VEcv")
  lambda3opt[i] <- rfkrigerfidwcv1
}
```

The optimal value and the maximal predictive accuracy are

```
lambda3.opt <- which (lambda3opt == max(lambda3opt))
lambda3[lambda3.opt]
```

```
[1] 0
```

```
round(max(lambda3opt), 2)
```

```
[1] 45.81
```

The value estimated for `lambda` suggests that the hybrid method resulted from `rfkrigeidwcv` is, in fact, *RFOK* according to Equation (10.7).

Similar to *RFkrigingIDW* with `hybrid.parameter = 2`, the parameters estimated or selected can also be used to optimize `lambda` for *RFkrigingIDW* with `hybrid.parameter = 3`. The results are in `lambda4opt`, and the optimal value and the maximal predictive accuracy are

```
lambda4.opt <- which (lambda4opt == max(lambda4opt))
lambda4[lambda4.opt]
```

```
[1] 0
```

```
round(max(lambda4opt), 2)
```

```
[1] 45.2
```

The value estimated for `lambda` suggests that the hybrid method resulted from `rfkrigeidwcv` is, in fact, *RFOK* but with $\lambda_k = 2/3$ according to Equation (10.7).

The predictive accuracy of the hybrid model resulted is lower than that of *RFkrigingIDW* with `hybrid.parameter = 2`. Hence, *RFkrigingIDW* with `hybrid.parameter = 2` and `lambda = 0`, that is, *RFOK*, should be used as a hybrid method for `species.richness`.

3. `gravel` in `petrel`

We will use the predictive variables selected for *RF* in `steprf3$steprfPredictorsAll[1][[1]]` `$variables.most.accurate` in Section 12.1.6 to optimize relevant parameters for `gravel`.

Estimation of `idp` and `nmaxidw`

```
idp3 <- (0:10) * 0.1
nmax3 <- c(0:20)
rfidwopt3 <- array(0, dim = c(length(idp3), length(nmax3)))

for (i in 1:length(idp3)) {
  for (j in 1:length(nmax3)) {
    set.seed(1234)
    rfidwcv1 <- rfidwcv(petrel[, 1:2], petrel[, c(1:2, 6:8)], petrel[, 5], idp =
        idp[i], nmax = nmax[j], predacc = "VEcv")
    rfidwopt3[i, j] <- rfidwcv1
  }
}
```

The optimal `idp` and `nmaxidw` are

```
para.rfidwopt3 <- which (rfidwopt3 == max(rfidwopt3), arr.ind = T)
idp3[para.rfidwopt3[, 1]]
```

```
[1] 0
```

```
nmax3[para.rfidwopt3[, 2]]
```

```
[1] 10
```

Since the value estimated for `idp` is 0, the method resulted is *RFKNN*. These parameters will be used for *RFkrigingIDW* (or *RFOKKNN* in this case).

Estimation of `nmaxkrige` and selection of `vgm.args`

```
nmax.ok <- c(5:40)
vgm.args <- c("Exp", "Gau", "Sph", "Exc", "Mat", "Ste", "Lin")
rfokopt3 <- matrix(0, length(nmax.ok), length(vgm.args))

for (i in 1:length(nmax.ok)) {
  for (j in 1:length(vgm.args)) {
    set.seed(1234)
    rfokcv1 <- rfokcv(petrel[, 1:2], petrel[, c(1:2, 6:8)], petrel[, 5], nmax =
        nmax.ok[i], vgm.args = vgm.args[j], predacc = "VEcv")
    rfokopt3[i, j] <- rfokcv1
  }
}
```

The optimal `nmaxkrige` and `vgm.args` are

```
para.rfokopt3 <- which(rfokopt3 == max(rfokopt3), arr.ind = T)
nmax.ok[para.rfokopt3[, 1]]
```

```
[1] 10
```

```
vgm.args[para.rfokopt3[, 2]]
```

```
[1] "Exp"
```

These parameters will be used for *RFOKKNN*.

Estimation of `lambda`

We will use the optimal `idp`, `nmaxidw`, `namxkrige`, and `vgm.args` to estimate `lambda` for *RFOKKNN* with `hybrid.parameter = 2`.

```
longlat <- petrel[, 1:2]
y = petrel[, 5]
trainx = petrel[, c(1:2, 6:8)]
```

```
lambda5 <- c(0:100)*0.02
lambda5opt <- NULL
```

```
for (i in 1:length(lambda5)) {
  set.seed(1234)
  rfkrigerfidwcv1 <- rfkrigeidwcv(longlat = longlat, trainx = trainx, trainy = y,
      transformation = "none", formula = res1 ~ 1, vgm.args = "Exp", nmaxkrige =
      10, idp = 0, nmaxidw = 10, lambda = lambda5[i], hybrid.parameter = 2,
      validation = "CV",  predacc = "VEcv")
  lambda5opt[i] <- rfkrigerfidwcv1
}
```

The optimal value and the maximal predictive accuracy are

```
lambda5.opt <- which (lambda5opt == max(lambda5opt))
lambda5[lambda5.opt]
```

```
[1] 2
```

```
round(max(lambda5opt), 2)
```

```
[1] 35.78
```

The value estimated for `lambda` suggests that the hybrid method resulted from `rfkrigeidwcv` is, in fact, *RFKNN* according to Equation (10.6).

Similar to *RFOKKNN* with `hybrid.parameter = 2`, the optimal parameters can also be used to optimize `lambda` for *RFOKKNN* with `hybrid.parameter = 3`. The results are in `lambda6opt`, and the optimal value and the maximal predictive accuracy are

```
lambda6.opt <- which (lambda6opt == max(lambda6opt))
lambda6[lambda6.opt]
```

```
[1] 2
```

```
round(max(lambda6opt), 2)
```

```
[1] 35.59
```

The value estimated for `lambda` suggests that the hybrid method resulted from `rfkrigeidwcv` is, in fact, *RFKNN* but with $\lambda_m = 2/3$ according to Equation (10.6).

The predictive accuracy of the hybrid model resulted is lower than that of *RFOKKNN* with `hybrid.parameter = 2`. Hence, *RFOKKNN* with `hybrid.parameter = 2` and `lambda = 2`, that is, *RFKNN*, should be used as a hybrid method for `gravel`.

12.1.8 *GBM*

For *GBM*, both variable selection and parameter optimization need to be conducted. We will use `stepgbm` to select predictive variables and `gbmcv` to optimize relevant parameters.

1. zinc in meuse

```
library(stepgbm)

zn3 <- meuse[, -c(3:5, 9)]
zn3 <- na.omit(zn3)
logzn3 <- log(zn3$zinc)

set.seed(1234)

stepgbm1 <- stepgbm(trainx = zn3[, -3], trainy = logzn3,
method = "KIRVI", family = "gaussian", rpt = 2, cv.fold = 5, predacc = "VEcv", min
    .n.var = 2, n.cores = 8, delta.predacc = 0.01)
```

The predictive variables selected are

```
stepgbm1$stepgbmPredictorsAll[[1]]$variables.most.accurate

[1] "x"        "elev"    "dist"    "ffreq"    "dist.m"
```

Estimation of `interaction.depth` *and* `train.fraction`

We can estimate the optimal `interaction.depth` and `train.fraction` for *GBM* predictive models. This will be demonstrated based on the above `stepgbm1`.

```
library(spm)
interact.depth1 <- c(1:10)
train.fract1 <- c(5:10) * 0.1
gbmopt1 <- matrix(0, length(interact.depth1), length(train.fract1))

for (i in 1:length(interact.depth1)) {
  for (j in 1:length(train.fract1)) {
    set.seed(1234)
    gbmcv1 <- gbmcv(trainx = zn3[, stepgbm1$stepgbmPredictorsAll[[1]]$variables.
        most.accurate], trainy = logzn3, family = "gaussian", learning.rate =
        0.001, interaction.depth = interact.depth1[i], train.fraction = train.
        fract1[j], n.minobsinnode = 5, cv.fold = 5, predacc = "VEcv", n.cores = 8)
    gbmopt1[i, j] <- gbmcv1
  }
}
```

The optimal `interaction.depth`, `train.fraction`, and the maximal predictive accuracy are

```
gbm.opt1 <- which (gbmopt1 == max(gbmopt1), arr.ind = T)
interact.depth1[gbm.opt1[, 1]]

[1] 8

train.fract1[gbm.opt1[, 2]]
```

```
[1] 0.9
```

```
round(max(gbmopt1), 2)
```

```
[1] 81.83
```

Estimation of `n.minobsinnode` *and further estimation of* `interaction.depth`

Given that the estimation of `train.fraction` is 0.9, we will use it to estimate the optimal `n.minobsinnode` and further estimate `interaction.depth` based on `stepgbm1`. The further estimation of `interaction.depth` is to examine if it changes with `n.minobsinnode`.

```
interact.depth1.2 <- c(8:15)
n.minobsinnode1 <- c(1:5)
gbmopt1.2 <- matrix(0, length(interact.depth1.2), length(n.minobsinnode1))

for (i in 1:length(interact.depth1.2)) {
  for (j in 1:length(n.minobsinnode1)) {
    set.seed(1234)
    gbmcv1 <- gbmcv(trainx = zn3[, stepgbm1$stepgbmPredictorsAll[[1]]$variables.
        most.accurate], trainy = logzn3, family = "gaussian", learning.rate =
        0.001, interaction.depth = interact.depth1.2[i], train.fraction = 0.9, n.
        minobsinnode = n.minobsinnode1[j], cv.fold = 5, predacc = "VEcv", n.cores
        = 8)
    gbmopt1.2[i, j] <- gbmcv1
  }
}
```

The optimal `n.minobsinnode`, `interaction.depth`, and the maximal predictive accuracy are

```
gbm.opt1.2 <- which (gbmopt1.2 == max(gbmopt1.2), arr.ind = T)
n.minobsinnode1[gbm.opt1.2[, 2]]
```

```
[1] 1
```

```
interact.depth1.2[gbm.opt1.2[, 1]]
```

```
[1] 14
```

```
round(max(gbmopt1.2), 2)
```

```
[1] 83.32
```

The results suggest that the optimal `interaction.depth` changes with the value of `n.minobsinnode`. The changes of `interaction.depth` with `n.minobsinnode` should be considered in parameter estimation for *GBM*. These optimal parameters result in a more accurate predictive model and should be used for *GBM*.

2. `species.richness` in `sponge2`

The variables for `species.richness` in `sponge2` have been selected in Section 8.3.2 and stored in `stepgbm2$stepgbmPredictorsAll[[2]]$variables.most.accurate`. They are

```
[1] "easting" "sand" "prof_cur_o" "tpi3" "relief_5" "homo3" "bathy_lmi7"
[8] "relief_3" "entro7"
```

According to Section 8.3.3, `interaction.depth = 5`, `train.fraction = 1`, and `n.minobsinnode = 8` are optimal for *GBM*.

3. gravel in petrel

```
set.seed(1234)

stepgbm3 <- stepgbm(trainx = petrel[, -c(3:5)], trainy = log(petrel[, 5] + 1),
    method = "KIRVI", family = "gaussian", rpt = 2, cv.fold = 5, predacc = "VEcv", min
    .n.var = 2, n.cores = 8, delta.predacc = 0.01)
```

The predictive variables selected are

```
stepgbm3$stepgbmPredictorsAll[[1]]$variables.most.accurate
```

```
[1] "long"  "lat"   "bathy" "dist"  "slope"
```

Estimation of interaction.depth *and* train.fraction

For train.fraction, the default value is 1; and the default value for interaction.depth is 2. In fact, we can estimate the optimal interaction.depth and train.fraction based on the above stepgbm3.

```
library(spm)
interact.depth3 <- c(10:15)
train.fract3 <- c(5:10) * 0.1
gbmopt3 <- matrix(0, length(interact.depth3), length(train.fract3))

for (i in 1:length(interact.depth3)) {
  for (j in 1:length(train.fract3)) {
    set.seed(1234)
    gbmcv1 <- gbmcv(trainx = petrel[, stepgbm3$stepgbmPredictorsAll[[1]]$variables
        .most.accurate], trainy = log(petrel[, 5] + 1), family = "gaussian",
        learning.rate = 0.001, interaction.depth = interact.depth3[i], train.
        fraction = train.fract3[j], n.minobsinnode = 5, cv.fold = 5, predacc = "
        VEcv", n.cores = 8)
    gbmopt3[i, j] <- gbmcv1
  }
}
```

The optimal interaction.depth, train.fraction, and the maximal predictive accuracy are

```
gbm.opt3 <- which (gbmopt3 == max(gbmopt3), arr.ind = T)
interact.depth3[gbm.opt3[, 1]]
```

```
[1] 14
```

```
train.fract3[gbm.opt3[, 2]]
```

```
[1] 1
```

```
round(max(gbmopt3), 2)
```

```
[1] 40.07
```

Estimation of n.minobsinnode *and further estimation of* interaction.depth

The further estimation of interaction.depth is to test if it changes with n.minobsinnode. We will use train.fraction = 1 to estimate the optimal n.minobsinnode together with interaction.depth.

```
interact.depth3.2 <- c(10:20)
n.minobsinnode <- c(1:10)
gbmopt3.2 <- matrix(0, length(interact.depth3.2), length(n.minobsinnode))

for (i in 1:length(interact.depth3.2)) {
```

```
for (j in 1:length(n.minobsinnode)) {
  set.seed(1234)
  gbmcv1 <- gbmcv(trainx = petrel[, stepgbm3$stepgbmPredictorsAll[[1]]$variables
      .most.accurate], trainy = log(petrel[, 5] + 1), family = "gaussian",
      learning.rate = 0.001, interaction.depth = interact.depth3.2[i], train.
      fraction = 1, n.minobsinnode = n.minobsinnode[j], cv.fold = 5, predacc = "
      VEcv", n.cores = 8)
  gbmopt3.2[i, j] <- gbmcv1
  }
}
```

The optimal `n.minobsinnode`, `interaction.depth`, and the maximal predictive accuracy are

```
gbm.opt3.2 <- which (gbmopt3.2 == max(gbmopt3.2), arr.ind = T)
n.minobsinnode[gbm.opt3.2[, 2]]
```

```
[1] 2
```

```
interact.depth3.2[gbm.opt3.2[, 1]]
```

```
[1] 18
```

```
round(max(gbmopt3.2), 2)
```

```
[1] 41.16
```

The results also suggest that the optimal `interaction.depth` changes with the value of `n.minobsinnode`. These optimal parameters result in a more accurate predictive model and should be used for *GBM*.

12.1.9 *GBMkrigingIDW*

We will use `gbmidwcv`, `gbmokcv`, and `gbmkrigeidwcv` to select the most accurate *GBMkrigingIDW* method (that can be any method in Table 11.2) for `zinc`, `species.richness`, and `gravel`. The parameters estimated and predictive variables selected for *GBM* will be used for *GBMkrigingIDW*.

1. `zinc` in `meuse`

We will use the predictive variables selected in `stepgbm1$stepgbmPredictorsAll[[1]]$variables.most.accurate` and the optimal parameters for *GBM* in Section 12.1.8 to further optimize relevant parameters for *GBMkrigingIDW*.

Estimation of `idp` and `nmaxidw`

```
idp1.gbm <- (20:30) * 0.2
nmax1.gbm <- c(90:100)
gbmidwopt1 <- array(0, dim = c(length(idp1.gbm), length(nmax1.gbm)))

for (i in 1:length(idp1.gbm)) {
  for (j in 1:length(nmax1.gbm)) {
    set.seed(1234)

    gbmidwcv1 <- gbmidwcv(longlat = meuse[, c(1:2)], trainx = meuse[, stepgbm1$
        stepgbmPredictorsAll[[1]]$variables.most.accurate], trainy = log(meuse$
        zinc), family = "gaussian", learning.rate = 0.001, interaction.depth = 14,
        train.fraction = 0.9, n.minobsinnode = 1, idp = idp1.gbm[i], nmax = nmax1
        .gbm[j], cv.fold = 10, predacc = "VEcv", n.cores = 8)
    gbmidwopt1[i, j] <- gbmidwcv1
  }
}
```

The optimal `idp` and `nmaxidw` are

```
para.gbmidwopt1 <- which (gbmidwopt1 == max(gbmidwopt1), arr.ind = T)
idp1.gbm[para.gbmidwopt1[, 1]]
```

```
[1] 4
```

```
nmax1.gbm[para.gbmidwopt1[, 2]]
```

```
[1] 97
```

Estimation of `nmaxkrige` and selection of `vgm.args`

```
nmax.ok.gbm <- c(5:40)
vgm.args.gbm <- c("Exp", "Gau", "Sph", "Exc", "Mat", "Ste", "Lin")
gbmokopt1 <- matrix(0, length(nmax.ok.gbm), length(vgm.args.gbm))

for (i in 1:length(nmax.ok.gbm)) {
  for (j in 1:length(vgm.args.gbm)) {
    set.seed(1234)

    gbmokcv1 <- gbmokcv(longlat = meuse[, c(1:2)], trainx = meuse[, stepgbm1$
        stepgbmPredictorsAll[[1]]$variables.most.accurate], trainy = log(meuse$
        zinc), family = "gaussian", learning.rate = 0.001, interaction.depth = 14,
        train.fraction = 0.9, n.minobsinnode = 1, nmax = nmax.ok[i], vgm.args =
        vgm.args[j], cv.fold = 10, predacc = "VEcv", n.cores = 8)
    gbmokopt1[i, j] <- gbmokcv1
  }
}
```

The optimal `nmaxkrige` and `vgm.args` are

```
para.gbmokopt1 <- which(gbmokopt1 == max(gbmokopt1), arr.ind = T)
nmax.ok.gbm[para.gbmokopt1[, 1]]
```

```
[1] 32
```

```
vgm.args.gbm[para.gbmokopt1[, 2]]
```

```
[1] "Sph"
```

Estimation of `lambda`

We will use the optimal `idp`, `nmaxidw`, `namxkrige`, and `vgm.args` to estimate `lambda` for *GBMkrigingIDW* with `hybrid.parameter = 2`.

```
longlat <- meuse[, 1:2]
y = log(meuse$zinc)
trainx = meuse[, c(1:2, 7:8, 10, 14)]

lambda1.gbm <- c(0:100)*0.02
lambda1.gbmopt <- NULL

for (i in 1:length(lambda1.gbm)) {
  set.seed(1234)

  gbmkrigegbmidwcv1 <- gbmkrigeidwcv(longlat = longlat, trainx = trainx, trainy =
      y, family = "gaussian", learning.rate = 0.001, interaction.depth = 14, train
      .fraction = 0.9, n.minobsinnode = 1, transformation = "none", formula = res1
      ~ 1, vgm.args = "Sph", nmaxkrige = 32, idp = 4, nmaxidw = 97, hybrid.
      parameter = 2, lambda = lambda1.gbm[i], validation = "CV", predacc = "VEcv"
      , n.cores = 8)

  lambda1.gbmopt[i] <- gbmkrigegbmidwcv1
}
```

The optimal value and the maximal predictive accuracy are

```
lambda1.gbm.opt <- which (lambda1.gbmopt == max(lambda1.gbmopt))
lambda1.gbm[lambda1.gbm.opt]
```

```
[1] 0.9
```

```
round(max(lambda1.gbmopt), 2)
```

```
[1] 83.9
```

Similar to *GBMkrigingIDW* with `hybrid.parameter = 2`, the parameters estimated or selected can also be used to optimize `lambda` for *GBMkrigingIDW* with `hybrid.parameter = 3`. The results are in `lambda2.gbmopt`, and the value estimated for `lambda` and the maximal predictive accuracy are

```
lambda2.gbm.opt <- which (lambda2.gbmopt == max(lambda2.gbmopt))
lambda2.gbm[lambda2.gbm.opt]
```

```
[1] 1.04
```

```
round(max(lambda2.gbmopt), 2)
```

```
[1] 83.82
```

The predictive accuracy of the hybrid model resulted is lower than that of *GBMkrigingIDW* with `hybrid.parameter = 2`. Hence, the *GBMkrigingIDW* (i.e., *GBMOKIDW*) with `hybrid.parameter = 2` and `lambda = 0.9` should be used for `zinc`.

2. `species.richness` in `sponge2`

We will use the predictive variables selected and the optimal parameters for *GBM* as shown in Section 12.1.8 to optimize relevant parameters for *GBMkrigingIDW*.

Estimation of `idp` and `nmaxidw`

```
idp2.gbm <- (0:20) * 0.2
nmax2.gbm <- c(5:20)
gbmidwopt2 <- array(0, dim = c(length(idp2.gbm), length(nmax2.gbm)))

for (i in 1:length(idp2.gbm)) {
  for (j in 1:length(nmax2.gbm)) {
    set.seed(1234)

    gbmidwcv1 <- gbmidwcv(longlat = sponge2[, c(1:2)], trainx = sponge2[, c(1, 5,
        45, 52, 62, 73, 57, 61, 72)], trainy = sponge2[, 3], family = "poisson",
        learning.rate = 0.001, interaction.depth = 5, train.fraction = 1, n.
        minobsinnode = 8, idp = idp2.gbm[i], nmax = nmax2.gbm[j], cv.fold = 10,
        predacc = "VEcv", n.cores = 8)

    gbmidwopt2[i, j] <- gbmidwcv1
  }
}
```

The optimal `idp` and `nmaxidw` are

```
para.gbmidwopt2 <- which (gbmidwopt2 == max(gbmidwopt2), arr.ind = T)
idp2.gbm[para.gbmidwopt2[, 1]]
```

```
[1] 0
```

```
nmax2.gbm[para.gbmidwopt2[, 2]]
```

```
[1] 7
```

Since `idp = 0`, *GBMIDW* is going to be *GBMKNN*.

Estimation of `nmaxkrige` and selection of `vgm.args`

```
nmax.ok.gbm <- c(5:40)
vgm.args.gbm <- c("Exp", "Gau", "Sph", "Exc", "Mat", "Ste", "Lin")
gbmokopt2 <- matrix(0, length(nmax.ok.gbm), length(vgm.args.gbm))

for (i in 1:length(nmax.ok.gbm)) {
  for (j in 1:length(vgm.args.gbm)) {
    set.seed(1234)

    gbmokcv1 <- gbmokcv(longlat = sponge2[, c(1:2)], trainx = sponge2[, c(1, 5,
        45, 52, 62, 73, 57, 61, 72)], trainy = sponge2[, 3], family = "poisson",
        learning.rate = 0.001, interaction.depth = 5, train.fraction = 1, n.
        minobsinnode = 8, nmax = nmax.ok.gbm[i], vgm.args = vgm.args.gbm[j], cv.
        fold = 10, predacc = "VEcv", n.cores = 8)

    gbmokopt2[i, j] <- gbmokcv1
  }
}
```

The optimal `nmaxkrige` and `vgm.args` are

```
para.gbmokopt2 <- which(gbmokopt2 == max(gbmokopt2), arr.ind = T)
nmax.ok.gbm[para.gbmokopt2[, 1]]
```

```
[1] 14
```

```
vgm.args.gbm[para.gbmokopt2[, 2]]
```

```
[1] "Gau"
```

Estimation of `lambda`

We will use the optimal `idp`, `nmaxidw`, `namxkrige`, and `vgm.args` to estimate `lambda` for *GBMOKKNN* with `hybrid.parameter = 2`.

```
longlat <- sponge2[, 1:2]
y = sponge2[, 3]
trainx = sponge2[, c(1, 5, 45, 52, 62, 73, 57, 61, 72)]

lambda3.gbm <- c(0:100)*0.02
lambda3.gbmopt <- NULL

for (i in 1:length(lambda3.gbm)) {
  set.seed(1234)

  gbmkrigegbmidwcv1 <- gbmkrigeidwcv(longlat = longlat, trainx = trainx, trainy =
      y, family = "poisson", interaction.depth = 5, train.fraction = 1, n.
      minobsinnode = 8, vgm.args = "Gau", nmaxkrige = 14, idp = 0, nmaxidw = 7,
      hybrid.parameter = 2, lambda = lambda3.gbm[i], validation = "CV", predacc =
      "VEcv", n.cores = 8)

  lambda3.gbmopt[i] <- gbmkrigegbmidwcv1
}
```

The optimal value and the maximal predictive accuracy are

```
lambda3.gbm.opt <- which (lambda3.gbmopt == max(lambda3.gbmopt))
lambda3.gbm[lambda3.gbm.opt]
```

```
[1] 1.64
```

```
round(max(lambda3.gbmopt), 2)
```

```
[1] 46.14
```

Similar to *GBMOKKNN* with `hybrid.parameter` = 2, the parameters estimated or selected can also be used to optimize `lambda` for *GBMOKKNN* with `hybrid.parameter` = 3. The results are in `lambda4.gbmopt`, and the optimal `lambda` and the maximal predictive accuracy are

```
lambda4.gbm.opt <- which (lambda4.gbmopt == max(lambda4.gbmopt))
lambda4.gbm[lambda4.gbm.opt]
```

```
[1] 1.64
```

```
round(max(lambda4.gbmopt), 2)
```

```
[1] 46.02
```

The predictive accuracy of the hybrid model resulted is lower than that of *GBMOKKNN* with `hybrid.parameter` = 2. Hence, the *GBMOKKNN* with `hybrid.parameter` = 2 and `lambda` = 1.64 should be used for `species.richness`.

3. `gravel` in `petrel`

We will use the predictive variables selected in `stepgbm3$stepgbmPredictorsAll[[1]]$variables`
`.most.accurate` and the optimal parameters for *GBM* in Section 12.1.8 to optimize relevant parameters for *GBMkrigingIDW*.

Estimation of `idp` and `nmaxidw`

```
longlat = petrel[, c(1:2)]
trainx = petrel[, stepgbm3$stepgbmPredictorsAll[[1]]$variables.most.accurate]
trainy = log(petrel[, 5] + 1)

idp3.gbm <- (0:20) * 0.2
nmax3.gbm <- c(5:20)
gbmidwopt3 <- array(0, dim = c(length(idp3.gbm), length(nmax3.gbm)))

for (i in 1:length(idp3.gbm)) {
  for (j in 1:length(nmax3.gbm)) {
    set.seed(1234)

    gbmidwcv1 <- gbmidwcv(longlat = longlat, trainx = trainx, trainy = trainy,
        family = "gaussian", interaction.depth = 18, train.fraction = 1, n.
        minobsinnode = 2, idp = idp3.gbm[i], nmax = nmax3.gbm[j], cv.fold = 10,
        predacc = "VEcv", n.cores = 8)

    gbmidwopt3[i, j] <- gbmidwcv1
  }
}
```

The optimal `idp` and `nmaxidw` are

```
para.gbmidwopt3 <- which (gbmidwopt3 == max(gbmidwopt3), arr.ind = T)
idp3.gbm[para.gbmidwopt3[, 1]]
```

```
[1] 0.2
```

```
nmax3.gbm[para.gbmidwopt3[, 2]]
```

```
[1] 15
```

Estimation of nmaxkrige and selection of 'vgm.args

```
longlat = petrel[, c(1:2)]
trainx = petrel[, stepgbm3$stepgbmPredictorsAll[[1]]$variables.most.accurate]
trainy = log(petrel[, 5] + 1)
```

```
nmax.ok.gbm2 <- c(5:20)
vgm.args.gbm <- c("Exp", "Gau", "Sph", "Exc", "Mat", "Ste", "Lin")
gbmokopt3 <- matrix(0, length(nmax.ok.gbm2), length(vgm.args.gbm))
```

```
for (i in 1:length(nmax.ok.gbm2)) {
  for (j in 1:length(vgm.args.gbm)) {
    set.seed(1234)

    gbmokcv1 <- gbmokcv(longlat = longlat, trainx = trainx, trainy = trainy,
        family = "gaussian", interaction.depth = 18, train.fraction = 1, n.
        minobsinnode = 2, nmax = nmax.ok.gbm2[i], vgm.args = vgm.args.gbm[j], cv.
        fold = 10, predacc = "VEcv", n.cores = 8)

    gbmokopt3[i, j] <- gbmokcv1
  }
}
```

The optimal nmaxkrige and vgm.args are

```
para.gbmokopt3 <- which(gbmokopt3 == max(gbmokopt3), arr.ind = T)
nmax.ok.gbm2[para.gbmokopt3[, 1]]
```

```
[1] 15
```

```
vgm.args.gbm[para.gbmokopt3[, 2]]
```

```
[1] "Lin"
```

Estimation of lambda

We will use the optimal idp, nmaxidw, namxkrige, and vgm.args to estimate lambda for *GBMkrigingIDW* with hybrid.parameter = 2.

```
longlat = petrel[, c(1:2)]
trainx = petrel[, stepgbm3$stepgbmPredictorsAll[[1]]$variables.most.accurate]
trainy = log(petrel[, 5] + 1)
```

```
lambda5.gbm <- c(0:100)*0.02
lambda5.gbmopt <- NULL
```

```
for (i in 1:length(lambda5.gbm)) {
  set.seed(1234)

  gbmkrigegbmidwcv1 <- gbmkrigeidwcv(longlat = longlat, trainx = trainx, trainy =
      trainy, family = "gaussian", interaction.depth = 18, train.fraction = 1, n.
      minobsinnode = 2, vgm.args = "Lin", nmaxkrige = 15, idp = 0.2, nmaxidw = 15,
       hybrid.parameter = 2, lambda = lambda5.gbm[i], validation = "CV",  predacc
      = "VEcv", n.cores = 8)

  lambda5.gbmopt[i] <- gbmkrigegbmidwcv1
}
```

The optimal value and the maximal predictive accuracy are

```
lambda5.gbm.opt <- which (lambda5.gbmopt == max(lambda5.gbmopt))
lambda5.gbm[lambda5.gbm.opt]
```

```
[1] 0
```

```
round(max(lambda5.gbmopt), 2)
```

```
[1] 43.91
```

The value estimated for `lambda` suggests that the hybrid method resulted from `gbmkrigeidwcv` is, in fact, *GBMOK* according to Equation (10.7).

Similar to *GBMkrigingIDW* with `hybrid.parameter = 2`, the parameters estimated or selected can also be used to optimize `lambda` for *GBMkrigingIDW* with `hybrid.parameter = 3`. The results are in `lambda6.gbmopt`, and the optimal `lambda` and the maximal predictive accuracy are

```
lambda6.gbm.opt <- which (lambda6.gbmopt == max(lambda6.gbmopt))
lambda6.gbm[lambda6.gbm.opt]
```

```
[1] 0
```

```
round(max(lambda6.gbmopt), 2)
```

```
[1] 43.96
```

The value estimated for `lambda` also suggests that the hybrid method resulted from `gbmkrigeidwcv` is, in fact, *GBMOK* but with $\lambda_k = 2/3$ according to Equation (10.7).

The predictive accuracy of the hybrid model resulted is higher than that of *GBMkrigingIDW* with `hybrid.parameter = 2`. Hence, the *GBMkrigingIDW* with `hybrid.parameter = 3` and `lambda = 0`, that is, *GBMOK* with $\lambda_k = 2/3$, should be used for `gravel`.

12.1.10 *SVM*

For *SVM*, both variable selection and parameter optimization need to be conducted. We will use `rfe` to select the predictive variables, and `svmcv` to optimize `type`, `kernel`, `cost`, and `gamma`.

1. `zinc` in `meuse`

```
zn3 <- meuse[, -c(3:5, 9:13)]
```

```
library(caret)
set.seed(1234)
```

```
svmrfe.zn <- rfe(zn3[, -c(3)], zn3[, 3], sizes = c(4:10, 15, 20, 30), metric = "
    RMSE", rfeControl = rfeControl(functions = caretFuncs, method = "repeatedcv",
    repeats = 5), method = "svmRadial", tuneLength = 12, trControl =
    trainControl(method = "cv"), verbose = FALSE)
```

The variables selected are

```
svmrfe.zn$optVariables
```

```
[1] "dist"   "dist.m" "elev"   "y"
```

We will use the variables selected in `svmrfe.zn$optVariables` for the parameter optimization.

Selection of type and kernel

```
library(spm2)

type <- c("eps-regression", "nu-regression")
kernel <- c("linear", "polynomial", "radial", "sigmoid")
typeopt1 <- matrix(0, length(type), length(kernel))

for (i in 1:length(type)) {
  for (j in 1:length(kernel)) {
    set.seed(1234)
    svmcv1 <- svmcv(zinc ~ ., trainxy = zn3[, -1], y = zn3[, 3], type = type[i],
        kernel = kernel[j], validation = "CV", predacc = "VEcv")
    typeopt1[i, j] <- svmcv1
  }
}
```

```
type.opt1 <- which (typeopt1 == max(typeopt1), arr.ind = T)
type[type.opt1[, 1]]
```

```
[1] "eps-regression"
```

```
kernel[type.opt1[, 2]]
```

```
[1] "radial"
```

The results show that type = eps-regression and kernel = radial (i.e., the default settings) result in the most accurate *SVM* regression model.

Estimation of cost and gamma

```
library(spm2)
```

```
cost1 = c(1:10)*0.5
gamma1 = c(1:40)*0.001
svm1opt <- matrix(0, length(cost1), length(gamma1))

for (i in 1:length(cost1)) {
  for (j in 1:length(gamma1)) {
    set.seed(1234)
    svmcv1 <- svmcv(zinc ~ ., trainxy = zn3[, -1], y = zn3[, 3], cost = cost1[i],
        gamma = gamma1[j], validation = "CV", predacc = "VEcv")
    svm1opt[i, j] <- svmcv1
  }
}
```

The optimal cost and gamma are

```
svm1.opt <- which (svm1opt == max(svm1opt), arr.ind = T)
cost1[svm1.opt[, 1]]
```

```
[1] 4
```

```
gamma1[svm1.opt[, 2]]
```

```
[1] 0.04
```

2. species.richness in sponge2

We will use the variables selected (i.e., all predictive variables) in Section 9.2 and the optimal parameters (i.e., `type = eps-regression`, `cost = 4`, and `gamma = 0.009`) in Section 9.3 for `species .richness`. The `kernel` argument is assumed to be optimal with the default setting, although one may test to see if this assumption is correct.

3. `gravel` in `petrel`

```
set.seed(1234)

svmrfe.gravel <- rfe(petrel[, -c(3:5)], petrel[, 5], sizes = c(4:10, 15, 20, 30),
    metric = "RMSE", rfeControl = rfeControl(functions = caretFuncs, method = "
    repeatedcv", repeats = 5), method = "svmRadial", tuneLength = 12,  trControl
    = trainControl(method = "cv"), verbose = FALSE)
```

The variables selected are

```
svmrfe.gravel$optVariables
```

```
[1] "lat"    "dist"   "bathy" "slope" "long"
```

We will use the variables selected in `svmrfe.gravel$optVariables` for the parameter optimization.

Selection of `type` *and* `kernel`

```
type <- c("eps-regression", "nu-regression")
kernel <- c("linear", "polynomial", "radial", "sigmoid")
typeopt2 <- matrix(0, length(type), length(kernel))

for (i in 1:length(type)) {
  for (j in 1:length(kernel)) {
    set.seed(1234)
    svmcv1 <- svmcv(gravel ~ ., trainxy =  petrel[, -c(3, 4, 8)], y = petrel[, 5],
        type = type[i], kernel = kernel[j], validation = "CV", predacc = "VEcv")
    typeopt2[i, j] <- svmcv1
  }
}
```

```
type.opt2 <- which (typeopt2 == max(typeopt2), arr.ind = T)
type[type.opt2[, 1]]
```

```
[1] "nu-regression"
```

```
kernel[type.opt2[, 2]]
```

```
[1] "radial"
```

The results show that `type = nu-regression` and `kernel = radial` result in the most accurate *SVM* regression model.

Estimation of `cost` *and* `gamma`

```
library(spm2)
```

```
cost2 = c(1:20)*0.5
gamma2 = c(100:200)*0.01
svm2opt <- matrix(0, length(cost2), length(gamma2))

for (i in 1:length(cost2)) {
  for (j in 1:length(gamma2)) {
    set.seed(1234)
    svmcv1 <- svmcv(gravel ~ ., trainxy =  petrel[, -c(3, 4, 8)], y = petrel[, 5],
        type = "nu-regression", cost = cost2[i], gamma = gamma2[j], validation =
        "CV", predacc = "VEcv")
```

```
      svm2opt[i, j] <- svmcv1
  }
}
```

The optimal cost and gamma are

```
svm2.opt <- which (svm2opt == max(svm2opt), arr.ind = T)
cost2[svm2.opt[, 1]]
```

```
[1] 4.5
```

```
gamma2[svm2.opt[, 2]]
```

```
[1] 1.98
```

12.1.11 *SVMkrigingIDW*

We will use svmidwcv, svmkrigecv, and svmkrigeidwcv to select the most accurate *SVMkrigingIDW* method (that can be any method in Table 11.3) for zinc, species.richness , and gravel. The parameters and predictive variables selected for *SVM* will be used for *SVMkrigingIDW*.

1. zinc **in** meuse

We will use the predictive variables selected in svmrfe.zn$optVariables and the optimal parameters for *SVM* in Section 12.1.10 to perform the parameter optimization for *SVM-krigingIDW*.

Estimation of idp ***and*** nmaxidw

```
zn3 <- meuse[, -c(3:5, 9:13)]
```

```
model <- zinc ~ .
longlat <- zn3[, 1:2]
```

```
idp1.svm <- (0:20) * 0.2
nmax1.svm <- c(80:100)
svmidwopt1 <- array(0, dim = c(length(idp1.svm), length(nmax1.svm)))
```

```
for (i in 1:length(idp1.svm)) {
  for (j in 1:length(nmax1.svm)) {
    set.seed(1234)
    svmidwcv1 <- svmidwcv(formula = model, longlat = longlat, trainxy = zn3[, -1],
 y = zn3[, 3], gamma = 0.04, cost = 4, scale = TRUE, idp = idp1.svm[i], nmaxidw =
      nmax1.svm[j],
 validation = "CV", predacc = "VEcv")
    svmidwopt1[i, j] <- svmidwcv1
  }
}
```

The optimal idp and nmaxidw are

```
para.svmidwopt1 <- which (svmidwopt1 == max(svmidwopt1), arr.ind = T)
idp1.svm[para.svmidwopt1[, 1]]
```

```
[1] 2.8
```

```
nmax1.svm[para.svmidwopt1[, 2]]
```

```
[1] 97
```

Estimation of `nmaxkrige` and selection of `vgm.args`

```
zn3 <- meuse[, -c(3:5, 9:13)]

model <- zinc ~ .
longlat <- zn3[, 1:2]

nmax.svmok1 <- c(5:40)
vgm.args <- c("Exp", "Gau", "Sph", "Exc", "Mat", "Ste", "Lin")
svmokopt1 <- matrix(0, length(nmax.svmok1), length(vgm.args))

for (i in 1:length(nmax.svmok1)) {
  for (j in 1:length(vgm.args)) {
    set.seed(1234)
    svmkrigecv1 <- svmkrigecv(formula.svm = model, longlat = longlat, trainxy =
        zn3[, -1], y = zn3[, 3], gamma = 0.04, cost = 4, formula.krige = res1 ~ 1,
        nmaxkrige = nmax.svmok1[i], vgm.args = vgm.args[j], validation = "CV",
        predacc = "VEcv")
    svmokopt1[i, j] <- svmkrigecv1
 }
}
```

The optimal `nmaxkrige` and `vgm.args` are

```
svmok.opt1 <- which(svmokopt1 == max(svmokopt1), arr.ind = T)
nmax.svmok1[svmok.opt1[, 1]]
```

```
[1] 8
```

```
vgm.args[svmok.opt1[, 2]]
```

```
[1] "Exp"
```

Estimation of `lambda`

We will use the optimal `idp`, `nmaxidw`, `namxkrige`, and `vgm.args` to estimate `lambda` for *SVMkrigingIDW* with `hybrid.parameter = 2`.

```
zn3 <- meuse[, -c(3:5, 9:13)]

model <- zinc ~ .
longlat <- zn3[, 1:2]

lambda7 <- c(0:100)*0.02
lambda7opt <- NULL

for (i in 1:length(lambda7)) {
  set.seed(1234)

  svmkrigesvmidwcv1 <- svmkrigeidwcv(formula.svm = model, longlat = longlat,
      trainxy = zn3[, -1], y = zn3[, 3], gamma = 0.04, cost = 4, formula.krige =
      res1 ~ 1, vgm.args = "Exp", nmaxkrige = 8, idp = 2.8, nmaxidw = 97, lambda =
      lambda7[i], hybrid.parameter = 2, validation = "CV", predacc = "VEcv")

  lambda7opt[i] <- svmkrigesvmidwcv1
}
```

The optimal `lambda` and the maximal predictive accuracy are

```
lambda7.opt <- which (lambda7opt == max(lambda7opt))
lambda7[lambda7.opt]
```

```
[1] 0.24
```

```
round(max(lambda7opt), 2)
```

```
[1] 75.49
```

Similar to *SVMkrigingIDW* with `hybrid.parameter = 2`, the parameters estimated or selected can also be used to optimize `lambda` for *SVMkrigingIDW* with `hybrid.parameter = 3`. The results are in `lambda8.gbmopt`, and the optimal `lambda` and the maximal predictive accuracy are

```
lambda8.opt <- which (lambda8opt == max(lambda8opt))
lambda8[lambda8.opt]
```

```
[1] 0
```

```
round(max(lambda8opt), 2)
```

```
[1] 73.93
```

The value estimated for `lambda` suggests that the hybrid method resulted from `svmkrigeidwcv` with `hybrid.parameter = 3` is, in fact, *SVMOK* but with $\lambda_k = 2/3$ according to Equation (10.7).

The predictive accuracy of the hybrid model resulted is lower than that of *SVMkrigingIDW* with `hybrid.parameter = 2`. Hence, the *SVMkrigingIDW* (i.e., *SVMOKIDW*) with `hybrid.parameter = 2` and `lambda = 0.24` should be used for `zinc`.

2. `species.richness` in `sponge2`

We will use all predictive variables in `sponge2` and the optimal parameters for *SVM* in Section 12.1.10 to perform the parameter optimization for *SVMkrigingIDW* for `species.richness`.

Estimation of `idp` and `nmaxidw`

```
model <- species.richness ~ .
longlat <- sponge2[, 1:2]

idp2.svm <- (0:20) * 0.2
nmax2.svm <- c(5:20)
svmidwopt2 <- array(0, dim = c(length(idp2.svm), length(nmax2.svm)))

for (i in 1:length(idp2.svm)) {
  for (j in 1:length(nmax2.svm)) {
    set.seed(1234)
    svmidwcv1 <- svmidwcv(formula = model, longlat = longlat, trainxy = sponge2[,
        -4], y = sponge2[, 3], gamma = 0.009, cost = 4, scale = TRUE, idp = idp2.
        svm[i], nmaxidw = nmax2.svm[j], validation = "CV", predacc = "VEcv")
    svmidwopt2[i, j] <- svmidwcv1
  }
}
```

The optimal `idp` and `nmaxidw` are

```
para.svmidwopt2 <- which (svmidwopt2 == max(svmidwopt2), arr.ind = T)
idp2.svm[para.svmidwopt2[, 1]]
```

```
[1] 1
```

```
nmax2.svm[para.svmidwopt2[, 2]]
```

```
[1] 11
```

Estimation of `nmaxkrige` and selection of `vgm.args`

```
model <- species.richness ~ .
longlat <- sponge2[, 1:2]

nmax.svmok2 <- c(5:40)
vgm.args <- c("Exp", "Gau", "Sph", "Exc", "Mat", "Ste", "Lin")
svmokopt2 <- matrix(0, length(nmax.svmok2), length(vgm.args))

for (i in 1:length(nmax.svmok2)) {
  for (j in 1:length(vgm.args)) {
    set.seed(1234)
    svmkrigecv1 <- svmkrigecv(formula.svm = model, longlat = longlat, trainxy =
        sponge2[, -4], y = sponge2[, 3], gamma = 0.009, cost = 4, formula.krige =
        res1 ~ 1, nmaxkrige = nmax.svmok2[i], vgm.args = vgm.args[j], validation =
        "CV", predacc = "VEcv")
    svmokopt2[i, j] <- svmkrigecv1
  }
}
```

The optimal `nmaxkrige` and `vgm.args` are

```
svmok.opt2 <- which(svmokopt2 == max(svmokopt2), arr.ind = T)
nmax.svmok2[svmok.opt2[, 1]]
```

```
[1] 6
```

```
vgm.args[svmok.opt2[, 2]]
```

```
[1] "Exc"
```

Estimation of `lambda`

We will use the optimal `idp`, `nmaxidw`, `namxkrige`, and `vgm.args` to estimate `lambda` for *SVMkrigingIDW* with `hybrid.parameter = 2`.

```
model <- species.richness ~ .
longlat <- sponge2[, 1:2]

lambda9 <- c(0:100)*0.02
lambda9opt <- NULL

for (i in 1:length(lambda9)) {
  set.seed(1234)

  svmkrigesvmidwcv1 <- svmkrigeidwcv(formula.svm = model, longlat = longlat,
      trainxy = sponge2[, -4], y = sponge2[, 3], gamma = 0.009, cost = 4, formula.
      krige = res1 ~ 1, vgm.args = "Exc", nmaxkrige = 6, idp = 1, nmaxidw = 11,
      lambda = lambda9[i], hybrid.parameter = 2, validation = "CV", predacc = "
      VEcv")

  lambda9opt[i] <- svmkrigesvmidwcv1
}
```

The optimal `lambda` and the maximal predictive accuracy are

```
lambda9.opt <- which (lambda9opt == max(lambda9opt))
lambda9[lambda9.opt]
```

```
[1] 0.7
```

```
round(max(lambda9opt), 2)
```

```
[1] 50.95
```

Similar to *SVMkrigingIDW* with `hybrid.parameter` = 2, the parameters estimated or selected can also be used to optimize `lambda` for *SVMkrigingIDW* with `hybrid.parameter` = 3. The results are in `lambda10.gbmopt`, and the optimal `lambda` and the maximal predictive accuracy are

```
lambda10.opt <- which (lambda10opt == max(lambda10opt))
lambda10[lambda10.opt]
```

```
[1] 0
```

```
round(max(lambda10opt), 2)
```

```
[1] 50.8
```

The value estimated for `lambda` suggests that the hybrid method resulted from `svmkrigeidwcv` with `hybrid.parameter` = 3 is, in fact, *SVMOK* but with $\lambda_k = 2/3$ according to Equation (10.7).

The predictive accuracy of the hybrid model resulted is lower than that of *SVMkrigingIDW* with `hybrid.parameter` = 2. Hence, the *SVMkrigingIDW* (i.e., *SVMOKIDW*) with `hybrid.parameter` = 2 and `lambda` = 0.7 is going to be used for `species.richness`.

3. `gravel` **in** `petrel`

We will use the variables selected in `svmrfe.gravel$optVariables` and the parameters optimized for *SVM* in Section 12.1.10 to perform the parameter optimization for *SVMkrigingIDW* for `gravel`.

Estimation of `idp` *and* `nmaxkidw`

```
model <- gravel ~ .
longlat <- petrel[, 1:2]

idp3.svm <- (0:10) * 0.2
nmax3.svm <- c(20:40)
svmidwopt3 <- array(0, dim = c(length(idp3.svm), length(nmax3.svm)))

for (i in 1:length(idp3.svm)) {
  for (j in 1:length(nmax3.svm)) {
    set.seed(1234)
    svmidwcv1 <- svmidwcv(formula = model, longlat = longlat, trainxy =  petrel[,
        -c(3, 4, 8)], y = petrel[, 5], type = "nu-regression", cost = 4.5, gamma =
        1.98, scale = TRUE, idp = idp3.svm[i], nmaxidw = nmax3.svm[j], validation
        = "CV", predacc = "VEcv")
    svmidwopt3[i, j] <- svmidwcv1
  }
}
```

The optimal `idp` and `nmaxidw` are

```
para.svmidwopt3 <- which (svmidwopt3 == max(svmidwopt3), arr.ind = T)
idp3.svm[para.svmidwopt3[, 1]]
```

```
[1] 0
```

```
nmax3.svm[para.svmidwopt3[, 2]]
```

```
[1] 31
```

Since `idp` = 0, *SVMIDW* is going to be *SVMKNN*.

Estimation of `nmaxkrige` and selection of `vgm.args`

```
model <- gravel ~ .
longlat <- petrel[, 1:2]

nmax.svmok3 <- c(5:40)
vgm.args <- c("Exp", "Gau", "Sph", "Exc", "Mat", "Ste", "Lin")
svmokopt3 <- matrix(0, length(nmax.svmok3), length(vgm.args))

for (i in 1:length(nmax.svmok3)) {
  for (j in 1:length(vgm.args)) {
    set.seed(1234)
    svmkrigecv1 <- svmkrigecv(formula = model, longlat = longlat, trainxy =
        petrel[, -c(3, 4, 8)], y = petrel[, 5], type = "nu-regression", cost =
        4.5, gamma = 1.98, formula.krige = res1 ~ 1, nmaxkrige = nmax.svmok3[i],
        vgm.args = vgm.args[j], validation = "CV", predacc = "VEcv")

    svmokopt3[i, j] <- svmkrigecv1
 }
}
```

The optimal `nmaxkrige` and `vgm.args` are

```
svmok.opt3 <- which(svmokopt3 == max(svmokopt3), arr.ind = T)
nmax.svmok3[svmok.opt3[, 1]]
```

```
[1] 27
```

```
vgm.args[svmok.opt3[, 2]]
```

```
[1] "Lin"
```

Estimation of `lambda`

We will use the optimal `idp`, `nmaxidw`, `namxkrige`, and `vgm.args` to estimate `lambda` for *SVMkrigingIDW* with `hybrid.parameter = 2`.

```
model <- gravel ~ .
longlat <- petrel[, 1:2]

lambda11 <- c(0:100)*0.02
lambda11opt <- NULL

for (i in 1:length(lambda11)) {
  set.seed(1234)

  svmkrigesvmidwcv1 <- svmkrigeidwcv(formula.svm = model, longlat = longlat,
      trainxy =  petrel[, -c(3, 4, 8)], y = petrel[, 5], type = "nu-regression",
      cost = 4.5, gamma = 1.98, formula.krige = res1 ~ 1, vgm.args = "Lin",
      nmaxkrige = 27, idp = 0, nmaxidw = 31, lambda = lambda11[i], hybrid.
      parameter = 2, validation = "CV",  predacc = "VEcv")

  lambda11opt[i] <- svmkrigesvmidwcv1
}
```

The optimal `lambda` and the maximal predictive accuracy are

```
lambda11.opt <- which (lambda11opt == max(lambda11opt))
lambda11[lambda11.opt]
```

```
[1] 2
```

```
round(max(lambda11opt), 2)
```

```
[1] 44.6
```

The value estimated for `lambda` suggests that the hybrid method resulted from `svmkrigeidwcv` with `hybrid.parameter = 2` is, in fact, *SVMKNN*.

Similar to *SVMkrigingIDW* with `hybrid.parameter = 2`, the parameters estimated or selected can also be used to optimize `lambda` for *SVMkrigingIDW* with `hybrid.parameter = 3`. The results are in `lambda12.gbmopt`, and the optimal `lambda` and the maximal predictive accuracy are

```
lambda12.opt <- which (lambda12opt == max(lambda12opt))
lambda12[lambda12.opt]
```

```
[1] 2
```

```
round(max(lambda12opt), 2)
```

```
[1] 44.83
```

The value estimated for `lambda` suggests that the hybrid method resulted from `svmkrigeidwcv` with `hybrid.parameter = 3` is also *SVMKNN* but with $\lambda_i = 2/3$ according to Equation (10.6).

The predictive accuracy of the hybrid model resulted is higher than that of *SVMkrigingIDW* with `hybrid.parameter = 2`. Hence, the *SVMKNN* with $\lambda_i = 2/3$ is going to be used for `gravel`.

12.2   Predictive accuracy of spatial predictive methods

To get a *stabilized accuracy* estimation, we repeat the cross-validation 100 times with the variables selected and parameters optimized for each of 11 models developed in Sections 12.1.1 to 12.1.11 for `zinc`, `species.richness`, and `gravel`, respectively, by following the example provided in Chapter 3.

The predictive accuracy of 11 models developed for `zinc`, `species.richness`, and `gravel` are compared in Figures 12.1, 12.2 and 12.3, respectively, with the median and range of the predictive accuracy based on 100 repetitions of 10-fold cross-validation.

It should be noted that:
(1) Hybrid methods are usually more accurate than their components;
(2) Machine learning methods are more accurate than the non-machine learning methods;
(3) When good predictive variables (e.g., the predictive variables that are highly correlated with the dependent variable) are available (i.e., the predictive variables in the `meuse` data set for `zinc`), non-machine learning methods such as *KED* and *GLMkrigingIDW* perform equally well as the machine learning methods, their predictive accuracy are mostly much higher than the average accuracy of the predictive models published in the environmental sciences (Li 2016; Li, Alvarez, et al. 2017), and their predictive performances are mostly good or even excellent (Li 2016);
(4) When good predictive variables are unavailable (e.g., the predictive variables in the `sponge2` data set for `species.richness`), the machine learning methods out-perform the non-machine learning methods, the predictive accuracy of the machine learning methods and the hybrid methods are equivalent to or slightly higher than the average accuracy of the predictive models published in the environmental sciences (Li 2016; Li, Alvarez, et al. 2017), and their predictive performances are mostly at average level (Li 2016); and

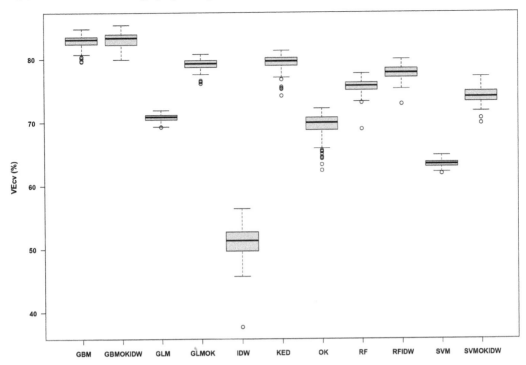

FIGURE 12.1: Predictive accuracy of 11 models developed from various spatial predictive methods for `zinc`. (1) *GBM*, (2) *GBMOKIDW* with `hybrid.parameter = 2` and `lambda = 0.9`, (3) *GLM*, (4) *GLMOK*, (5) *IDW*, (6) *KED*, (7) *OK*, (8) *RF*, (9) *RFIDW*, (10) *SVM*, and (11) *SVMOKIDW* with `hybrid.parameter = 2` and `lambda = 0.24`.

(5) The optimal predictive model changes with response variables, suggesting that "no free lunch theorems" for optimization (Wolpert and Macready 1997) are applicable to spatial predictive modeling and spatial predictive model needs to be selected and optimized for each response variable (Li 2013c).

It should also be noted that the performance of various methods may be improved if predictive variables and relevant parameters are further optimized. For instance, the default values for certain arguments, such as `transformation` and `anis`, are assumed to be optimal and used for the hybrid methods in comparison. The optimization of the values for these arguments may further increase the predictive accuracy of the hybrid methods despite the fact that they are already more accurate than their components.

Moreover, one may be interested in comparing the predictions of each methods. Relevant functions, such as `idwpred`, `krigepred`, `gbmkrigeidwpred`, `rfkrigeidwpred`, and `svmkrigeidwpred`, which have been introduced in previous chapters, can be used to generate spatial predictions. Then relevant tools, such as `spplot` (see Section 5.2.4 for further information), can be used to depict the predictions.

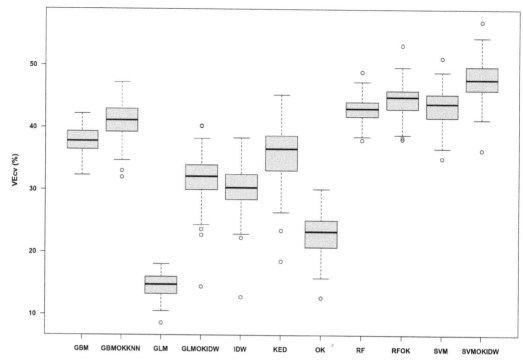

FIGURE 12.2: Predictive accuracy of 11 models developed from various spatial predictive methods for `species.richness`. (1) *GBM*, (2) *GBMOKKNN* with `hybrid.parameter = 2` and `lambda = 1.64`, (3) *GLM*, (4) *GLMOKIDW* with `hybrid.parameter = 2` and `lambda = 1.88`, (5) *IDW*, (6) *KED*, (7) *OK*, (8) *RF*, (9) *RFOK*, (10) *SVM*, and (11) *SVMOKIDW* with `hybrid .parameter = 2` and `lambda = 0.7`.

12.3   Notes on prediction uncertainty

Issues associated with prediction *uncertainty* for kriging methods, *LM*, *GLM* and *RF*, have been briefly addressed in Chapters 4 to 8. As shown above, for any of the spatial predictive methods, predictive errors based on validation can be produced for a predictive model developed, which leads to only one error value for the predictive model. If it is used as an uncertainty measurement, all predictions would have the same uncertainty value (Bishop, Minasny, and McBratney 2006). The prediction uncertainty produced so far is either not measuring prediction uncertainty, or depending on various factors as discussed by Li (2019a). Consequently, how to assess prediction uncertainty needs further study and any uncertainty measures based on the information of predictive accuracy are worth further investigation and recommended for future research (Li 2019a).

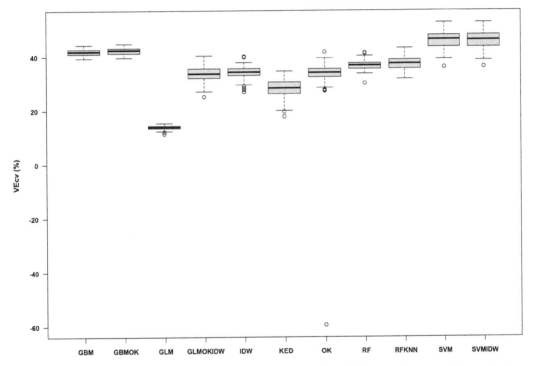

FIGURE 12.3: Predictive accuracy of 11 models developed from various spatial predictive methods for `gravel` data. (1) *GBM*, (2) *GBMOK* with `hybrid.parameter` = 3, (3) *GLM*, (4) *GLMOKIDW* with `hybrid.parameter` = 2 and `lambda` = 0.32, (5) *IDW*, (6) *KED*, (7) *OK*, (8) *RF*, (9) *RFKNN*, (10) *SVM*, and (11) *SVMKNN* with `hybrid.parameter` = 3.

A

Data sets used in this book

A.1 Location data for Petrel area

A.1.1 Data source

The data source for the Petrel area is at [http://pid.geoscience.gov.au/dataset/ga/75121] or [https://ecat.ga.gov.au/geonetwork/srv/eng/catalog.search#/metadata/75121], which provides seabed sediments data in the Petrel sub-basin in the Australian continental Exclusive Economic Zone (EEZ) and detailed descriptions of the data. The data can also be retrieved from [https://d28rz98at9flks.cloudfront.net/75121/75121_csv.zip].

A.1.2 Location data

```
# load data into R
petrel.mud <- read.csv('./data/Seabed mud content in the Petrel sub-basin in the
    Australian continental EEZ 2012.csv')

# to show variable names in the data set
names(petrel.mud)

[1] "long" "lat"  "mud"

# remove mud variable and keep the location information
petrel.mud2 <- petrel.mud[, -3]

summary(petrel.mud2)

      long              lat
 Min.   :125.5    Min.    :-15.540
 1st Qu.:127.3    1st Qu.:-13.265
 Median :128.6    Median :-12.260
 Mean   :128.5    Mean    :-12.260
 3rd Qu.:129.7    3rd Qu.:-11.285
 Max.   :131.5    Max.    : -9.655

# export lat and long data for Petrel area
write.csv(petrel.mud2, "./data/latlong.Petrel.csv", row.names = FALSE)
```

DOI: 10.1201/9781003091776-A

351

A.2   Bathymetry data for Petrel area

A.2.1   Data source

The data source for bathymetry is at [https://data.gov.au/data/dataset/australian-bathymetry-and-topography-grid-june-2009/resource/f5901db4-c075-491e-a6f9-b9c6392a87de], which provides bathymetry data for the Australian EEZ and detailed descriptions of the data. The data can also be retrieved from [https://d28rz98at9flks.cloudfront.net/67703/ESRI_Grid.zip].

A.2.2   Bathymetry data for Australian EEZ

The bathymetry data at a resolution of 0.0025 decimal degrees can be loaded into R by

```
library(raster)
bathy.aust <- raster('./data/ESRI_Grid/ESRI_Grid/ausbath_09_v4/w001001.adf')
bathy.aust[bathy.aust>0] <- NA
bathy.aust
```

A.2.3   Bathymetry data for Petrel area

Crop the `bathy.aust` for the Petrel area.

```
cropbox1 <- extent(127.1410, 130.7, -14.831, -11.272) # extract a square from
    Petrel region
bathy.c <- crop(bathy.aust, cropbox1)
bathy.c
```

Export the `bathy` data for the Petrel area

```
writeRaster(bathy.c,"./data/bathy_250m_Petrel.tif","GTiff", overwrite=TRUE)
```

Aggregate the data to 1 km resolution

```
bathy.c.aggr <- aggregate(bathy.c, fact=4)

bathy.c.aggr

Loading required package: raster

Loading required package: sp

class      : RasterLayer
dimensions : 356, 356, 126736  (nrow, ncol, ncell)
resolution : 0.01, 0.01  (x, y)
extent     : 127.1401, 130.7001, -14.83244, -11.27244  (xmin, xmax, ymin, ymax)
crs        : +proj=longlat +datum=WGS84 +no_defs +ellps=WGS84 +towgs84=0,0,0
source     : memory
names      : layer
values     : -177, 0  (min, max)
```

Export the aggregated `bathy` data

```
writeRaster(bathy.c.aggr,"./data/bathy_1km_Petrel.tif","GTiff", overwrite=TRUE)
```

A.3 Sample data for Petrel area

Load and subset sample data for the Petrel area:

```
library(spm)
data("petrel")
names(petrel)


[1] "long"    "lat"     "mud"     "sand"    "gravel" "bathy"  "dist"    "relief"
[9] "slope"


pp1 <- subset(petrel, long>= 127.141 & long <= 130.7 & lat >= -14.831 & lat <=
    -11.272)
dim(pp1)


[1] 114    9


summary(pp1[, c(1,2)]) # produce summary data for long, lat


      long              lat
 Min.   :127.3   Min.    :-14.78
 1st Qu.:128.9   1st Qu.:-13.91
 Median :129.8   Median :-12.38
 Mean   :129.5   Mean    :-12.79
 3rd Qu.:130.0   3rd Qu.:-12.29
 Max.   :130.5   Max.    :-11.40
```

Export the location information of sample data for the Petrel area.

```
write.csv(pp1[, c(1,2)], "./data/Sample for Petrel.csv", row.names = FALSE)
```

A.4 Bee count data

A.4.1 Data source

The data source is at [https://doi.org/10.25919/5f17b34638cca] or [https://data.csiro.au/collections/collection/CIcsiro:45533], which provides bee count data and relevant predictive variables along with a brief description of the data. The detailed descriptions of the data are available in Arthur et al. (2020).

A.4.2 Bee count data and relevant predictive variables

```
## load data into R
hbee1 <- read.csv("./data/bee1sub.csv", sep = ",", header = T)
dim(hbee1)


[1] 212   61


names(hbee1)
```

```
[1] "transid"               "transsurv"            "plotsurv"
[4] "paddock"               "plot"                 "obs"
[7] "hbee"                  "nbee"                 "hover"
[10] "date"                 "sx"                   "fx"
[13] "sy"                   "fy"                   "loc"
[16] "pair"                 "inf"                  "rankinf"
[19] "dupl"                 "temp"                 "windspeed"
[22] "winddir"              "cloudc"               "disttoedgecalc"
[25] "disttoedgemeasured"   "w100"                 "w200"
[28] "w300"                 "w400"                 "w500"
[31] "w600"                 "w700"                 "w800"
[34] "w900"                 "w1000"                "w1500"
[37] "w2000"                "c100"                 "c200"
[40] "c300"                 "c400"                 "c500"
[43] "c1000"                "c1500"                "area"
[46] "perimeter"            "gyration"             "paratio"
[49] "shape"                "fractaldimention"     "circumscircle"
[52] "contiguity"           "links100"             "links200"
[55] "links300"             "links400"             "links500"
[58] "links1000"            "links1500"            "links2000"
[61] "windspeed2"
```

A.5 Hard data

A.5.1 Data source

The data source is at [http://www.rdatamining.com/books/dmar/code/Data-for-DMAR-book.zip?attredirects=0&d=1], which provides hard data, `Appendix_A_hardness.csv`. The detailed descriptions of the data are available in Li, Siwabessy, Tran, et al. (2014).

A.5.2 Hard data and relevant predictive variables

Hard data and relevant predictive variables are retrieved and saved as `hard2` below.

```
hard2 <- read.csv("./data/Appendix_A_hardness.csv", sep = ",", header = T)
  # since a `hard` data set exists in the `spm` package, we call this data set `
     hard2`.
dim(hard2)
```

```
[1] 140   17
```

```
names(hard2)
```

```
[1] "Area"          "easting"       "northing"      "prock"         "bathy"
[6] "bs"            "bathy.moran"   "planar.curv"   "profile.curv"  "relief"
[11] "slope"        "surface"       "tpi"           "homogeneity"   "bs.moran"
[16] "variance"     "hardness"
```

A.6 Sponge data

A.6.1 Data Source

The data source is at [http://dx.doi.org/10.1016/j.envsoft.2017.07.016], which provides sponge data, `1-s2.0-S1364815217301615-mmc2.csv` and the detailed descriptions of the data. Further information are available in Li, Alvarez, et al. (2017).

A.6.2 Sponge data and relevant predictive variables

Sponge data and relevant predictive variables are retrieved as follows.

```
sponge2 <- read.csv("./data/1-s2.0-S1364815217301615-mmc2.csv", sep = ",", header
    = T)
  # since a `sponge` data set exists in the `spm` package, we call this data set `
      sponge2`.
dim(sponge2)
```

```
[1] 77 81
```

```
names(sponge2)
```

```
 [1] "long.sample"    "lat.sample"    "species.richness"  "mud"
 [5] "sand"           "gravel"        "bathy"             "bs25"
 [9] "bs10"           "bs11"          "bs12"              "bs13"
[13] "bs14"           "bs15"          "bs16"              "bs17"
[17] "bs18"           "bs19"          "bs20"              "bs21"
[21] "bs22"           "bs23"          "bs24"              "bs26"
[25] "bs27"           "bs28"          "bs29"              "bs30"
[29] "bs31"           "bs32"          "bs33"              "bs34"
[33] "bs35"           "bs36"          "bs_o"              "bs_homo_o"
[37] "bs_entro_o"     "bs_var_o"      "bs_lmi_o"          "bathy_o"
[41] "bathy_lmi_o"    "tpi_o"         "slope_o"           "plan_cur_o"
[45] "prof_cur_o"     "relief_o"      "rugosity_o"        "dist.coast"
[49] "rugosity3"      "rugosity5"     "rugosity7"         "tpi3"
[53] "tpi5"           "tpi7"          "bathy_lmi3"        "bathy_lmi5"
[57] "bathy_lmi7"     "plan_curv3"    "plan_curv5"        "plan_curv7"
[61] "relief_3"       "relief_5"      "relief_7"          "slope3"
[65] "slope5"         "slope7"        "prof_cur3"         "prof_cur5"
[69] "prof_cur7"      "entro3"        "entro5"            "entro7"
[73] "homo3"          "homo5"         "homo7"             "var3"
[77] "var5"           "var7"          "bs_lmi3"           "bs_lmi5"
[81] "bs_lmi7"
```

Given that spatial information of the `sponge.grid` data set is stored in `utm`, we need to reproject `sponge2` from `longitude` and `latitude` in WGS84 to `easting` and `northing` in utm zone 52 south and save it as below.

```
library(sf)
```

```
Linking to GEOS 3.8.0, GDAL 3.0.4, PROJ 6.3.1
```

```
spng2 <- st_as_sf(sponge2, coords=c("long.sample", "lat.sample"))
st_crs(spng2) <- 4326
# crs(spng2)
spng2.utm <- st_transform(spng2, st_crs(32752))
class(spng2.utm)
```

```
[1] "sf"                "data.frame"
```

```
crs(spng2.utm)
```

```
CRS arguments:
 +proj=utm +zone=52 +south +datum=WGS84 +units=m +no_defs
```

```
spng3.utm <- cbind((st_coordinates(spng2.utm)), st_set_geometry(spng2.utm, NULL))
class(spng3.utm)
```

```
[1] "data.frame"
```

```
colnames(spng3.utm)[c(1:2)] <- c("easting",  "northing")
```

The data set is saved as `sponge2`.

```
write.csv(spng3.utm, "./data/sponge2.csv", row.names = FALSE)
```

References

Adler, D., and D. Murdoch. 2020. *Rgl: 3D Visualization Using Opengl.* https://r-forge.r-project.org/projects/rgl/.

Allouche, O., A. Tsoar, and R. Kadmon. 2006. "Assessing the Accuracy of Species Distribution Models: Prevalence, Kappa and True Skill Statistic (Tss)." *Journal of Applied Ecology* 43: 1223–32.

Appelhans, T., E. Mwangomo, D. R. Hardy, A. Hemp, and T. Nauss. 2015. "Evaluating Machine Learning Approaches for the Interpolation of Monthly Air Temperature at Mt. Kilimanjaro, Tanzania." *Spatial Statistics* 14: 91–113.

Arthur, A. D., J. Li, S. Henry, and S. A. Cunningham. 2010. "Influence of Woody Vegetation on Pollinator Densities in Oilseed Brassica Fields in an Australian Temperate Landscape." *Basic and Applied Ecology* 11: 406–14.

———. 2020. *Influence of Woody Vegetation on Pollinator Densities in Oilseed Brassica Fields in an Australian Temperate Landscape.* Dataset. v2 ed. Canberra: CSIRO. https://doi.org/10.25919/5f17b34638cca.

Austin, M. 2007. "Species Distribution Models and Ecological Theory: A Critical Assessment and Some Possible New Approaches." *Ecological Modelling* 200: 1–19.

Bakar, K. S., and S. K. Sahu. 2015. "spTimer: Spatio-Temporal Bayesian Modeling Using R." *Journal of Statistical Software* 63 (15): 1–32. http://www.jstatsoft.org/v63/i15.

Becker, N., W. Werft, and A. Benner. 2018. *PenalizedSVM: Feature Selection Svm Using Penalty Functions.* https://CRAN.R-project.org/package=penalizedSVM.

Benedetti, R., and F. Piersimoni. 2017. "A Spatially Balanced Design with Probability Function Proportional to the Within Sample Distance." *Biometrical Journal* 59: 1067–84.

Benedetti, R., F. Piersimoni, and P. Postigione. 2017. "Spatially Balanced Sampling: A Review and a Reappraisal." *International Statistical Review* 85 (3): 439–54. https://doi.org/DOI:10.1111/insr.12216.

Bennett, N. D., B. F. W. Croke, G. Guariso, J. H. A. Guillaume, S. H. Hamilton, A. J. Jakeman, S. Marsili-Libelli, et al. 2013. "Characterising Performance of Environmental Models." *Environmental Modelling & Software* 40: 1–20.

Bishop, T. F. A., B. Minasny, and A. B. McBratney. 2006. "Uncertainty Analysis for Soil-Terrain Models." *International Journal of Geographical Information Science* 20 (2): 117–34.

Bivand, R., T. Keitt, and B. Rowlingson. 2019. *Rgdal: Bindings for the 'Geospatial' Data Abstraction Library.* https://CRAN.R-project.org/package=rgdal.

Bivand, R. S., E. Pebesma, and V. Gomez-Rubio. 2013. *Applied Spatial Data Analysis with R, Second Edition.* Springer, NY. https://asdar-book.org/.

Borcard, D., F. Gillet, and P. Legendre. 2011. *Numerical Ecology with R*. New York: Springer.

Breiman, L. 2001. "Random Forests." *Machine Learning* 45: 5–32.

Breiman, L., A. Cutler, A. Liaw, and M. Wiener. 2018. *RandomForest: Breiman and Cutler's Random Forests for Classification and Regression*. https://www.stat.berkeley.edu/~breiman/RandomForests/.

Breiman, L., J. H. Friedman, R. A. Olshen, and C. J. Stone. 1984. *Classification and Regression Trees*. Belmont: Wadsworth.

Burrough, P. A., and R. A. McDonnell. 1998. *Principles of Geographical Information Systems*. Oxford: Oxford University Press.

Cai, L., and Y. Zhu. 2015. "The Challenges of Data Quality and Data Quality Assessment in the Big Data Era." *Data Science Journal* 14 (2): 1–10. https://dx.doi.org/10.5334/dsj-2015-002.

Chambers, J. M., and T. J. Hastie. 1992. *Statistical Models in S*. Pacific Grove, CA: Wadsworth; Brooks/Cole Advanced Books; Software.

Chen, J., M.-C. Li, and W. Wang. 2012. "Statistical Uncertainty Estimation Using Random Forests and Its Application to Drought Forecast." *Mathematical Problems in Engineering* 2012: ID 915053. https://doi.org/doi:10.1155/2012/915053.

Chen, T., T. He, M. Benesty, V. Khotilovich, Y. Tang, H. Cho, K. Chen, et al. 2020. *Xgboost: Extreme Gradient Boosting*. https://github.com/dmlc/xgboost.

Chorti, A., and D. T. Hristopulos. 2008. "Non-Parametric Identification of Anisotropic (Elliptic) Correlations in Spatially Distributed Data Sets." *IEEE Transactions on Signal Processing* 56 (10): 4738–51.

Clark, I., and W. V. Harper. 2001. *Practical Geostatistics 2000*. Geostokos (Ecosse) Limited.

Collins, F. C., and P. V. Bolstad. 1985. "A Comparison of Spatial Interpolation Techniques in Temperature Estimation." In *Proceedings, Third International Conference/Workshop on Integrating Gis and Environmental Modeling, Santa Fe, Nm*. Santa Barbara, CA: National Center for Geographic Information; Analysis. http://www.ncgia.ucsb.edu/conf/SANTA_FE_CD-ROM/main.html.

Cortes, C., and V. Vapnik. 1995. "Support-Vector Networks." *Machine Learning* 20: 273–97.

Coulston, J. W., C. E. Blinn, V. A. Thomas, and R. H. Wynne. 2016. "Approximating Prediction Uncertainty for Random Forest Regression Models." *Photogrammetric Engineering & Remote Sensing* 82 (3): 189–97.

Crawley, M. J. 2007. *The R Book*. Chichester: John Wiley & Sons, Ltd.

Cutler, D. R., T. C.Jr. Edwards, K. H. Beard, A. Cutler, K. T. Hess, J. Gibson, and J. J. Lawler. 2007. "Random Forests for Classification in Ecology." *Ecography* 88 (11): 2783–92.

Das, P. 2020. *SigFeature: Significant Feature Selection Using Svm-Rfe & T-Statistic*.

Diaz-Uriarte, R., and S. A. de Andres. 2006. "Gene Selection and Classification of Microarray Data Using Random Forest." *BMC Bioinformatics* 7 (3): 1–13.

Diesing, M., T. Thorsnes, and L. R. Bjarnadóttir. 2021. "Organic Carbon Densities and Accumulation Rates in Surface Sediments of the North Sea and Skagerrak." *Biogeosciences* 18 (6): 2139–60. https://doi.org/10.5194/bg-18-2139-2021.

Diggle, P. J., and P. J. Ribeiro Jr. 2010. *Model-Based Geostatistics*. New York: Springer.

Dormann, C. F., J. Elith, S. Bacher, C. Buchanan, G. Carl, G. Carré, J. R. García Marquéz, et al. 2013. "Collinearity: A Review of Methods to Deal with It and a Simulation Study Evaluating Their Performance." *Ecography* 36: 27–46.

Drucker, H., C. J. C. Burges, L. Kaufman, A. J. Smola, and V. Vapnik. 1996. "Support Vector Regression Machines." In *Advances in Neural Information Processing Systems 9, Neural Information Processing Systems 1996*, edited by M. C. Mozer, M. I. M. I. Jordan, and T. Petsche, 155–61. MIT Press.

Elith, J., C. H. Graham, R. P. Anderson, M. Dulik, S. Ferrier, A. Guisan, R. J. Hijmans, et al. 2006. "Novel Methods Improve Prediction of Species' Distributions from Occurrence Data." *Ecography* 29: 129–51.

Elith, J., and J. Leathwick. 2009. "Species Distribution Models: Ecological Explanation and Prediction Across Space and Time." *Annual Review of Ecology and Evolution, and Systematics* 40: 677–97.

Erdely, A., and I. Castillo. 2017. *Cumstats: Cumulative Descriptive Statistics*. https://CRAN.R-project.org/package=cumstats.

Fielding, A. H., and J. F. Bell. 1997. "A Review of Methods for the Assessment of Prediction Errors in Conservation Presence/Absence Models." *Environmental Conservation* 24 (1): 38–49. %3CGo%20to%20ISI%3E://A1997XV97800008.

Foster, S. D. 2019. *MBHdesign: Spatial Designs for Ecological and Environmental Surveys*. https://CRAN.R-project.org/package=MBHdesign.

Foster, S. D., G. R. Hosack, E. Lawrence, R. Przeslawski, P. Hedge, M. J. Caley, N. S. Barrett, et al. 2017. "Spatially Balanced Designs That Incorporate Legacy Sites." *Methods in Ecology and Evolution*, 1–10. https://doi.org/DOI:10.1111/2041-210X.12782.

Fox, E. W., J. M. Ver Hoef, and A. R. Olsen. 2020. "Comparing Spatial Regression to Random Forests for Large Environmental Data Sets." *PLOS ONE* 15(3): e0229509. https://doi.org/10.1371/journal.pone.0229509.

Friedman, J., T. Hastie, R. Tibshirani, B. Narasimhan, K. Tay, and N. Simon. 2020. *Glmnet: Lasso and Elastic-Net Regularized Generalized Linear Models*. https://CRAN.R-project.org/package=glmnet.

Furnival, G. M., and R. W. Wilson. 1974. "Regressions by Leaps and Bounds." *Technometrics* 16: 499–511.

Genuer, R., J. Poggi, and C. Tuleau-Malot. 2019. *VSURF: Variable Selection Using Random Forests*. https://github.com/robingenuer/VSURF.

Glatzer, E. 2015. *Vardiag: Variogram Diagnostics*. https://CRAN.R-project.org/package=vardiag.

Goovaerts, P. 1997. *Geostatistics for Natural Resources Evaluation*. New York: Oxford University Press.

Gräler, B., E. Pebesma, and G. Heuvelink. 2016. "Spatio-Temporal Interpolation Using Gstat." *The R Journal* 8 (1): 204–18. https://journal.r-project.org/archive/2016/RJ-2016-014/index.html.

Greenwell, B., B. Boehmke, J. Cunningham, and GBM Developers. 2020. *Gbm: Generalized Boosted Regression Models*. https://github.com/gbm-developers/gbm.

Han, J., and M. Kamber. 2006. *Data Mining: Concept and Techniques.* 2nd ed. Amsterdam: Elsevier.

Harrell Jr, F. E. 2001. *Regression Modelling Strategies: With Applications to Linear Models, Logistic Regression, and Survival Analysis.* New York: Springer-Verlag.

Hastie, T., R. Tibshirani, and J. Friedman. 2009. *The Elements of Statistical Learning: Data Mining, Inference, and Prediction.* 2nd ed. New York: Springer.

Helleputte, T., P. Gramme, and J. Paul. 2017. *LiblineaR: Linear Predictive Models Based on the Liblinear c/C++ Library.* http://dnalytics.com/liblinear/.

Helwig, N. E. 2018. *Bigsplines: Smoothing Splines for Large Samples.* https://CRAN.R-project.org/package=bigsplines.

Hengl, T., G. B. M. Heuvelink, B. Kempen, J. G. B. Leenaars, M. G. Walsh, K. D. Shepherd, A. Sila, et al. 2015. "Mapping Soil Properties of Africa at 250 M Resolution: Random Forests Significantly Improve Current Predictions." *PLOS ONE* 10 (6): e0125814.

Hengl, T., E. van Loon, H. Sierdsema, and W. Bouten. 2008. "Advanced Spatio-Temporal Analysis of Ecological Data: Examples in R." *ICCSA* 1: 692–707.

Henningsen, A., and O. Toomet. 2019. *MiscTools: Miscellaneous Tools and Utilities.* https://CRAN.R-project.org/package=miscTools.

Hiemstra, P. 2013. *Automap: Automatic Interpolation Package.* https://CRAN.R-project.org/package=automap.

Hijmans, R. J. 2020. *Raster: Geographic Data Analysis and Modeling.* https://CRAN.R-project.org/package=raster.

Hijmans, R. J., S. Phillips, J. Leathwick, and J. Elith. 2017. *Dismo: Species Distribution Modeling.* http://rspatial.org/sdm/.

Hinz, J., I. Grigoryev, and A. Novikov. 2020. "An Application of High-Dimensional Statistics to Predictive Modeling of Grade Variability." *Geosciences* 10 (4): 116. https://www.mdpi.com/2076-3263/10/4/116.

Huston, M. A. 1997. "Hidden Treatments in Ecological Experiments: Re-Evaluating the Ecosystem Function of Biodiversity." *Oecologia* 110: 449–60.

Hutchinson, M. F. 1995. "Interpolating Mean Rainfall Using Thin Plate Smoothing Splines." *International Journal of Geographical Information Systems* 9 (4): 385–403.

Isaaks, E. H., and R. M. Srivastava. 1989. *Applied Geostatistics.* New York: Oxford University Press.

James, G., D. Witten, T. Hastie, and R. Tibshirani. 2017. *An Introduction to Statistical Learning with Applications in R.* Springer Texts in Statistics. New York: Springer.

Jiang, W., and J. Li. 2013. "Are Spatial Modelling Methods Sensitive to Spatial Reference Systems for Predicting Marine Environmental Variables." In *20th International Congress on Modelling and Simulation*, 387–93.

———. 2014. "The Effects of Spatial Reference Systems on the Predictive Accuracy of Spatial Interpolation Methods." Record 2014/01. Geoscience Australia: Canberra, pp33.

Karatzoglou, A., A. Smola, and K. Hornik. 2019. *Kernlab: Kernel-Based Machine Learning Lab.* https://CRAN.R-project.org/package=kernlab.

Kincaid, T., T. Olsen, and M. Weber. 2020. *Spsurvey: Spatial Survey Design and Analysis.* https://CRAN.R-project.org/package=spsurvey.

Knotters, M., D. J. Brus, and J. H. Oude Voshaar. 1995. "A Comparison of Kriging, Co-Kriging and Kriging Combined with Regression for Spatial Interpolation of Horizon Depth with Censored Observations." *Geoderma* 67: 227–46.

Kohavi, R. 1995. "A Study of Cross-Validation and Bootstrap for Accuracy Estimation and Model Selection." In *International Joint Conference on Artificial Intelligence (Ijcai), Morgan Kaufmann*, 1137–43.

Krige, D. G. 1951. "A Statistical Approach to Some Mine Valuations Problems at the Witwatersrand." *Journal of the Chemical, Metallurgical and Mining Society of South Africa* 52: 119–39.

Kuhn, M. 2020. *Caret: Classification and Regression Training.* https://github.com/topepo/caret/.

Kuhn, M., and K. Johnson. 2013. *Applied Predictive Modeling.* New York: Springer.

Kursa, M. B., and W. R. Rudnicki. 2010. "Feature Selection with the Boruta Package." *Journal of Statistical Software* 36 (11): 1–13. http://www.jstatsoft.org/v36/i11/.

———. 2020. *Boruta: Wrapper Algorithm for All Relevant Feature Selection.* https://gitlab.com/mbq/Boruta/.

Leathwick, J. R., J. Elith, and T. Hastie. 2006. "Comparative Performance of Generalised Additive Models and Multivariate Adaptive Regression Splines for Statistical Modelling of Species Distributions." *Ecological Modelling* 199: 188–96.

Leek, J. T., and R. D. Peng. 2015. "What Is the Question?" *Science* 347: 1314–5.

Legates, D. R., and G. J. McCabe. 2013. "A Refined Index of Model Performance: A Rejoinder." *International Journal of Climatology* 33: 1053–6.

Li, J. 2008. "Statistical Modelling and Prediction: Coping with Data Noise, Small N Large P, and the Value of Ecological Knowledge." In *Spatial and Statistical Modelling Workshop, Geoscience Australia.*

———. 2011. "Novel Spatial Interpolation Methods for Environmental Properties: Using Point Samples of Mud Content as an Example." *The Survey Statistician: The Newsletter of the International Association of Survey Statisticians* No. 63: 15–16.

———. 2013a. "Assessing the Accuracy of Spatial Predictive Models in the Environmental Sciences." In *Australian Statistical Conference.*

———. 2013b. "Predicting the Spatial Distribution of Seabed Gravel Content Using Random Forest, Spatial Interpolation Methods and Their Hybrid Methods." In *The International Congress on Modelling and Simulation (Modsim) 2013*, 394–400.

———. 2013c. "Predictive Modelling Using Random Forest and Its Hybrid Methods with Geostatistical Techniques in Marine Environmental Geosciences." In *The Proceedings of the Eleventh Australasian Data Mining Conference (Ausdm 2013), Canberra, Australia, 13-15 November 2013*, edited by P. Christen, P. Kennedy, L. Liu, K-L. Ong, A. Stranieri, and Y. Zhao. Vol. 146. Conferences in Research; Practice in Information Technology, Vol. 146.

———. 2016. "Assessing Spatial Predictive Models in the Environmental Sciences: Accuracy Measures, Data Variation and Variance Explained." *Environmental Modelling & Software* 80: 1–8. https://doi.org/10.1016/j.envsoft.2016.02.004.

———. 2017. "Assessing the Accuracy of Predictive Models for Numerical Data: Not R nor R2, Why Not? Then What?" *PLOS ONE* 12 (8): e0183250. https://doi.org/10.1371/journal.pone.0183250.

———. 2018a. "A New R Package for Spatial Predictive Modelling: Spm." In *UseR! 2018*.

———. 2018b. "Guidelines for Spatial Predictive Modelling – with a Case Study Using Spm in R." Geoscience Australia, Canberra (Unpublished).

———. 2019a. "A Critical Review of Spatial Predictive Modeling Process in Environmental Sciences with Reproducible Examples in R." *Applied Sciences* 9: 2048. https://doi.org/10.3390/app9102048.

———. 2019b. *Spm: Spatial Predictive Modeling*. https://CRAN.R-project.org/package=spm.

———. 2021a. *Spm2: Spatial Predictive Modeling*. https://CRAN.R-project.org/package=spm2.

———. 2021b. *Stepgbm: Stepwise Variable Selection for Generalized Boosted Regression Modeling*. https://CRAN.R-project.org/package=stepgbm.

———. 2021c. *Steprf: Stepwise Predictive Variable Selection for Random Forest*. https://CRAN.R-project.org/package=steprf.

Li, J., B. Alvarez, J. Siwabessy, M. Tran, Z. Huang, R. Przeslawski, L. Radke, F. Howard, and S. Nichol. 2017. "Application of Random Forest, Generalised Linear Model and Their Hybrid Methods with Geostatistical Techniques to Count Data: Predicting Sponge Species Richness." *Environmental Modelling and Software* 97: 112–29. https://doi.org/10.1016/j.envsoft.2017.07.016.

Li, J., and A. Heap. 2008. "A Review of Spatial Interpolation Methods for Environmental Scientists." Geoscience Australia, Record 2008/23, 137pp.

———. 2011. "A Review of Comparative Studies of Spatial Interpolation Methods in Environmental Sciences: Performance and Impact Factors." *Ecological Informatics* 6: 228–41. https://doi.org/10.1016/j.ecoinf.2010.12.003.

Li, J., and A. D. Heap. 2014. "Spatial Interpolation Methods Applied in the Environmental Sciences: A Review." *Environmental Modelling & Software* 53: 173–89. https://doi.org/10.1016/j.envsoft.2013.12.008.

Li, J., A. D. Heap, A. Potter, and J. Daniell. 2011a. "Application of Machine Learning Methods to Spatial Interpolation of Environmental Variables." *Environmental Modelling & Software* 26: 1647–59.

Li, J., A. D. Heap, A. Potter, Z. Huang, and J. Daniell. 2011. "Can We Improve the Spatial Predictions of Seabed Sediments? A Case Study of Spatial Interpolation of Mud Content Across the Southwest Australian Margin." *Continental Shelf Research* 31: 1365–76. https://doi.org/10.1016/j.csr.2011.05.015.

Li, J., A. Heap, A. Potter, and J. Daniell. 2010. "Can Machine Learning Methods Be Applied for Spatial Predictions of Environmental Properties?" In *Australian Statistical Conference*.

Li, J., A. Heap, A. Potter, and J. J. Daniell. 2011b. "Predicting Seabed Mud Content Across the Australian Margin Ii: Performance of Machine Learning Methods and Their Combination with Ordinary Kriging and Inverse Distance Squared." Geoscience Australia, Record 2011/07, 69pp.

Li, J., D. W. Hilbert, T. Parker, and S. Williams. 2009. "How Do Species Respond to Climate Change Along an Elevation Gradient? A Case Study of the Grey-Headed Robin (Heteromyias Albispecularis)." *Global Change Biology* 15: 255–67.

Li, J., A. Potter, Z. Huang, J. J. Daniell, and A. Heap. 2010. "Predicting Seabed Mud Content Across the Australian Margin: Comparison of Statistical and Mathematical Techniques Using a Simulation Experiment." Record. Geoscience Australia, Record 2010 /11, 146pp.

Li, J., A. Potter, Z. Huang, and A. Heap. 2012a. "Advances in Spatial Modelling: Hybrids of Machine Learning Methods and the Existing Spatial Interpolation Methods and Their Application to Marine Environmental Variables." In *Australian Statistical Conference*.

———. 2012b. "Predicting Seabed Sand Content Across the Australian Margin Using Machine Learning and Geostatistical Methods." Geoscience Australia, Record 2012/48, 115pp.

Li, J., J. Siwabessy, Z. Huang, and S. Nichol. 2019. "Developing an Optimal Spatial Predictive Model for Seabed Sand Content Using Machine Learning, Geostatistics and Their Hybrid Methods." *Geosciences* 9 (4):180. https://doi.org/10.3390/geosciences9040180.

Li, J., J. Siwabessy, M. Tran, Z. Huang, and A. Heap. 2014. "Predicting Seabed Hardness Using Random Forest in R." In *Data Mining Applications with R*, edited by Y. Zhao and Y. Cen, 299–329. Amsterdam: Elsevier.

Li, J., M. Tran, and J. Siwabessy. 2016. "Selecting Optimal Random Forest Predictive Models: A Case Study on Predicting the Spatial Distribution of Seabed Hardness." *PLOS ONE* 11 (2): e0149089. https://doi.org/10.1371/journal.pone.0149089.

Liaw, A., and M. Wiener. 2002. "Classification and Regression by randomForest." *R News* 2 (3): 18–22.

Ließ, M., J. Schmidt, and B. Glaser. 2016. "Improving the Spatial Prediction of Soil Organic Carbon Stocks in a Complex Tropical Mountain Landscape by Methodological Specifications in Machine Learning Approaches." *PLoS One* 11(4): e0153673. https://doi.org/doi: 10.1371/journal.pone.0153673[1].

Lobo, J. M., A. Jiménez-Valverde, and R. Real. 2008. "AUC: A Misleading Measure of the Performance of Predictive Distribution Models." *Global Ecology and Biogeography* 7: 145–51.

Lumley, T. 2020. *Leaps: Regression Subset Selection*. https://CRAN.R-project.org/package=leaps.

Maier, H. R., Z. Kapelan, J. Kasprzyk, J. Kollat, L. S. Matott, M. C. Cunha, G. C. Dandy, et al. 2014. "Evolutionary Algorithms and Other Metaheuristics in Water Resources: Current Status, Research Challenges and Future Directions." *Environmental Modelling & Software* 62 (0): 271–99. https://doi.org/10.1016/j.envsoft.2014.09.013.

Matheron, G. 1963. "Principles of Geostatistics." *Economic Geology* 58: 1246–66.

McArthur, M. A., B. P. Brooke, R. Przeslawski, D. A. Ryan, V. L. Lucieer, S. Nichol, A. W. McCallum, C. Mellin, I. D. Cresswell, and L. C. Radke. 2010. "On the Use of Abiotic Surrogates to Describe Marine Benthic Biodiversity." *Estuarine, Coastal and Shelf Science* 88: 21–32.

McCullagh, P., and J. A. Nelder. 1999. *Generalized Linear Models*. 2nd ed. Monographs on Statistics and Applied Probability 37. Boca Raton: Chapman & Hall/CRC.

---

[1]https://doi.org/doi:%2010.1371/journal.pone.0153673%20

McLeod, A. I., C. Xu, and Y. Lai. 2020. *Bestglm: Best Subset Glm and Regression Utilities.* https://CRAN.R-project.org/package=bestglm.

Mentch, L., and G. Hooker. 2016. "Quantifying Uncertainty in Random Forests via Confidence Intervals and Hypothesis Tests." *Journal of Machine Learning Research* 17: 1–41.

Meyer, D., E. Dimitriadou, K. Hornik, A. Weingessel, and F. Leisch. 2019. *E1071: Misc Functions of the Department of Statistics, Probability Theory Group (Formerly: E1071), Tu Wien.* https://CRAN.R-project.org/package=e1071.

Miller, K., M. Puotinen, R. Przeslawski, Z. Huang, P. Bouchet, B. Radford, J. Li, et al. 2016. "Ecosystem Understanding to Support Sustainable Use, Management and Monitoring of Marine Assets in the North and North-West Regions: Final Report for Nesp D1 2016e."

Mitasova, H., L. Mitas, W. M. Brown, D. P. Gerdes, I. Kosinovsky, and T. Baker. 1995. "Modelling Spatially and Temporally Distributed Phenomena: New Methods and Tools for Grass Gis." *International Journal of Geographical Information Systems* 9 (4): 433–46.

Morgan, J. A., and J. F. Tatar. 1972. "Calculation of the Residual Sum of Squares for All Possible Regressions." *Technometrics* 14: 317–25.

Moriasi, D. N., J. G. Arnold, M. W. Van Liew, R. L. Bingner, R. D. Harmel, and T. L. Veith. 2007. "Model Evaluation Guidelines for Systematic Quantification of Accuracy in Watershed Simulations." *American Society of Agricultural and Biological Engineers* 50 (3): 885–900.

Nychka, D., R. Furrer, J. Paige, S. Sain, F. Gerber, and M. Iverson. 2020. *Fields: Tools for Spatial Data.* https://CRAN.R-project.org/package=fields.

O'Brien, R. M. 2007. "A Caution Regarding Rules of Thumb for Variance Inflation Factors." *Quality & Quantity* 41: 673–90.

Pebesma, E. 2018. "Simple Features for R: Standardized Support for Spatial Vector Data." *The R Journal* 10 (1): 439–46. https://doi.org/10.32614/RJ-2018-009.

Pebesma, E., and R. Bivand. 2020. *Sp: Classes and Methods for Spatial Data.* https://CRAN.R-project.org/package=sp.

Pebesma, E. J. 2004. "Multivariable Geostatistics in S: The Gstat Package." *Computer & Geosciences* 30: 683–91.

Pebesma, E. J., and R. S. Bivand. 2005. "Classes and Methods for Spatial Data in R." *R News* 5 (2): 9–13. https://CRAN.R-project.org/doc/Rnews/.

Pebesma, E., J. O. Skoien with contributions from O. Baume, A. Chorti, D. T. Hristopulos, H. Kazianka, S. J. Melles, and G. Spiliopoulos. 2018. *Intamap: Procedures for Automated Interpolation.* https://CRAN.R-project.org/package=intamap.

Pinheiro, J., D. Bates, and R-core. 2020. *Nlme: Linear and Nonlinear Mixed Effects Models.* https://svn.r-project.org/R-packages/trunk/nlme/.

Pinheiro, J. C., and D. M. Bates. 2000. *Mixed-Effects Models in S and S-Plus.* New York: Springer.

Pipino, L. L., Y. W. Lee, and R. Y. Wang. 2002. "Data Quality Assessment." *Communications of the ACM* 45 (4): 211–18.

Przeslawski, R., J. Daniell, T. Anderson, J. Vaughn Barrie, A. Heap, M. Hughes, J. Li, et al. 2011. "Seabed Habitats and Hazards of the Joseph Bonaparte Gulf and Timor Sea, Northern Australia." Geoscience Australia, Record 2008/23, 69pp.

Radke, L. C., J. Li, G. Douglas, R. Przeslawski, S. Nichol, J. Siwabessy, Z. Huang, J. Trafford, T. Watson, and T. Whiteway. 2015. "Characterising Sediments for a Tropical Sediment-Starved Shelf Using Cluster Analysis of Physical and Geochemical Variables." *Environmental Chemistry* 12 (2): 204–26.

Radke, L., T. Nicholas, P. Thompson, J. Li, E. Raes, M. Carey, I Atkinson, Z. Huang, J. Trafford, and S. Nichol. 2017. "Baseline Biogeochemical Data from Australia's Continental Margin Links Seabed Sediments to Water Column Characteristics." *Marine and Freshwater Research* http://dx.doi.org/10.1071/MF16219.

R Core Team. 2020. *R: A Language and Environment for Statistical Computing*. Vienna, Austria: R Foundation for Statistical Computing. https://www.R-project.org/.

Ribeiro Jr, P. J., P. J. Diggle, M. Schlather, R. Bivand, and B. Ripley. 2020. *GeoR: Analysis of Geostatistical Data*. https://CRAN.R-project.org/package=geoR.

Ripley, B. 2019. *Tree: Classification and Regression Trees*. https://CRAN.R-project.org/package=tree.

———. 2020. *MASS: Support Functions and Datasets for Venables and Ripley's Mass*. https://CRAN.R-project.org/package=MASS.

Ripley, B. D. 1981. *Spatial Statistics*. New York: John Wiley & Sons.

Roudier, P. 2011. *Clhs: A R Package for Conditioned Latin Hypercube Sampling*.

———. 2020. *Clhs: Conditioned Latin Hypercube Sampling*. https://CRAN.R-project.org/package=clhs.

RStudio Team. 2020. *RStudio: Integrated Development Environment for R*. Boston, MA: RStudio, PBC. http://www.rstudio.com/.

Ruiz-Álvarez, M., F. Alonso-Sarria, and F. Gomariz-Castillo. 2019. "Interpolation of Instantaneous Air Temperature Using Geographical and Modis Derived Variables with Machine Learning Techniques." *International Journal of Geo-Information* 8 (382). https://doi.org/10.3390/ijgi8090382.

Sanabria, L. A., X. Qin, J. Li, R. P. Cechet, and C. Lucas. 2013. "Spatial Interpolation of Mcarthur's Forest Fire Danger Index Across Australia: Observational Study." *Environmental Modelling & Software* 50: 37–50.

Sarkar, D. 2008. *Lattice: Multivariate Data Visualization with R*. New York: Springer. http://lmdvr.r-forge.r-project.org.

Seo, Y., S. Kim, and V. P. Singh. 2015. "Estimating Spatial Precipitation Using Regression Kriging and Artificial Neural Network Residual Kriging (Rknnrk) Hybrid Approach." *Water Resources Management* 29: 2189–2204.

Sestelo, M., N. M. Villanueva, and Javier Roca-Pardinas. 2015. *FWDselect: Selecting Variables in Regression Models*. http://cran.r-project.org/package=FWDselect.

Slaets, J. I. F., H-P. Piepho, P. Schmitter, T. Hilger, and G. Cadisch. 2017. "Quantifying Uncertainty on Sediment Loads Using Bootstrap Confidence Intervals." *Hydrology and Earth System Sciences* 21: 571–88.

Smith, S. J., N. Ellis, and C. R. Pitcher. 2011. "Conditional Variable Importance in R Package extendedForest." *R Vignette <Http://Gradientforest.r-Forge.r-Project.org/Conditional-Importance.pdf>*. http://gradientforest.r-forge.r-project.org/Conditional-importance.pdf.

Specht, D. F. 1990. "Probabilistic Neural Networks." *Neural Networks* 3: 109–18.

Stein, A., M. Hoogerwerf, and J. Bouma. 1988. "Use of Soil Map Delineations to Improve (Co-)Kriging of Point Data on Moisture Deficits." *Geoderma* 43: 163–77.

Stephens, D., and M. Diesing. 2014. "A Comparison of Supervised Classification Methods for the Prediction of Substrate Type Using Multibeam Acoustic and Legacy Grain-Size Data." *PLOS ONE* 9 (4): e93950:14.

———. 2015. "Towards Quantitative Spatial Models of Seabed Sediment Composition." *PLOS ONE* 10 (11): e0142502.

Stevens, D. L., and A. R. Olsen. 2004. "Spatially Balanced Sampling of Natural Resources." *Journal of the American Statistical Association* 99 (465): 262–78.

Therneau, T., and B. Atkinson. 2019. *Rpart: Recursive Partitioning and Regression Trees.* https://CRAN.R-project.org/package=rpart.

Thibaud, E., B. Petitpierre, O. Broennimann, A. C. Davison, and A. Guisan. 2014. "Measuring the Relative Effect of Factors Affecting Species Distribution Model Predictions." *Methods in Ecology and Evolution* 5: 947–55.

Turner, A. J., J. Li, and W. Jiang. 2017. "Effects of Spatial Reference Systems on the Accuracy of Spatial Predictive Modelling Along a Latitudinal Gradient." In *22nd International Congress on Modelling and Simulation*, 106–12.

van Lieshout, M. N. M. 2019. *Theory for Spatial Statistics.* Chapman & Hall/CRC.

Venables, W. N., and B. D. Ripley. 2002. *Modern Applied Statistics with S-Plus.* 4th ed. New York: Springer-Verlag.

Verfaillie, E., V. Van Lancker, and M. Van Meirvenne. 2006. "Multivariate Geostatistics for the Predictive Modelling of the Surficial Sand Distribution in Shelf Seas." *Continental Shelf Research* 26: 2454–68.

Wackernagel, H. 2003. *Multivariate Geostatistics: An Introduction with Applications.* 3rd ed. Berlin: Springer.

Wager, S., T. Hastie, and B. Efron. 2014. "Confidence Intervals for Random Forests: The Jackknife and the Infinitesimal Jackknife." *Journal of Machine Learning Research* 15: 1625–51.

Wahba, G. 1990. *Spline Models for Observational Data.* CBMS-Nsf Regional Conference Series in Applied Mathematics. Philadelphia, Pennsylvania: Society for Industrial; Applied Mathematics.

Wahba, G., and J. Wendelberger. 1980. "Some New Mathematical Methods for Variational Objective Analysis Using Splines and Cross-Validation." *Monthly Weather Review* 108: 1122–45.

Walvoort, D., D. Brus, and J. de Gruijter. 2020. *Spcosa: Spatial Coverage Sampling and Random Sampling from Compact Geographical Strata.* https://CRAN.R-project.org/package=spcosa.

Wang, J-F., A. Stein, B-B. Gao, and Y. Ge. 2012. "A Review of Spatial Sampling." *Spatial Statistics* 2: 1–14. https://doi.org/10.1016/j.spasta.2012.08.001.

Webster, R., and M. Oliver. 2001. *Geostatistics for Environmental Scientists.* Chichester: John Wiley & Sons, Ltd.

Wikle, C. K., A. Zammit-Mangion, and N. Cressie. 2019. *Spatio-Temporal Statistics with R.* Chapman & Hall/CRC.

Wolpert, D., and W. Macready. 1997. "No Free Lunch Theorems for Optimization." *IEEE Transactions on Evolutionary Computation* 1: 67–82.

Wright, M. N., S. Wager, and P. Probst. 2020. *Ranger: A Fast Implementation of Random Forests.* https://github.com/imbs-hl/ranger.

Wright, M. N., and A. Ziegler. 2017. "Ranger: A Fast Implementation of Random Forests for High Dimensional Data in C++ and R." *Journal of Statistical Software* 77 (1): 1–17.

Xia, F., B. Hu, Y. Zhu, W. Ji, S. Chen, D. Xu, and Z. Shi. 2020. "Improved Mapping of Potentially Toxic Elements in Soil via Integration of Multiple Data Sources and Various Geostatistical Methods." *Remote Sensing* 12 (November): 3775. https://doi.org/10.3390/rs12223775.

Xie, Y. 2016. *Bookdown: Authoring Books and Technical Documents with R Markdown.* Boca Raton, Florida: Chapman & Hall/CRC. https://bookdown.org/yihui/bookdown.

Zhang, X., G. Liu, H. Wang, and X. Li. 2017. "Application of a Hybrid Interpolation Method Based on Support Vector Machine in the Precipitation Spatial Interpolation of Basins." *Water* 9: w9100760.

Zuur, A., E. N. Leno, and C. S. Elphick. 2010. "A Protocol for Data Exploration to Avoid Common Statistical Problems." *Methods in Ecology and Evolution* 1 (1): 3–14.

Index